Learning Materials in Biosciences

Learning Materials in Biosciences textbooks compactly and concisely discuss a specific biological, biomedical, biochemical, bioengineering or cell biologic topic. The textbooks in this series are based on lectures for upper-level undergraduates, master's and graduate students, presented and written by authoritative figures in the field at leading universities around the globe.

The titles are organized to guide the reader to a deeper understanding of the concepts covered.

Each textbook provides readers with fundamental insights into the subject and prepares them to independently pursue further thinking and research on the topic. Colored figures, step-by-step protocols and take-home messages offer an accessible approach to learning and understanding.

In addition to being designed to benefit students, Learning Materials textbooks represent a valuable tool for lecturers and teachers, helping them to prepare their own respective coursework.

Anke Brüning-Richardson • Sabine Knipp
Editors

Cell Migration in Development, Health and Disease

Springer

Editors
Anke Brüning-Richardson
University of Huddersfield
Huddersfield, UK

Sabine Knipp
Core Facility Imaging, Zentrum fuer Medizinische Forschung (Centre for Medical Research)
Johannes Kepler University Linz
Linz, Austria

ISSN 2509-6125 ISSN 2509-6133 (electronic)
Learning Materials in Biosciences
ISBN 978-3-031-64531-0 ISBN 978-3-031-64532-7 (eBook)
https://doi.org/10.1007/978-3-031-64532-7

This Springer imprint is published by the registered company Springer Nature Switzerland AG
The registered company address is: Gewerbestrasse 11, 6330 Cham, Switzerland

Contents

Part I

Introduction to Cell Migration

Introduction to Cell Migration

A. Brüning-Richardson

What You Will Learn in This Chapter
Cell migration is a fundamental biological process that supports and drives vital developments in humans. It drives embryonic growth and plays an important role in our ability to respond to damages as observed in wound healing or to external threats by foreign invaders as part of our immune responses. It follows that defective cell migration associated with embryonic growth, wound healing and immune responses may lead to developmental disorders or disease. The study of the cellular mechanisms and underlying structural components of cell migration is becoming increasingly of interest to scientists and clinicians alike to tackle the effects of deregulated migratory activities. This chapter gives an overview of the topics addressed in detail in the following chapters. This includes an introduction to the history of cell migration, why and how cells migrate and modes of cell migration and disease, allowing a brief overview of the chapters within this book. Also discussed briefly is how cell migration can be targeted to combat disease.

A. Brüning-Richardson (✉)
Department of Biological and Geographical Sciences, University of Huddersfield, Huddersfield, UK
e-mail: a.bruning-richardson@hud.ac.uk

A. Brüning-Richardson, S. Knipp (eds.), *Cell Migration in Development, Health and Disease*, Learning Materials in Biosciences,
https://doi.org/10.1007/978-3-031-64532-7_1

1 History

Descriptions of cell migration prior to the invention of microscopy are lacking, but findings of and references to diseases associated with defective cell migration, such as cancers, do exist and reach as far back as the Egyptian dynasties of 2000 BC. Modern technologies, such as computerised tomography (CT) scans, uncovered cancers in Egyptian mummies, including brain tumours and multiple myeloma [1], and attempts to carry out basic surgery are now known to have occurred as far back as the fourth century BC [2], as documented by Hippocratic physicians. The Greek physician Hippocrates (ca. 460–377) first gave the disease its name based on the resemblance of cancerous infiltrative outgrowths to a crab; 'carcinos' is the Greek word for cancer. Surgeries were described by Archigenes of Apamea (born ca. 54 AD), a Roman medical author and practitioner, to allow excision of cancerous growths [2].

In terms of the discovery of the structures associated with the outgrowths observed in cancers, the scientist and architect Robert Hooke (1635–1703) was the first to identify and describe cells in his work *Micrographia* in 1665 while analysing a thin specimen of cork, which he viewed with an early, primitive version of a compound light microscope [3]. Following on, the scientists Mathias Jacob Schleiden (1804–1881) and Theodor Schwann (1810–1882) developed the classical cell theory and stipulated that '1. All living things are made up of cells. 2. Cells are the smallest units of living things 3. Cells only rise from existing cells' (1838) [4, 5]. In 1885, the first attempts were made by Wilhelm Roux (1850–1924), a German zoologist, to keep cells alive in cell cultures. His interest in cells, cell structure and function led to the design of a special vessel for maintaining the cells ('Roux bottle'), which is still used today [6, 7]. Roux also discovered that it was possible to keep tissue extracted from chick embryos alive in warmed saline solutions. His studies and discoveries made him known as the founder of experimental embryology. The scientist Antonie van Leeuwenhoek (1632–1723) developed the first standard microscope, consisting of a single lens with a magnification capacity of 275×, which allowed him to view single-celled animals or 'animalcules' (ca. 1673/1674) and to identify cells, noting their particular shape [8, 9]. In 1676, van Leeuwenhoek described his observations of cells by microscopy in a publication by the Royal Society. The scientist and pathologist Rudolf Virchow (1821–1902) realised that cells form tissues; this work was presented in 'omnis cellula e cellula', which can be translated as 'one living cell is derived from another' [10, 11]. The first links of tissues to the development of tumours were established. An important milestone alongside the development of microscopy was the creation of synthetic fluorescent dyes in the late 1880s, which found diverse application in fluorescence microscopy and live cell imaging [12, 13]. By the early twentieth century, the first micro-cinematographic constructions were introduced in Europe, which also found use in the scientific and medical fields. Basic microscopy and cinematography were succeeded by more complex designs, such as phase contrast microscopy in the 1930s, to improve specimen resolution. In addition, the discovery in the 1980s of green fluorescent protein (GFP) in jellyfish [14] allowed enhanced resolution and examination of cellular features within a specimen. The first confocal microscopes with pinhole technology were developed in the 1980s, and most recently we now make use of microscope systems such as light-sheet microscopy (LSM) for live three-dimensional (3D) imaging, Lattice or super-resolution microscopy (SRM), such as structured illumination microscopy (SIM), to allow high resolution of subcellular structures. Other upcoming techniques include stimulated emission depletion (STED) microscopy and single-molecule localisation microscopy (SMLM) [15–18].

With the development of cameras and live cell imaging, it became clear that the cells discovered and studied by microscopy previously were in many cases motile and able to move from one location to another. This ability to move/migrate was then finally associated with the various functions a cell can have, from being part of the immune system to allowing the closure of a wound and, at the other extreme, enabling a primary tumour to disseminate and reseed as a metastatic tumour. One of the most memorable movies generated by David Rogers in the 1950s is that of a neutrophil chasing and finally engulfing bacteria (https://www.youtube.com/watch?v=JnlULOjUhSQ); this fascinating movie clip is one of the first examples to demonstrate chemotactic cell migration. Continued development of microscopy-related technologies now enables us to study cell migration at the subcellular level even within living organisms. Other scientific discoveries gathering momentum are emerging, from basic to highly sophisticated, in vitro and in vivo assays to assess cell migration. Simple live cell imaging to determine cell velocity and directionality can be as informative as highly elaborate migration chambers for cells to follow external cues and stimuli and invasion assays for the investigation of migration in a 3D environment. Lastly, there are now opportunities to bioengineer such environments by 3D bioprinting, which may one day replace in vivo models of cell migration. Another field that has evolved with the study of cell migration is mathematical modelling to predict outcome of manipulation of migratory activities. This includes models based on differential equations or agent-based modelling commonly used to support research targeting cell migration [19, 20]. Software tools such as ImageJ or Fiji, which are both freely accessible, have been developed to help with the analysis and interpretation of generated data [21, 22]. Novel approaches also include the use of biological tools such as oncolytic viruses that utilise cellular structures such as microtubules to move around their cellular host to reach, for example, the site of viral replication. As such, alongside advances in technologies for the study of cell migration in great detail, there are also now enhanced activities in relation to the development of medicines. These also include novel inhibitors, which improve our ability to target rogue cells such as cancer cells or dysfunctional, migration-associated proteins that promote developmental defects. Cell migration, its study at the cellular level in vivo and in vitro, mathematical modelling, software development and bioengineering will be discussed in depth in the following chapters. The cornerstones of cell migration can thus be summarised by the emerging and continued interest in cells and cellular activity, the developments in the field of microscopy, our continued quest to understand diseases driven by migratory activities and our ability

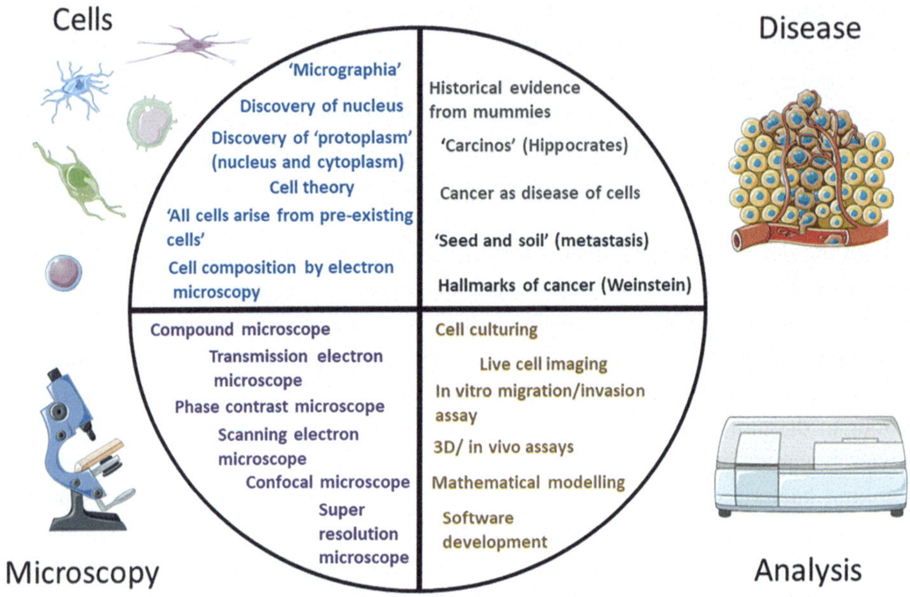

Fig. 1 The four 'cornerstones' of the study of cell migration in health and disease discussed in this chapter. The study of cell migration in health and disease is made possible by milestone discoveries in cell biology, advances in microscopy, the relationship between diseases such as cancer and cell migration and the development of techniques and technologies allowing data generation and interpretation. Highlighted in different colours here are important events for the four cornerstones. This figure was created with https://smart.servier.com/

to study cell migration in vitro and in vivo. Figure 1 depicts the cornerstones of the study of cell migration.

2 Why Do Cells Migrate

The discovery of cells and the notion that cells eventually make up tissues and organs organised in specialised formations and arrangements with specific functions also led to thorough investigations of cell structure and why cells migrate. We now know that cell migration has various functions: firstly, during embryonic development, it drives the generation of specific tissues within a set time frame (perhaps even preceded by the propelling abilities of sperm cells to move); secondly, during repair such as tissue regeneration and wound healing, it allows cells to replace or repair and move into their allocated position; thirdly, during immune responses to harmful intruders in our bodies, it enables immune cells to reach sites of potential threat swiftly; and fourthly, in a disease context, cell migration promotes the dissemination of cancer cells within the body or may be the basis of developmental disorders.

2.1 Cell Migration and Development

Cell migration is a driver of embryonic development. Human embryonic development may be described as the prenatal development up to the eighth week of the developing foetus. During this time frame, many cellular processes are taking place that are associated with the foundations of organ development. (Please note: for a detailed account of embryonic development, please access textbooks such as Alberts' *Molecular Biology of the Cell*, 7th edition.) After implantation into the endometrial lining of the uterus, the fertilised egg consists of the embryoblast and trophoblast. Whereas the embryoblast will develop into different embryonic body structures, the trophoblast is responsible for the development of prenatal support systems. Coinciding with the beginning of the second week, the embryoblast differentiates into the epiblast and hypoblast, leading to the creation of the bilaminar disc. At the same time, the amniotic cavity and yolk sac, as well as the extra-embryonic mesoderm and later chorionic cavity, are formed. Formation of the amniotic cavity is achieved by migration of the epiblast downwards away from the trophoblast. Migration of hypoblast cells leads to the creation of an extracellular matrix and the extra-embryonic mesoderm. In the third and fourth weeks, the bilaminar disc differentiates into the trilaminar disc in the context of gastrulation. From the three overlapping discs (entoderm, mesoderm and ectoderm), all structures and organs of the body are formed. The nervous system is also formed in these weeks. During 'neurulation', the neural plate and neural tube are created from the ectoderm after stimulation with signalling molecules of the Chorda dorsalis. Cell migration of neural crest cells allows the movement away from the neuroepithelium to different tissues (Fig. 2). Bilateral symmetry is established at the same time as gastrulation, and the single-layered embryo adopts a 3D structure by craniocaudal and lateral folding. Weeks 5–8 are characterised by the establishment of missing organs and further differentiation of the embryonic tissue [23, 24]. As such, embryonic development is a tightly regulated process requiring strict adherence to cellular processes, which includes cell migration. This is particularly striking during brain development. Brain development in the embryo is an overly complex process. Cells have to migrate to specific sites where they eventually fulfil a particular function. There is now evidence that strings of nerve fibres may function as a type of tract along which cells can migrate to reach their final destination. A study by Martinez-Chavez et al. [25] in mouse embryos focussed on neuronal progenitor cells, which are formed 10 days after fertilisation in the hindbrain. These cells finally differentiate into nerve cells that migrate into other regions of the hindbrain, similar to processes in the development of the human brain. To investigate the mechanisms of cell migration, the researchers studied a protein called GL13. Knockdown of the gene in mice led to greatly reduced cell migration in the nerve cells; however, after knockdown in only the developing nerve cells, these cells were still able to migrate to their final destination. The research suggested that this ability to still migrate is based on following certain tracts. It is known that outside the brain and the spinal cord (central nervous system), animals also possess the peripheral nervous system, which consists of nerve cells and nerve tracts important for the transmission of

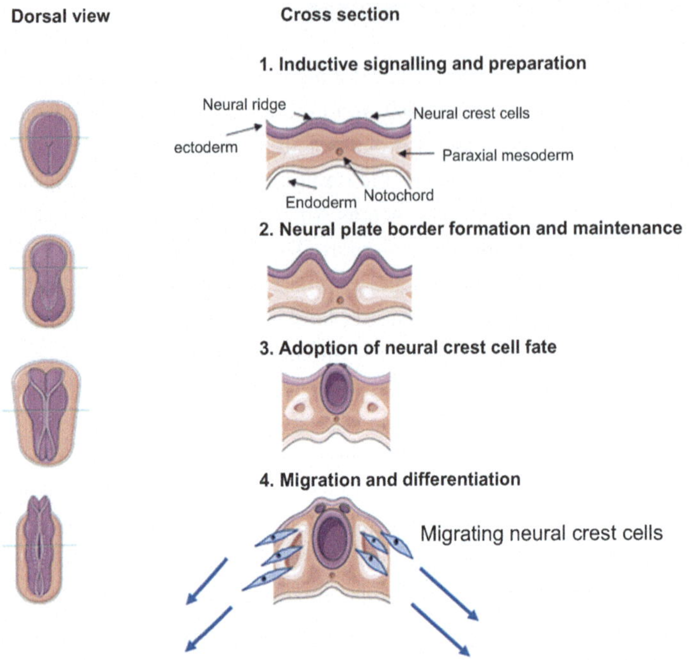

Fig. 2 The steps of neurulation. (1) Differentiation from the ectoderm leads to the formation and thickening of the neuroectodermal tissue into the neural plate. The ectoderm is separated from the neural plate by the neural plate border. (2) Dorsal folding of the neural plate leads to the joining of the ends at the neural plate borders, known as the neural crest. (3) The neural crest is disconnected from the epidermis by the closure of the neural tube. Differentiation of neural crest cells leads to the creation of most of the peripheral nervous system. (4) Degeneration of the notochord leads to the appearance of the nucleus pulposus of the intervertebral discs. Differentiation of mesoderm cells into skeletal muscle, the somites and precursors of the axial skeleton begins. Neural crest cells (NCC) undergo epithelial-mesenchymal transition (EMT) and migrate away from the neuroepithelium to different regions where they are involved in the formation of different tissues. This figure was created with https://smart.servier.com/

electrical impulses to the musculature, digestive tract and glands. The researchers could show that the migratory cells use the nerve tracts for orientation as they migrate along them. Only mouse embryo cells that expressed GL13 were able to move along the tracts; in the knockdown animals, the nerve tracts were much looser and appeared ‘fuzzy’ under the microscope. It was concluded that the change in morphology led to the loss of function in the tracts. Follow-up research aims to determine the role of GL13 in this interaction between nerve cells and nerve tracts. The several stages of neurulation, including inductive signalling, neural plate formation and maintenance, adoption of neural crest cell fate and migration and differentiation, are depicted in Fig. 2.

2.2 Cell Migration and Wound Healing/Regeneration

Wound healing is a multistep process to close a wound in the epithelium by filling it with granulation tissue. This is characteristic in wounds associated with the skin. During the exudative (*homeostatic*) phase, the wound opening is filled by exudation with blood components, with coagulated blood and fibrin (Fig. 3, top-left panel). Inflammation begins when circulating platelets converge on the wound site to form a fibrin clot that seals the wound. Neutrophils and macrophages likewise converge at the wound to clear debris and kill invasive cellular species, mitigating infection (Fig. 3, top-right panel). Each of these cellular types migrates through the circulatory system. During the resorptive phase, increased reabsorption of this blood coagulate occurs via the dermal connective tissue by circulating 'resolving' macrophages. At the same time, migration of basal epithelial

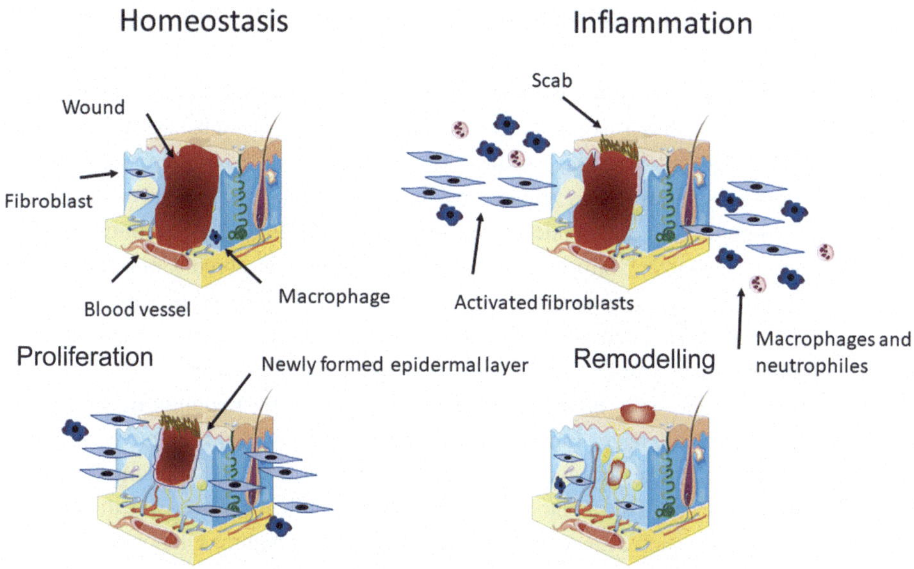

Fig. 3 The steps of wound healing. (1) The first step of wound healing is homeostasis. This is aimed at preventing blood loss by vasoconstriction and the creation of blood clots. Platelets initiate fibrin clot formation and also release chemoattractants that attract macrophages and fibroblasts to the site of the wound (indicated by the dark brown structure). (2) Inflammation. Inflammatory cells, including neutrophiles and macrophages, move into the blood clot to induce inflammatory responses. Macrophage activity supports angiogenesis and re-epithelisation and guides fibroblasts towards the wound. (3) Proliferation. Granulation tissue is formed after migration and proliferation of fibroblasts and endothelial cells. A new matrix, rich in collagen, is formed to replace the provisional matrix. (4) Remodelling. Activated fibroblasts have laid the foundations for a newly formed epidermal layer that had been compromised. This is an ongoing process until the layer is fully formed and seals off the original wound. As a result, the wound is closed with the granulation tissue replaced by the creation of fresh, collagen-rich scar tissue. This figure was created with https://smart.servier.com/

cells occurs between the initial scab and granulation tissue. During the reparative phase, scar tissue is created by the granulation tissue. Concomitantly, proliferation of basal epithelial cells leads to renewal of the epithelial cells on the surface (Fig. 3, lower-left panel). In the final phase, the newly created scar on the surface consists of differentiated epithelial cells, whereas the deeper parts are devoid of vasculature. As melanocytes are not regenerated in this area, this scar assumes a white appearance (Fig. 3, lower-right panel). Therefore, wound healing can be divided into three distinct phases: inflammation, proliferation and remodelling. Since our body and various physiological processes need to maintain a general state of homeostasis to fully function, tissue damage is a major threat and is first tackled by inflammatory responses as part of the re-equilibrium process [26, 27]. The stages of wound healing, including homeostasis, inflammation, proliferation and remodelling, are illustrated in a schematic overview in Fig. 3.

2.3 Cell Migration and the Immune System

The immune system is an important line of defence against substances identified as foreign, which pose a threat to our body. It can be activated by substances recognised as antigens. These include proteins on the surfaces of bacteria, fungi, viruses, parasites or other substances such as toxins released by bacteria. If antigens interact with specific receptors on immune cells, a cascade of cell responses is triggered. After initial contact with an external 'challenger', information about this encounter is stored. Upon subsequent exposure, recognition and response are rapid due to instant recognition. Our own cells also possess surface-associated proteins. If these proteins are erroneously recognised by our own immune system as foreign, this triggers autoimmune responses.

Our body relies on both innate and adaptive (or acquired) immunity to combat the aforementioned challenges. Innate (non-specific) immunity and adaptive (specific) immunity are tightly connected and complement each other should a challenge by an intruder or irritant occur. The innate immunity targets challengers non-specifically and relies on immune cells such as macrophages, neutrophils and natural killer cells. The key role of this part of the immune system is to combat toxins and damaging microorganisms, e.g. those that enter the body via the skin or the digestive system. The adaptive (specific) immune response is based on the generation of antibodies targeted specifically at certain pathogens to which the body has been previously exposed. Therefore, this is also known as a 'learnt' or specific immune response. As the specific response is adaptable and able to 'remember', the body can fight off bacteria and viruses even after they have mutated, leading to new variants.

Cells of the immune system, such as white blood cells (neutrophils), macrophages and lymphocytes, are chemotactic cells, meaning they migrate towards concentration gradients of specific chemicals. Migration is triggered by chemokines, a type of cytokine, as well as other chemoattractants as chemotactic factors. When cells migrate towards an increasing chemical concentration, it is known as positive chemotaxis; when they follow a decreasing concentration, it is termed negative chemotaxis. Fascinating work by Weber et al. [28] details this behaviour. It is understood that immune cells are constantly ‘patrolling’ our body to target pathogens. To accomplish this, they leave the bloodstream to reach tissues and organs and subsequently re-enter the bloodstream via the lymphatic system. It is believed that cells can recognise the chemical composition of their environment either by physically interacting with structural molecules, such as components of connective tissue, or by sensing soluble signal molecules. Since the concentration of these signals increases as cells approach their source, cells can orient themselves along the concentration gradient of these molecules. Both principles of tasting and smelling have been proven in vitro cell culture experiments to guide cells in their migration. However, the mechanisms behind cell orientation in real tissue were not well understood. Weber et al. [28] demonstrated that dendritic cells in the skin rely on both strategies. They follow concentration gradients of signal molecules that are not soluble but are glycan-anchored to the connective tissue. One specific signal molecule, the chemokine CCL21, produced exclusively by lymph nodes, was found by the researchers to enable the dendritic cells to navigate through living tissue. Released by the lymph nodes, CCL21 diffuses into the surrounding tissue and establishes a concentration gradient by binding to glycans. The researchers further established that a cell can find its way to the next lymphatic vessel by comparing the concentration of CCL21 along its surface and then following the direction of the higher concentration. For this to happen, the cell must have a certain size to sense the gradient. A cell that is too small might become trapped at a local concentration peak, unable to perceive that a higher concentration peak is nearby. This was demonstrated by flooding the tissue with an excess of chemokine, which caused the cells to lose their ability to locate lymphatic vessels. Additionally, when the chemokine was unanchored from the tissue, the cells also failed to find their way, indicating that the gradient is not soluble but tied to the tissue [28].

2.4 Cell Migration and Cancer Metastasis

Once assigned to a particular site or organ within the body, a cell becomes differentiated and performs its particular function. However, during cancer development and progression, a normal, differentiated cell can undergo dedifferentiation. Through multiple mutational events, it acquires advantageous traits such as resistance to apoptosis, enhanced and unlimited cell proliferation and the ability to migrate away from its original site. This process involves intricate morphological changes and alterations in how cells interact with their environment (Fig. 4, left panel). It involves the change in orientation of the cell and its ability to form contacts with its neighbouring cells. Cell-to-cell interactions via cellular adhesion are lost and replaced with the reorientation of the cell, detachment and then egress from the original site within the tissue. Adoption of a mostly mesenchymal phenotype allows directed cell migration away from the original tissue (Fig. 4, middle panel), transvasination, invasion into healthy tissues and finally re-establishment of the

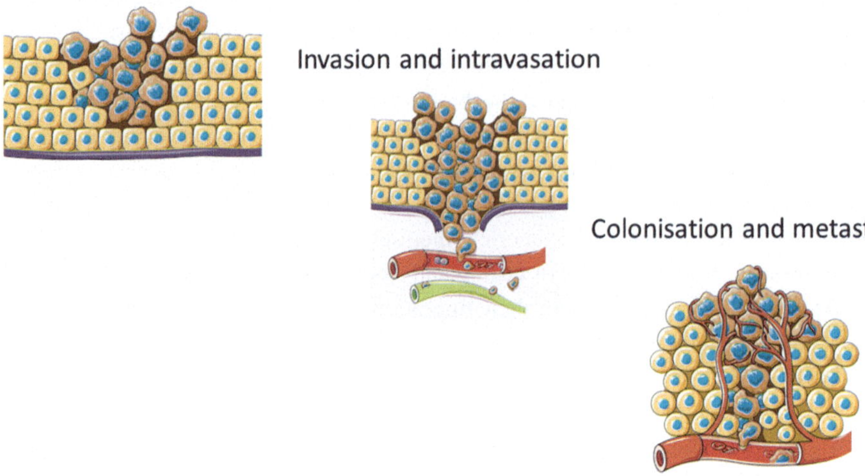

Fig. 4 The steps of tumour dissemination. (1) A primary tumour is located in its original site. Growth of the tumour and interaction with its microenvironment drives tumour progression. Cancer cells lose their contact with their neighbouring cells and re-orientate themselves. (2) They adopt an elongated, mesenchymal morphology and are able to invade normal tissue, intravasate, breaking down surrounding tissues and entering the blood stream. Cells are disseminated via the bloodstream to local lymph nodes and then to a new site of colonisation where they exhibit extravasation and a reversion to their epithelial morphology. (3) Colonisation of the tissue by the cells and growth of a secondary tumour is initiated. This figure was created with https://smart.servier.com/

tumour in a different site (Fig. 4, right panel) [29, 30]. Figure 4 depicts the three main stages of cancer metastasis.

3 Ways of Cell Migration

3.1 How Do Cells Migrate

According to their specific cellular functions, cells are defined by their particular morphology and their association with other cells. Cells that make up a tissue or organ are arranged in a particular order and orientation, possessing a morphology that allows them to be part of this tissue/organ and fulfil their specific function. An example of this is seen in the skin, which consists of three main layers: the epidermis consisting mainly of keratinocytes, the dermis and the hypodermis. Keratinocytes produce keratin, which is essential for hair, nails and the skin. Keratinocytes are characterised by their larger size among epidermal cells, intercellular bridges and abundant cytoplasm. Their structure is closely linked to their protective function, forming tight barriers to prevent invasion from the outside and anchoring other cells firmly in their position [31]. During wound healing, the migratory ability of keratinocytes is essential within the first 24 h post-injury, as they migrate from the wound's edges to facilitate wound closure as part of re-epithelialisation. Migrating cells also adopt specific morphologies and may associate with other cells depending on their background, function and the environment they encounter. Cells can be rounded, amoeboid or elongated and mesenchymal-like. They may migrate individually or in groups, with the most connected type forming a migrating cell sheet. Additionally, cells may transition from one phenotype to another.

3.2 Cell Motility

For a cell to migrate, it requires the power, direction and the ability to interact with its environment. Moreover, for a cell to move, it must adapt and alter its shape. Cell motility is enabled by the coordinated interaction of three key cellular components: the actin microfilaments, the microtubule network and intermediate filaments. The actin filament consists of two strands of globular molecules arranged in a helical shape. Apart from its role in cell migration as a force generator through filament assembly and disassembly dynamics, actin is crucial for regulating cell shape and providing mechanical support. Microtubules are composed of alpha and beta tubulin subunits and exhibit dynamic instability, constantly growing or shrinking. Their polarity within the network is essential for sustained directed migration [32]. Intermediate filaments are structural components of the cytoskeleton assembled from fibrous proteins. They primarily protect cells from mechanical stress and help maintain cell shape [33, 34]. Intermediate filaments also interact with and regulate focal adhesions [34], which are contact points between the cell and the extracellular matrix (ECM). Focal adhesions feature integrin receptors on the cell surface that connect to the actin cytoskeleton inside the cell. Polarisation of the cell marks the initial stage of cell migration, establishing a front-rear axis characterised by a prominent leading edge and trailing edge. This process is initiated by the activity of several proteins, notably small guanosine triphosphatases (GTPases) such as cell division control protein 42 (Cdc42), Rac and Rat sarcoma homolog (Rho). Cytoskeletal reorganisation and control of direction sensing of the cell are directly regulated by these key proteins. Movement is aided by polymerisation of the actin filament, which makes up the characteristic broad, flat lamellipodium and the long, thin filopodia of the leading edge under the control of Rac and Cdc42. Protrusions and actin dynamics and rearrangements are believed to be influenced by the activity of the microtubules, which are also in control of the cell's directionality. Interactions between the cell and its environment are made possible via the actin cytoskeleton and its points of contact with the ECM, the focal adhesions and integrin receptors. Contact with the substrate is established, and a traction force is generated during actomyosin contraction [35–43]. Cell migration is therefore a repetitive event of the generation of protrusions, adhesion to a substrate, then contraction and finally retraction, made possible by an interplay among the actin filament, microtubules and focal adhesions. The main components of cell migration, including the

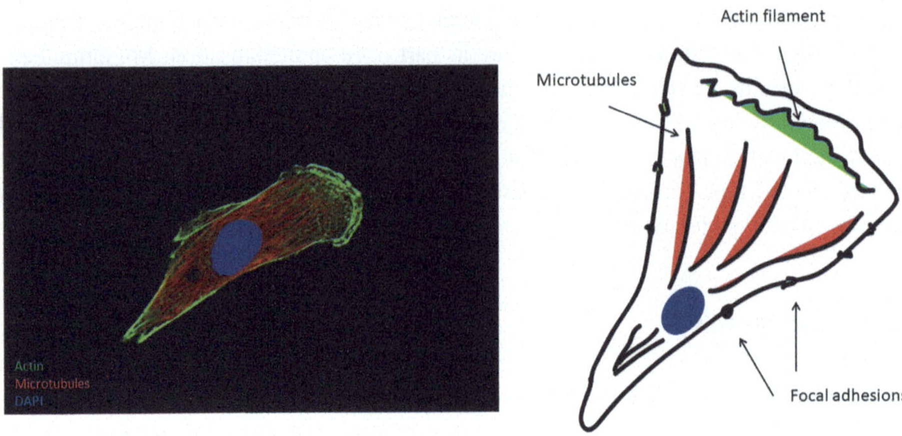

Fig. 5 Cell migration is dependent on the interplay of the actin filament, the microtubules and intermediate filament such as the focal adhesions. Image of a single migrating U251 glioma cell. The cell has been fluorescently labelled to highlight key features of cellular structures involved in cell migration. Here, the actin filament is highlighted in green, whereas the microtubules are highlighted in red. The cell nucleus appears blue. In addition, a cartoon of the same cell also highlighting the presence of the focal adhesions is depicted. Approximate length of cell: 20 μM. Image and cartoon: author's own. This figure was created with https://smart.servier.com/

actin filament, microtubules and focal adhesions, are shown in their specific locations in Fig. 5.

3.3 Single-Cell Migration

Single cells may adopt two major morphologies to migrate: an elongated, mesenchymal morphology or a rounded, amoeboid morphology. Apart from the visual appearance, there are clear differences between the two migration types. Amoeboid migration is fast (recorded cell speeds are around 10 μm/min [44]) and displays only a weak adhesive interaction with the environment. Cells are characterised by a well-defined front and rear. Such migrating cells are known to be pathfinders as they squeeze through pores and gaps in their environment. Contractility of the cell is achieved by cortical actomyosin, and the cell front is enriched with a dendritic actin network. On the other hand, mesenchymal cell migration is relatively slow (at less than 1 μm/min [44]), and a strong adhesive interaction with the environment/EMC is generated via focal adhesion points on the lamellipodia. These migrating cells are path-layers as they degrade the ECM via several enzymes, including the matrix metalloproteinases (MMPs). The lamellipodia are enriched with dendritic F-actin. Contractility of the cell is enabled by actomyosin filaments located behind the leading edge and the generation of thick stress fibres interacting with focal adhesions [45].

Single migrating cells are mostly found among the cells of the immune response and in migrating cancer cells. As mentioned earlier, the immune response consists of T cells and B cells as part of the innate and adaptive immune responses. T cells (such as T-helper cells and cytotoxic T cells) must be able to respond quickly to an intruder or externally induced threat and therefore adopt an amoeboid shape and polarised polarity [46]. They are adapted in such a way that they can cross any tissue within the body and are guided by either chemoattractants or random migration to reach their target and combat the potential threat [47]. It is now known that this migration in vivo is mainly a non-adherent confined migration. This means that the migratory activity is based on the retrograde flow of myosin II-dependent actin, allowing the cells to propel themselves forward by friction rather than relying on specific interactions between receptors and ligands. Immune cells are flexible in terms of cell speed and can alternate from migrating slowly to fast based on certain actin pools. This optimises the way the cells can explore their space and the environment they find themselves in. As mentioned previously, the cells are guided by certain chemical and physical cues in their environment, which are detected and responded to by extracellular/transmembrane receptors and transmitted to the actin cytoskeleton. Cellular activities recorded estimate that immune cells such as neutrophils can travel at a speed of 9 μm/min [44] and are also able to cross various cellular structures within their environments, such as tissues. Therefore, these cells can change their morphology based on the particular microenvironment they find themselves in.

3.4 Collective Cell Migration

On the other hand, many cells, such as border cells from an epithelium, can migrate in a collective and interactive manner [48, 49]. Collective cell migration is characterised by the coordinated movement of multiple cells that maintain cell-to-cell contacts while undergoing migratory processes. Through this connectivity, the cells move as a single structural unit, experiencing either passive or active translocation. This migration pattern is observed in morphogenesis, wound healing and invasive processes by cancer cells. Cells migrate in clusters, strands or sheets, adopting similar migratory morphologies driven by actin- or myosin-mediated protrusion formations, and they respond to chemotactic cues in their environment. In collective migration, leader cells at the forefront of the migrating cell collective generate necessary pulling forces and modify the ECM, whereas follower cells communicate among themselves and with the leaders. This communication aids in force generation and the establishment of chemotactic gradients [50]. The mechanical coupling of closely adjacent cells is believed to synchronise intercellular communication and cytoskeletal arrangements during collective cell migration. Signals from the microenvironment efficiently propagate throughout the cell sheet [49].

An important example of collective cell migration is again observed in wound healing. The ability to close wounds is vital to prevent blood loss, shock and potentially death. Re-epithelisation, which involves the movement of cells to the site of damaged tissue to cover the wound, relies on the increased motility of epithelial cells. Eventually, proliferation across the wound site leads to the creation of a new, permanent cellular structure and epidermis. During 'epithelialisation', keratinocytes at the wound edge must detach from the basement membrane via desmosomes to migrate, proliferate and differentiate. Cellular activities aimed at reducing adhesive properties, such as activation of PKCα or activity by the transcription factor Slug, are initiated. Cell migration is promoted by cytokines and growth factors. Simultaneously, as keratinocytes at the wound edge migrate, the cells behind this leading front begin to proliferate. Once a single layer of cells forms across the wound and the two epithelial tongues meet, a basement membrane forms. Proliferation ceases, and stratification of the keratinocytes allows for the renewal of a differentiated epidermis [26, 27].

4 Cell Migration and Disease

As previously discussed, cell migration is a finely tuned biological activity that promotes development, defence against intruders and healing processes. All these processes can be affected by gene mutations, gene deregulation or errors in post-translational processes, leading to dysfunction, gain of function or loss of function in cells orchestrating migratory events. During development, defects in the ability of cells to migrate can have devastating consequences, leading to the embryo death, birth defects or disorders affecting the central nervous system, heart or physical and mental abilities. Knockout studies in mice targeting key players in cell migration have revealed lethal effects in most cases. For instance, loss of actin resulted in early embryonic death, while targeting vinculin, a component of focal adhesion complexes, also led to lethality, with embryos showing defects in the neural tube and incorrect development of the heart, somites and limbs [51].

T-cell and B-cell migratory activity is vital for our defence against external threats but can also be affected by structural defects or loss of expression. For example, the Wiskott-Aldrich syndrome protein (WASp) is exclusively expressed in cells of the haematopoietic system, and defective protein function in most immune cell lineages has been associated with enhanced susceptibility to infections [52]. This is because the protein plays a crucial role in regulating the actin cytoskeleton, promoting the transmission of actin-regulating signals and controlling morphological changes [53, 54]. The disorder associated with reduced or loss of expression of this protein is known as Wiskott-Aldrich syndrome (WAS), which leads, among other defects, to immune deficiency [55].

Perhaps the most studied effect of deregulated cellular migratory activity is the role of cell migration in cancer. Cell migration and metastasis are included by Hanahan and Weinberg as one of the hallmarks of cancer [29]. This means that cells that are normally part of a tissue or organ lose their characteristics associated with their particular role during tumourigenesis. They then gain the ability to detach from their original site, travel through microenvironments and drive the dissemination of a primary tumour. Indeed, 90% of deaths in cancer patients are caused by metastasis. It is now well established that for cancer cells to detach, they need to lose their interaction with neighbouring cells at cell junctions via adhesion complexes such as members of the cadherins. For epithelial-derived cancer cells to detach, they must downregulate proteins such as E-cadherin and upregulate N-cadherin after the induction of TGF-β and the Snail, Slug and Twist transcription factors [56, 57]. Following this, the cells also lose their polarity and are able to 'break free', and depending on their external environment, they follow chemical or mechanical cues to move away from their original site. As an example of cell migration in cancer cells, in brain tumours, cancer cells dissociate from the original tumour and travel within the brain following neuronal tracts and blood vessels, finally reseeding the tumour in the surroundings in most cases of the original tumour.

In contrast to the diffusely disseminating type seen in high-grade brain tumours, an example of solid tumours is epithelial ovarian cancers. Here, the primary tumour develops in the ovaries and grows until conditions become unfavourable for further expansion. Outgrowth formation is initiated via ECM remodelling and loss of apical-basal cell polarity. The cancer cells are released from their original site as clusters, which then collect in ascitic fluids in the peritoneum. Here, the cells may remain, grow and survive by adhering to the ECM and inactivating pro-apoptotic factors. Supported by soluble immune-stimulating molecules such as epidermal growth factors (EGFs), secreted by tumour-associated macrophages (TAMs), the cells display enhanced cell mobility. They finally adhere to the mesothelium through interactions via surface receptors and create a metastatic tumour environment by reprogramming non-malignant cells including fibroblasts and neutrophils and undergoing mesenchymal-epithelial transition (MET) [58].

5 Targeting Cell Migration

As discussed, the migratory ability of cells is necessary for many biological processes. When migration is deregulated, diseases may follow as a consequence. Thus, it seems logical that targeting the migratory process and the key genes involved could prevent disease development or at least slow disease progression. In the case of diseases arising from developmental deregulation, gene therapy may be successful in correcting the faulty gene and restoring it to normal activity.

In targeting cancer dissemination and metastasis, most recent treatment strategies now include using small molecule inhibitors to target signalling pathways or key regulators involved in cell migration. This strategy suggests combining treatments of tumours with common cytotoxic drugs and radiotherapy to kill cancer cells within the main bulk of the tumour and inhibit their ability to break away and metastasise. In our group, we have investigated targeting cell migration in glioblastoma (GBM) by inhibiting glycogen synthase kinase-3 (GSK-3). GSK-3, originally associated with glucose homeostasis, is upregulated in GBM and plays a role in regulating microtubules, the actin cytoskeleton and focal adhesions involved in cell migration. Our studies indicated that inhibiting GSK-3 caused morphological changes in glioma cells, shifting them from a mesenchymal to a more rounded, amoeboid phenotype (Brüning-Richardson et al., unpublished data). Similar behaviour was observed when cells were treated with another inhibitor, CCG-1423, which targets mesenchymal cell migration through a novel signalling mechanism. We found that cells adapted to targeting the mesenchymal migratory pathway by switching to an amoeboid phenotype, thereby compensating for the inhibition. When challenged with an inhibitor targeting amoeboid cell migration, such as Rhosin hydrochloride, in combination with CCG-1423, migration was reduced but still continued due to a third switch to collective cell migration [59, 60]. Combining this with the Rac1 inhibitor II to prevent collective cell migration had the most pronounced effect on the cells' migratory ability (author's own unpublished data). We utilise 3D invasion assays to assess these drug effects in a more tumour-relevant environment. Brain tumour cells expressing GFP stably can track migration paths, demonstrating inhibitor activity on cell migration. Figure 6 illustrates an experimental setup in 3D, tracing migratory paths of cancer cells moving away from the original cancer spheroid embedded in a brain-like matrix. Different migration tracks are differentiated by colour. Targeting the migratory ability of cancer cells is an exciting frontier in cancer research that may reveal key regulators of tumour dissemination and metastasis.

Take-Home Message

This chapter provided a general introduction to important aspects of the fundamentals and study of cell migration:

- Cell migration is an expanding research field due to its critical role in processes such as development, wound healing, the immune response and disease.
- Cell migration in development is essential for distributing cells into specific architectures, such as organising the central nervous system and forming organs and tissues.
- Cell migration in wound healing drives tissue repair and renewal, maintaining tissue integrity and homeostasis.
- Cell migration in the immune response enables continuous surveillance and rapid immune cell responses.
- Defective cell migration can underlie disorders like chronic inflammations or facilitate the dissemination of cancers to distant sites in the body.

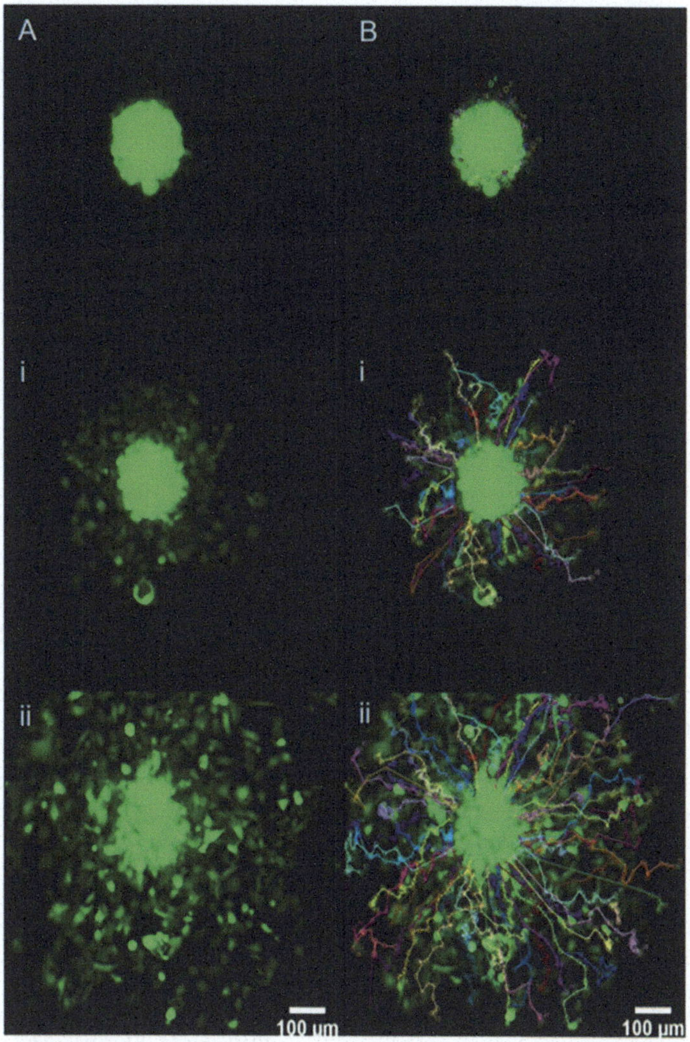

Fig. 6 Example of acquiring tracking data from a U251 glioma spheroid and migratory cells in 3D. The sequence of images was obtained from time stacks by confocal microscopy. (**a–aii**) Time stack of U251 cells treated with the inhibitor MI192. U251 was stably transfected to express GFP. (**b–bii**) Comparison of the same time series with tracking markers displayed. Markers are displayed at each time frame for the duration of cell migratory progression. For both sequences, time segments were 2, 24 and 72 h. Images: James Samarasekara

References

1. https://www.sciencedaily.com/releases/2017/12/171214101215.htm
2. https://www.ncbi.nlm.nih.gov/pmc/articles/PMC2820670/
3. Gest H. The discovery of microorganisms by Robert Hooke and Antoni van Leeuwenhoek, Fellows of The Royal Society. Notes Rec R Soc Lond. 2004;58:187–201.
4. Tavassoli M. The cell theory: a foundation to the edifice of biology. Am J Pathol. 1980;98(1):44. PMID: 6985772
5. https://hekint.org/2022/07/15/the-beginnings-of-cell-theory-schleiden-schwann-and-virchow/
6. https://scitechconnect.elsevier.com/the-beginning-of-tissue-culture/
7. Richter M, Piwocka O, Musielak M, Piotrowski I, Suchorska WM, Trzeciak T. From donor to the lab: a fascinating journey of primary cell lines. Front Cell Dev Biol. 2021;9:711381. https://doi.org/10.3389/fcell.2021.711381. PMID: 34395440; PMCID: PMC8356673
8. Wollman AJ, Nudd R, Hedlund EG, Leake MC. From animaculum to single molecules: 300 years of the light microscope. Open Biol. 2015;5(4):150019. https://doi.org/10.1098/rsob.150019. PMID: 25924631; PMCID: PMC4422127
9. Davis IM. Antoni van Leeuwenhoek and measuring the invisible: The context of 16th and 17th century micrometry. Stud Hist Philos Sci. 2020;83:75–85. https://doi.org/10.1016/j.shpsa.2020.03.004. Epub 2020 Mar 24. PMID: 32958283
10. Ribatti D. Rudolf Virchow, the founder of cellular pathology. Rom J Morphol Embryol. 2019;60(4):1381–2. PMID: 32239122
11. Breathnach CS. Rudolf Virchow (1821-1902) and Die Cellularpathologie (1858). J Ir Coll Physicians Surg. 2002;31(1):43–6. PMID: 11908520
12. Specht EA, Braselmann E, Palmer AE. A critical and comparative review of fluorescent tools for live-cell imaging. Annu Rev Physiol. 2017;79:93–117. https://doi.org/10.1146/annurev-physiol-022516-034055. Epub 2016 Nov 16. PMID: 27860833
13. Lavis LD. Teaching old dyes new tricks: biological probes built from fluoresceins and rhodamines. Annu Rev Biochem. 2017;86:825–43. https://doi.org/10.1146/annurev-biochem-061516-044839. Epub 2017 Apr 7. PMID: 28399656
14. Shimomura O. Discovery of green fluorescent protein (GFP) (Nobel lecture). Angew Chem Int Ed Engl. 2009;48(31):5590–602. https://doi.org/10.1002/anie.200902240. PMID: 19579247
15. Chen J, Liu W, Xu Z. [Comparison and progress review of various super-resolution fluorescence imaging techniques]. Se Pu. 2021;39(10):1055–64. Chinese. https://doi.org/10.3724/SP.J.1123.2021.06015. PMID: 34505427.
16. Wan Y, McDole K, Keller PJ. Light-sheet microscopy and its potential for understanding developmental processes. Annu Rev Cell Dev Biol. 2019;35:655–81. https://doi.org/10.1146/annurev-cellbio-100818-125311. Epub 2019 July 12. PMID: 31299171
17. Garini Y, Vermolen BJ, Young IT. From micro to nano: recent advances in high-resolution microscopy. Curr Opin Biotechnol. 2005;16(1):3–12. https://doi.org/10.1016/j.copbio.2005.01.003. PMID: 15722009
18. Knight AE, Peckham M. Recent innovations in super-resolution microscopy. Methods. 2015;88:1–2. https://doi.org/10.1016/j.ymeth.2015.08.004. Epub 2015 Aug 6. PMID: 26255962
19. Carlsson AE, Sept D. Mathematical modeling of cell migration. Methods Cell Biol. 2008;84:911–37. https://doi.org/10.1016/S0091-679X(07)84029-5. PMID: 17964954
20. Butler G, Rudge J, Dash PR. Mathematical modelling of cell migration. Essays Biochem. 2019;63(5):631–7. https://doi.org/10.1042/EBC20190020. PMID: 31654055
21. Collins TJ. ImageJ for microscopy. Biotechniques. 2007;43(1 Suppl):25–30. https://doi.org/10.2144/000112517. PMID: 17936939

22. Schindelin J, Arganda-Carreras I, Frise E, Kaynig V, Longair M, Pietzsch T, Preibisch S, Rueden C, Saalfeld S, Schmid B, Tinevez JY, White DJ, Hartenstein V, Eliceiri K, Tomancak P, Cardona A. Fiji: an open-source platform for biological-image analysis. Nat Methods. 2012;9(7):676–82. https://doi.org/10.1038/nmeth.2019. PMID: 22743772; PMCID: PMC3855844
23. Zhai J, Xiao Z, Wang Y, Wang H. Human embryonic development: from peri-implantation to gastrulation. Trends Cell Biol. 2022;32(1):18–29. https://doi.org/10.1016/j.tcb.2021.07.008. Epub 2021 Aug 17. PMID: 34417090
24. Shahbazi MN. Mechanisms of human embryo development: from cell fate to tissue shape and back. Development. 2020;147(14):dev190629. https://doi.org/10.1242/dev.190629. PMID: 32680920; PMCID: PMC7375473
25. Martinez-Chavez E, Scheerer C, Wizenmann A, Blaess S. The zinc-finger transcription factor GLI3 is a regulator of precerebellar neuronal migration. Development. 2018;145(24):dev166033. https://doi.org/10.1242/dev.166033. PMID: 30470704
26. Wilkinson HN, Hardman MJ. Wound healing: cellular mechanisms and pathological outcomes. Open Biol. 2020;10(9):200223. https://doi.org/10.1098/rsob.200223. Epub 2020 Sept 30. PMID: 32993416; PMCID: PMC7536089.
27. Sorg H, Tilkorn DJ, Hager S, Hauser J, Mirastschijski U. Skin wound healing: an update on the current knowledge and concepts. Eur Surg Res. 2017;58(1–2):81–94. https://doi.org/10.1159/000454919. Epub 2016 Dec 15. PMID: 27974711
28. Weber M, Hauschild R, Schwarz J, Moussion C, de Vries I, Legler DF, Luther SA, Bollenbach T, Sixt M. Interstitial dendritic cell guidance by haptotactic chemokine gradients. Science. 2013;339(6117):328–32. https://doi.org/10.1126/science.1228456. PMID: 23329049
29. Hanahan D, Weinberg RA. Hallmarks of cancer: the next generation. Cell. 2011;144(5):646–74. https://doi.org/10.1016/j.cell.2011.02.013. PMID: 21376230
30. Hanahan D. Hallmarks of cancer: new dimensions. Cancer Discov. 2022;12(1):31–46. https://doi.org/10.1158/2159-8290.CD-21-1059. PMID: 35022204
31. Schröder JM. The role of keratinocytes in defense against infection. Curr Opin Infect Dis. 2010;23(2):106–10. https://doi.org/10.1097/QCO.0b013e328335b004. PMID: 20010101
32. Etienne-Manneville S. Microtubules in cell migration. Annu Rev Cell Dev Biol. 2013;29:471–99. https://doi.org/10.1146/annurev-cellbio-101011-155711. Epub 2013 July 12. PMID: 23875648
33. Herrmann H, Bär H, Kreplak L, et al. Intermediate filaments: from cell architecture to nanomechanics. Nat Rev Mol Cell Biol. 2007;8:562–73. https://doi.org/10.1038/nrm2197.
34. Leube RE, Moch M, Windoffer R. Intermediate filaments and the regulation of focal adhesion. Curr Opin Cell Biol. 2015;32:13–20. https://doi.org/10.1016/j.ceb.2014.09.011. Epub 2014 Oct 17. PMID: 25460777
35. Akhshi TK, Wernike D, Piekny A. Microtubules and actin crosstalk in cell migration and division. Cytoskeleton (Hoboken). 2014;71(1):1–23. https://doi.org/10.1002/cm.21150. Epub 2013 Oct 29. PMID: 24127246
36. Trepat X, Chen Z, Jacobson K. Cell migration. Compr Physiol. 2012;2(4):2369–92. https://doi.org/10.1002/cphy.c110012. PMID: 23720251; PMCID: PMC4457291
37. Wong M, Gilmour D. Going your own way: Self-guidance mechanisms in cell migration. Curr Opin Cell Biol. 2021;72:116–23. https://doi.org/10.1016/j.ceb.2021.07.004. Epub 2021 Aug 14. PMID: 34403875
38. Friedl P, Sahai E, Weiss S, Yamada KM. New dimensions in cell migration. Nat Rev Mol Cell Biol. 2012;13(11):743–7. https://doi.org/10.1038/nrm3459. Epub 2012 Oct 17. PMID: 23072889

39. Kirfel G, Rigort A, Borm B, Herzog V. Cell migration: mechanisms of rear detachment and the formation of migration tracks. Eur J Cell Biol. 2004;83(11–12):717–24. https://doi.org/10.1078/0171-9335-00421. PMID: 15679116
40. Ridley AJ. Rho GTPases and cell migration. J Cell Sci. 2001;114(Pt 15):2713–22. https://doi.org/10.1242/jcs.114.15.2713. PMID: 11683406
41. Petrie RJ, Yamada KM. At the leading edge of three-dimensional cell migration. J Cell Sci. 2012;125(Pt 24):5917–26. https://doi.org/10.1242/jcs.093732. Epub 2013 Feb 1. PMID: 23378019; PMCID: PMC4067260
42. Petrie RJ, Doyle AD, Yamada KM. Random versus directionally persistent cell migration. Nat Rev Mol Cell Biol. 2009;10(8):538–49. https://doi.org/10.1038/nrm2729. Epub 2009 Jul 15. PMID: 19603038; PMCID: PMC2752299
43. Wehrle-Haller B, Imhof BA. Actin, microtubules and focal adhesion dynamics during cell migration. Int J Biochem Cell Biol. 2003;35(1):39–50. https://doi.org/10.1016/s1357-2725(02)00071-7. PMID: 12467646
44. http://book.bionumbers.org/how-fast-do-cells-move/
45. Bear JE, Haugh JM. Directed migration of mesenchymal cells: where signaling and the cytoskeleton meet. Curr Opin Cell Biol. 2014;30:74–82. https://doi.org/10.1016/j.ceb.2014.06.005. Epub 2014 July 5. PMID: 24999834; PMCID: PMC4177959
46. Vesperini D, Montalvo G, Qu B, et al. Characterization of immune cell migration using microfabrication. Biophys Rev. 2021;13:185–202. https://doi.org/10.1007/s12551-021-00787-9.
47. Galeano Niño JL, Pageon SV, Tay SS, Colakoglu F, Kempe D, Hywood J, Mazalo JK, Cremasco J, Govendir MA, Dagley LF, Hsu K, Rizzetto S, Zieba J, Rice G, Prior V, O'Neill GM, Williams RJ, Nisbet DR, Kramer B, Webb AI, Luciani F, Read MN, Biro M. Cytotoxic T cells swarm by homotypic chemokine signalling. Elife. 2020;9:e56554. https://doi.org/10.7554/eLife.56554. PMID: 33046212; PMCID: PMC7669268
48. Friedl P, Gilmour D. Collective cell migration in morphogenesis, regeneration and cancer. Nat Rev Mol Cell Biol. 2009;10:445–57. https://doi.org/10.1038/nrm2720.
49. Saraiva JE, Barriga EH. Chapter one—The basics of collective cell migration: unity makes strength. In: Pajic-Lijakovic I, Barriga E, editors. Viscoelasticity and collective cell migration. An interdisciplinary perspective across levels of organization. London: Academic; 2021. p. 1–19.
50. Mayor R, Etienne-Manneville S. The front and rear of collective cell migration. Nat Rev Mol Cell Biol. 2016;17:97–109. https://doi.org/10.1038/nrm.2015.14.
51. Kurosaka S, Kashina A. Cell biology of embryonic migration. Birth Defects Res C Embryo Today. 2008;84(2):102–22. https://doi.org/10.1002/bdrc.20125. PMID: 18546335; PMCID: PMC2542983
52. Bouma G, Burns SO, Thrasher AJ. Wiskott-Aldrich syndrome: Immunodeficiency resulting from defective cell migration and impaired immunostimulatory activation. Immunobiology. 2009;214(9–10):778–90. https://doi.org/10.1016/j.imbio.2009.06.009. Epub 2009 July 22. PMID: 19628299; PMCID: PMC2738782
53. Burns S, Cory GO, Vainchenker W, Thrasher AJ. Mechanisms of WASp-mediated hematologic and immunologic disease. Blood. 2004;104(12):3454–62. https://doi.org/10.1182/blood-2004-04-1678. Epub 2004 Aug 12. PMID: 15308573
54. Thrasher AJ, Burns SO. WASP: a key immunological multitasker. Nat Rev Immunol. 2010;10(3):182–92. https://doi.org/10.1038/nri2724. PMID: 20182458
55. Ochs HD, Thrasher AJ. The Wiskott-Aldrich syndrome. J Allergy Clin Immunol. 2006;117(4):725–38.; Quiz 739. PMID: 16630926. https://doi.org/10.1016/j.jaci.2006.02.005.

56. Wu JS, Jiang J, Chen BJ, Wang K, Tang YL, Liang XH. Plasticity of cancer cell invasion: Patterns and mechanisms. Transl Oncol. 2021;14(1):100899. https://doi.org/10.1016/j.tranon.2020.100899. Epub 2020 Oct 17. PMID: 33080522; PMCID: PMC7573380
57. Knights AJ, Funnell AP, Crossley M, Pearson RC. Holding tight: cell junctions and cancer spread. Trends Cancer Res. 2012;8:61–9. PMID: 23450077; PMCID: PMC3582402
58. Farsinejad S, Cattabiani T, Muranen T, Iwanicki M. Ovarian cancer dissemination—a cell biologist's perspective. Cancers (Basel). 2019;11(12):1957. https://doi.org/10.3390/cancers11121957. PMID: 31817625; PMCID: PMC6966436
59. Ketchen S, Rohwedder A, Knipp S, Esteves F, Struve N, Peckham M, Ladbury JE, Curd A, Short SC, Brüning-Richardson A. A novel workflow for three-dimensional analysis of tumour cell migration. Interface Focus. 2020;10(2):20190070. https://doi.org/10.1098/rsfs.2019.0070. Epub 2020 Feb 14. PMID: 32194931; PMCID: PMC7061943
60. Ketchen SE, Gamboa-Esteves FO, Lawler SE, Nowicki MO, Rohwedder A, Knipp S, Prior S, Short SC, Ladbury JE, Brüning-Richardson A. Drug resistance in glioma cells induced by a mesenchymal-amoeboid migratory switch. Biomedicines. 2021;10(1):9. https://doi.org/10.3390/biomedicines10010009. PMID: 35052688; PMCID: PMC8773151

Cell Migration in Development

Sabine Knipp

What You Will Learn in This Chapter

This chapter gives you a general overview on the pivotal role of cell migration during developmental morphogenesis. It highlights general mechanisms and molecular regulation pathways that are evolutionarily conserved throughout the animal kingdom. Representative developmental processes will be described to explain general migrational processes, for example epithelial-to-mesenchymal transition (EMT) or the role of proteins of the family of RhoGTPases for migration. It aims to help the reader to gain an insight into the complexity researchers face when studying migration during developmental processes and wants to draw attention on how studying developmental cell migration might help researchers to better understand cell migration during disease or regeneration.

S. Knipp (✉)
Core Facility Imaging, Zentrum fuer Medizinische Forschung (Centre for Medical Research), Johannes Kepler University Linz, Linz, Austria
e-mail: sabine.knipp@jku.at

A. Brüning-Richardson, S. Knipp (eds.), *Cell Migration in Development, Health and Disease*, Learning Materials in Biosciences,
https://doi.org/10.1007/978-3-031-64532-7_2

1 Introduction

Cell migration is a complex, evolutionarily conserved mechanism with its basic processes and major components preserved in organisms ranging from protozoa to mammals. It is a pivotal mechanism not only during development, but also for events like immune responses, cancer metastases, or wound healing (see other chapters of this book for more details). During development, cell migration is crucial for the morphogenesis of a single cell into a multicellular organism. Cells are often born in one region of the embryo and have to travel either individually or collectively as clusters, sheets, or loose chains over remarkable distances to their final destination. For example, neural crest cells migrate long distances throughout the embryo to form melanocytes and vascular smooth muscle cells or to establish the enteric nervous system. Many more essential developmental events, like the formation of the third germ layer in gastrulation, primordial germ cell migration, tracheal branching in insects, posterior lateral line formation in aquatic vertebrates, blood vessel sprouting, neural cell and axon migration, etc., rely on cell migration events. Thus, it is no surprise that disturbances of cellular migration during development are the basis of various developmental disorders, such as skeletal malformations (e.g. cleft palate, limb deformities), congenital diseases (e.g. Hirschsprung disease), brain disorders (e.g. Lissencephaly), neuroblastoma, and in the worst case even embryonic lethality [1–10].

To enable directed cell migration, cells need to interact and communicate with the substrate and surrounding cells and have to be able to respond to various types of environmental signals. They have to be able to transmit these extracellular cues to their cytoskeletal machinery and implement the information into changes of cytoskeletal structures, cell polarisation, and adhesion complexes. However, different cell types can not only adapt different modes of migration; it is important to note that they can also switch between migratory states in a highly dynamic and adaptable manner (reviewed e.g. in [1]). On their way, cells may be guided towards a particular place by soluble, gaseous, or matrix-associated signals, or even mechanical constraints, and sometimes their migration may appear random. In all modes and phases of cell migration, members of the family of RhoGTPases play an important role. Rho protein activation is due to guanine exchange factors (GEFs), which induce the exchange of GDP to GTP, while they are inactivated by GTPase-activating proteins (GAPs), which catalyse the hydrolysis of GTP to GDP [3, 9, 10–13].

This chapter intends to introduce general mechanisms shared between various developmental events, many of which are evolutionarily conserved throughout the animal kingdom.

2 The Journey Starts: Gastrulation, Neural Crest Cell Migration, and Epithelial-to-Mesenchymal Transition

For migration in a developing organism, first of all a signal must be generated to induce migration. In most cases this is due to extracellular cues, e.g. a diffusible chemical signal secreted by other cells; however, recent research supports the notion that in some cases cells are able to initiate migration by intrinsic signalling [1]. The cells must be able to perceive, interpret, and translate the signals intracellularly to set a complex and highly coordinated molecular machinery in motion, driving these cells to move. Furthermore, in case of epithelial origin, when cells are strongly interconnected and have a dedicated apico-basal polarity, migratory cells first need to detach from the surrounding tissue and neighbouring non-motile cells, i.e. cut cell–cell adhesion and change their polarity. This cell state conversion is known as epithelial-to-mesenchymal transition (EMT) [1, 5, 8, 11, 14]. In metazoan development EMT constitutes a crucial and evolutionary conserved process. The transition is accomplished through multiple mechanisms in different morphogenic contexts and is also implicated in processes like cancer metastasis or wound healing (Fig. 1). It is not irreversible, and cells can indeed cycle repeatedly between epithelial and mesenchymal states via EMT and the reverse process, mesenchymal-to-epithelial transition (MET) [14, 15]. The first descriptions of distinct phenotypes of mesenchymal and epithelial cells originate from the late nineteenth century. Although EMT was already recognised in the late nineteenth century and described in detail in 1908 by Frank Lillie, it only gained the interest of developmental biologists

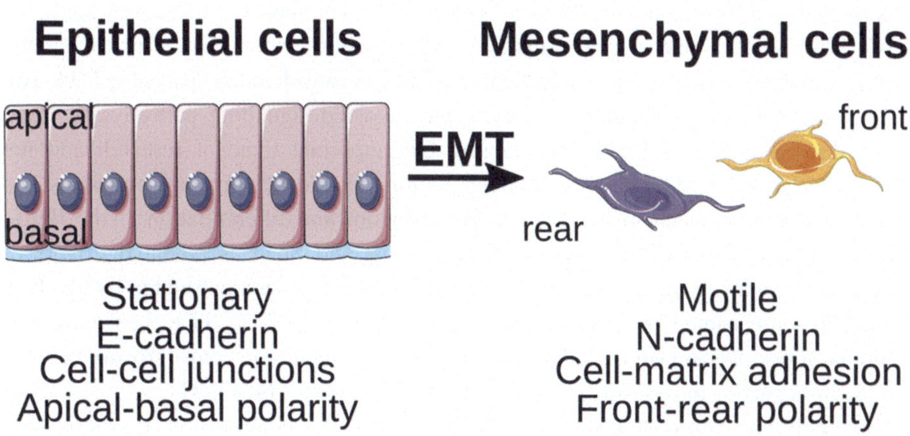

Fig. 1 Epithelial-to-mesenchymal transition (EMT). Stationary cuboid epithelial cells are organised in an epithelial sheet. During EMT cells remodel their polarity, adhesion properties, and migratory capacity to detach from the epithelial tissue and adapt a motile mesenchymal phenotype. Parts of this figure were drawn by using pictures from Servier Medical Art. Servier Medical Art by Servier is licensed under a Creative Commons Attribution 3.0 Unported License (https://creativecommons.org/licenses/by/3.0/)

as a distinct cell process in the second half of the twentieth century. In her pioneering works, Elizabeth Hay laid the basis for studying EMT from descriptions of epithelial to mesenchymal phenotype changes in the primitive streak of chick embryos during gastrulation (reviewed in [15–17]).

Epithelial cells are adherent cells closely attached to neighbouring cells by intercellular junctions in their lateral membranes (adherens and/or gap junctions, desmosomes, cell–extracellular matrix (ECM) interactions via integrins, etc.) forming an epithelial tissue that provides physiologically controlled subdomains within the organism. The cells are typically apico-basally polarised with a basally localised basement membrane, separating the epithelial tissue from other tissues. In contrast to this, mesenchymal cells lack intercellular junctions and are highly motile spindle-shaped cells with a front-rear polarisation (Fig. 1). The transition from an epithelial into a mesenchymal cell requires alterations in cell morphology, adhesion properties, and migration capacity accompanied by a dramatic cytoskeletal remodelling [7, 14, 16, 18]. EMT-inducing signals promote the disruption of the intercellular junctions and loss of apico-basal polarity in epithelial cells. Increased protease activity leads to the simultaneous breakdown of the basement membrane, and rearrangements of the cytoskeleton facilitate the cells to leave the epithelium and initiate migration, a process known as delamination. The former epithelial cells acquire a mesenchymal phenotype with the ability to migrate and invade during this process. At the same time cell adhesion properties change, e.g. marked by an increased expression of N-cadherin and downregulation of E-cadherin, and a mesenchymal front–rear cell polarity replaces the original apico-basal epithelial polarity. However, the exact criteria, i.e. a set of explicit unequivocal markers, to accurately distinguish epithelial from mesenchymal cells remains controversial and can vary between different contexts. Moreover, epithelial cells may only lose some of their traits and may present a combination of epithelial and mesenchymal cell characteristics, in which case EMT is considered as "partial" [7, 14–16].

To understand the molecular regulation and involved signalling pathways of EMT during development is a highly challenging but important topic of research and not yet fully achieved. EMT occurs embedded within other developmental processes and respective signalling mechanisms, e.g. cell fate induction and differentiation. In the 1990s, researchers identified the transcription factors Twist and Snail as potential drivers of EMT for Drosophila gastrulation, and a Snail ortholog was soon identified to be involved in chicken (here named Slug) and mouse gastrulation. However, since then numerous additional transcription factors, miRNAs, cell polarity proteins, receptor tyrosine kinases, and effector proteins have been shown to govern EMT events [14, 15, 17].

One common feature of EMT is a change in cadherin composition. Cadherins are transmembrane proteins whose homophilic interactions create cell–cell adherens with neighbouring cells. On the intracellular side, cadherins are tethered to the actin cytoskeleton by a multiprotein complex including alpha- and beta-catenins. The intercellular adhesion between cadherins proceeds in two steps: In the first step, cadherins on the surface of adjacent cells form homophilic adhesions, a process that happens within 30 s. During the second step, which can take up to 30 min, the initial interaction is strengthened.

This strengthening depends on Rac- and Cdc42-mediated actin polymerisation, anchoring the cell surface cadherins to the cytosol [12, 19, 20]. As mentioned earlier, during EMT E-cadherin levels are reduced, and concomitantly production of N-cadherin is increased. In this context, it is important to note that the force to separate adhesions between E-cadherins is reported to be four times stronger than between N-cadherins [19, 20].

Besides the reorganisation of adhesion properties, cells have to adapt their cytoskeleton into one that promotes a dynamic and motile phenotype during EMT and initiate the migratory cycle described in Sect. 3 of this chapter. Indeed, RhoGTPase plays a pivotal role in regulating cytoskeletal dynamics and rearrangement in this process. Rac and Cdc42 promote the formation of lamellipodia and filopodia by activating actin polymerisation at the leading edge, and in combination with Par complex proteins and PI3 kinase, RhoGTPases participate in the switch of cell polarity from apical-basal to front-rear. Rho meanwhile localises at the rear of the cell and promotes cell retraction and disassembly of adhesion, as well as promoting actin stress fibre formation. Also, reorganisation of the microtubule cytoskeleton is controlled by RhoGTPase, for example by Rac promoting microtubule polymerisation at the leading edge [15, 17].

The earliest developmental event of EMT in embryogenesis is the formation of the third germ layer, the mesoderm, during gastrulation. It is an evolutionarily conserved developmental process common to all metazoans, which establishes the three primary germ layers: ectoderm, mesoderm, and endoderm. Epithelial cells of the epiblast undergo EMT and form the mesoderm, either through invagination and collective cell migration (e.g. in *Xenopus*) or individual cell migration by ingression (e.g. in chicken embryos). The site of gastrulation is restricted by morphogenic gradients (such as those formed by members of the FGF family); thus, only a defined subpopulation of cells passes through EMT [7, 15, 16, 21]. In chick and mouse embryos gastrulation EMT is driven by canonical Wnt signalling and downstream of these proteins of the TGF-beta superfamily, while signalling through FGF receptors maintains the EMT regulatory network and mesenchymal cell migration. In response to these initial signals, cells activate transcription factors, e.g. members of the Snail family, to induce intracellular EMT signalling re-arranging cell adhesion properties (downregulation of E-cadherin for example) and cytoskeletal dynamics. Snail also represses expression of genes encoding tight junction components like claudins and occludins and genes important for apico-basal polarity like Discs large (Dlg) and Crumbs homolog 3 (Crb3). Since some proteins that need to be downregulated during EMT have a longer half-life than the time gastrulation takes (like E-cadherin), posttranscriptional regulatory mechanisms to promote rapid protein degradation and translocation have to be involved. For instance, beta-catenin tethering E-cadherin to the actin cytoskeleton is translocated from the cell membrane to the cytoplasm and either degraded or directed to the nucleus [7, 16, 22].

Another prominent developmental EMT event is the (evolutionarily newer) delamination of neural crest cells from the neural tube in vertebrates. Following gastrulation, premigratory neural crest cells form in the dorsal folds of the neural tube, situated at the border of the neural plate and non-neural ectoderm. Refined signals promote the expression of genes such as members of the Sox and Snail families and Foxd3 in these neural crest cells. This results in the cells delaminating from the neural folds and adapting a mesenchymal phenotype, i.e. to undergo EMT. Neural crest cells migrate in defined routes through the embryo before reaching their target sites where they differentiate into peripheral neurons and glia, melanocytes, cranial bones and muscles, odontoblasts, and many more [7, 16, 18]. The signalling pathways regulating EMT in neural crest cell delamination are similar to those acting to promote gastrulation. Again, a combination of Wnt, FGF, and BMP (bone morphogenetic protein, a member of the TGF-beta superfamily) is crucial for neural crest cell EMT induction. BMPs also seem important for the regulation of timing of neural crest cell migration, since the presence of Noggin, an inhibitor of BMP activity, prevents delamination until such time as Noggin is downregulated. Analogous to the situation during gastrulation, the breakdown of the basement membrane surrounding the neural tube is necessary to complete EMT of neural crest cells. This is achieved for instance by matrix metalloproteinases (MMPs) that degrade various extracellular matrix molecules, including collagens, laminin, and gelatins. MMPs can also target extracellular domains of E-cadherin and thus help to disrupt adherens junctions [7, 16].

3 Individual Cell Migration in Development

During organogenesis of the central nervous system (CNS), as an example, neurons often have to migrate away from their site of origin to the position they will finally function. Vertebrate cortical neurons are born in the ventricular zone of the telencephalon and must migrate towards the pial surface of the brain to build up the adult cerebral cortex. These neuronal precursors use radial glia cells as a scaffold to facilitate their migration. Other neurons migrate collectively in a process known as chain migration (e.g. olfactory interneuron precursors along the rostral migratory stream; see also Sect. 4). A third group of neurons migrates individually without close contact or formation of cell–cell adhesion to other cells, for instance populations of neural crest cells, but also subpopulations of neurons in the embryonic cortex [5, 23].

In general, individual cell migration, not only of neuronal cells but virtually all mesenchymal migrating cells, can be seen as a repeating cycle, and the first published proposal to view cell migration as a cyclic process originates from Michael Abercrombie in the 1970s (reviewed in [24]). To initiate migration, a cell needs first to polarise with a front and rear, then further extend protrusions in the direction of migration, and stabilise these by adhering to the substrate. This is followed by moving the cell body in the direction of established protrusions and detaching (i.e. disassemble cell adhesion) and retracting the cell rear before starting the cycle all over again (see Fig. 2) [13, 24].

On an organismal level, cell migration is generally initiated and perpetuated by extracellular signalling (e.g. growth factors, chemokines), with each individual cell translating these signals into intracellular regulation of the cytoskeleton, adhesion structures,

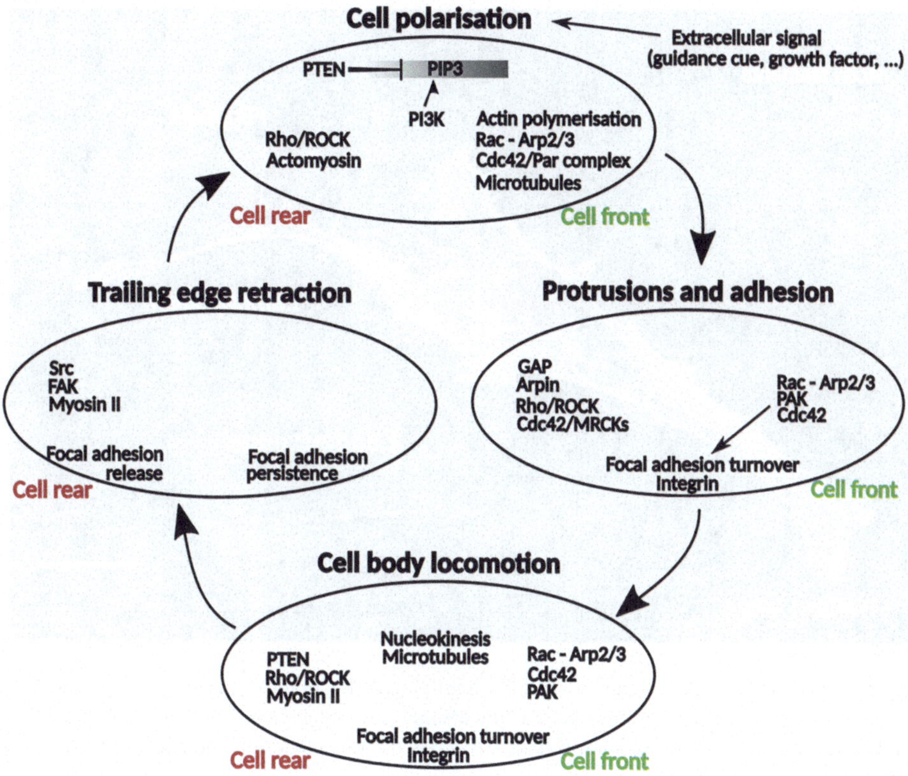

Fig. 2 Cell migration as a cyclic process. Cell migration can be subdivided into four steps which are constantly repeated. At first, the cell polarises, then extends protrusions in the direction of travel, and changes cell–substrate adhesion. In the third step, the cell body is moved forward, and in the final step, the cell rear is detached and retracted. See text for more details on the indicated signalling molecules

and plasma membrane turnover. Initiation of locomotion fundamentally depends on an asymmetric cell morphology with a defined leading and trailing edge. At the cell front, the leading edge, cells extend a sheet-like structure named lamellipodium and thin finger-like filopodia (Fig. 3). Both structures are highly dynamic and are pushed outwards from the cell by the assembly of actin filaments. Within the protrusions the organisation of actin filaments differs, while filaments form a branching, intertwined network in lamellipodia, actin filaments are organised in long, parallel running bundles in filopodia. Cellular polarisation depends strongly on intracellular differential activity of members of the family of small RhoGTPases, foremost Rac, RhoA, and Cdc42, which can regulate cytoskeletal dynamics, cell adhesion, and protrusion formation, but also Par proteins and atypical protein kinase (aPKC) [8, 12, 13, 24–26].

Cdc42 (cell division control protein 42) at the front of the cell acts through the Par polarity complex and microtubules to establish directionality and migratory persistence.

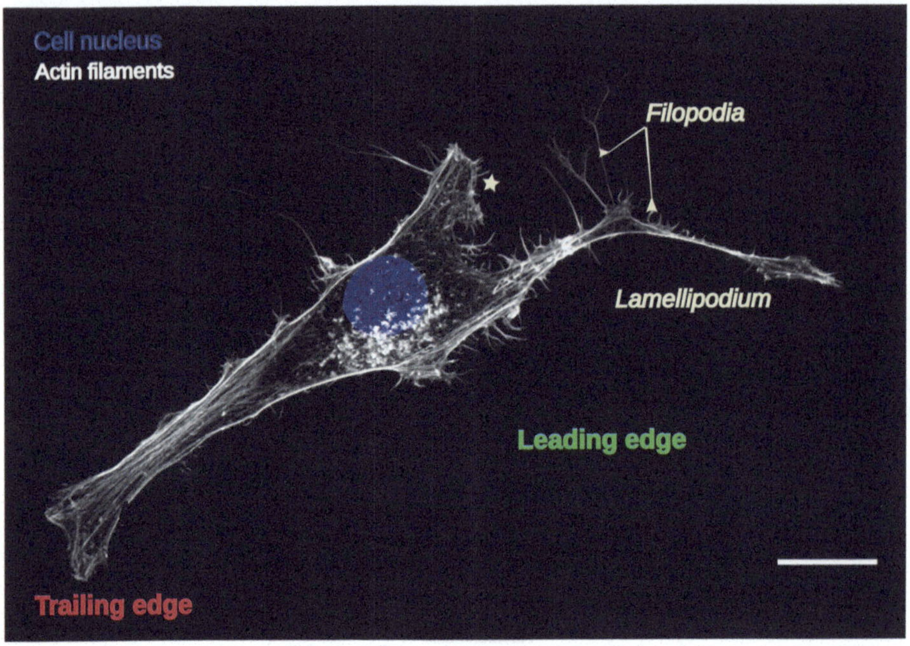

Fig. 3 Representative microscopic image of a migrating cell. Confocal image of a migrating U87 glioblastoma cell in 2D-cell culture. Actin filaments (grey) were labelled with fluorescent phalloidin and DNA in the cell nucleus with DAPI (blue). Asterisk marks a retracting lamellipodium. Scale bar represents 20 μm

In addition to RhoGTPAses, a gradient of phosphatidylinositol triphosphate (PIP3), decreasing from front to rear, contributes to cell polarisation. This gradient is produced by the action of phosphoinositide-3 kinase (PI3K) and regulated by PTEN, which is located mainly to the trailing edge where it lowers the level of PIP3. In cell regions away from the leading edge, lamellipodia formation is inhibited by feedback loops involving Rho/ROCK, the WAVE complex activating Arp2/3, and actomyosin contractility (Fig. 2). For example, RhoC activity enables lamellipodium extension at the leading edge in response to epidermal growth factor (EGF), by turning off cofilin activity in cell regions away from the leading edge. Active cofilin generates free actin filament ends that become accessible to G-actin and initiate Arp2/3-driven actin polymerisation in lamellipodia and thus would support lamellipodium protrusion [3, 8, 27].

Once a cell is polarised with a leading and trailing edge, intracellular signalling orients protrusions to the leading edge and thus directional migration can start [5, 25]. While actin filaments push protrusions of the cell membrane forward and establish lamellipodia and filopodia, lamellipodia are stabilised by microtubules and the formation of focal adhesions. Filamentous actin is already intrinsically polarised with a fast-growing "barbed" end and slow-growing "pointed" ends. This inherent polarity is also used by the cell to drive membrane protrusions. Actin filament disassembly of older filaments to generate free actin monomers for the barbed ends is assisted and regulated by a number of proteins, primarily profilin, ADF/cofilin, and RhoGTPases [12, 13]. A positive feedback loop with Cdc42 and integrins selectively at the leading edge stimulates Rac activation and integrin-mediated adhesion. Rac interacts here for instance with VASP and the WAVE complex to activate Arp2/3-mediated actin polymerisation. A large number of actin-binding proteins regulate leading edge actin dynamics, facilitating different stages of assembly, disassembly, sequestering, and crosslinking. Unravelling this complicated network constitutes an exciting field of ongoing research. Negative feedback loops that include the Arp2/3 inhibitor Arpin and GAPs for Rac restrict Rac activity outside the leading edge. Also, concentrated further to the cell rear, RhoC acts to downregulate cofilin activity via LIM kinase, thus decreasing actin polymerisation, and via ROCK stimulating actomyosin contractility [3, 8, 27].

Once a leading edge protrusion is established, it needs to stabilise and adhere to the substrate, allowing the cell to push itself forward. This is enabled primarily by the interplay between RhoGTPases (Rac/Cdc42) and PAK family kinases promoting integrin-based adhesion turnover. To allow meaningful cell migration, it needs a fine balance between cytoskeletal dynamics and cellular adhesion. In parallel with actin filament polymerisation at the leading edge, actomyosin contraction in the cell body and the rear of the cell is important to drive the cell body forward and the detachment of the trailing edge. This is mainly regulated by Rho and ROCK, but also Cdc42 here via its downstream serine/threonine protein kinase effectors MRCK [3, 8].

Formation of cellular protrusions also requires rapid insertion of new cellular components at the leading edge as well as turnover of cytosol. This allows the formation of new plasma membrane components and insertion of new cell adhesion points. Moving the whole cell body in the direction of the leading edge also involves intensive cytoplasmic and cell organelle rearrangement, processes which are not yet fully uncovered. Live cell imaging has revealed that the centrosome and Golgi apparatus reorient to the front of the nucleus and move towards the leading edge, accompanied by other cell organelles. Forces generated in the leading edge of the cell are transmitted to the centrosome, pulling it forward. As the centrosome is constantly linked with the nucleus via a microtubule network, the nucleus is thereby pulled towards the centrosome by dyneins associated with this microtubule cage. At the same time, actomyosin contraction at the rear of the cell pushes the nucleus forward. This process is also known as "nucleokinesis". In the cell rear, PTEN activity is increased and correlates with the activity of myosin II, both promoting trailing edge contractility and retraction [8, 24, 28].

To finally enable the whole cell to move forward, the trailing edge must detach and retract. While focal adhesions to the substratum maintain at the front of the cell, they are released at the cell rear. This is regulated by FAK (focal adhesion kinase) and Src signalling and accompanied by myosin II contractility. Release of focal adhesion at the cell rear in turn contributes to the maintenance of cell polarity and promotes leading edge activity, providing a positive feedback loop for the continued migration cycle [8].

It should be mentioned, though due to lack of space not further discussed, that lamellipodia are not essential for cell migration; in fact a number of cell types (e.g. melanoblasts and fibroblasts) predominantly use filopodia to promote migration, and cells that lack integrins or have upregulated Rho/ROCK activity tend to migrate using bleb-based motility (a form of motility seen for instance in amoeboid migration) (see [1] or [3] for more information). However, cells adapting mesenchymal morphology are responsible for most migratory events during development [8].

In the developing nervous system, neurite extension provides a special case of cell migration. Neurites protrude through tissues to reach their destinations and form connections with their effectors, leaving their cell body, i.e. the cell rear, behind. At the tip of each growing neurite is a special structure, the growth cone, first described by Ramón y Cajal in the 1890s. It can be subdivided into two main regions, known as the central and the peripheral zones. The central zone contains bundled microtubules, providing structural support, while the peripheral zone is characterised by thin, sheet-like protrusions, lamellipodia, and long, finger-like processes, filopodia. Indeed, the growth cone can be considered as a leading edge, which follows the same mechanics and is regulated by the same molecules as any mesenchymal cell migrating through tissue. And just like in migrating cells, motility depends on a highly dynamic actin cytoskeleton regulated by RhoGTPases [8, 23, 29]. Furthermore, the neuronal growth cone can not only be regarded as a special case of cell migration; numerous hallmarks for molecular guidance and regulation of migration, e.g. the role of cellular adhesion molecules, or of the second messengers cAMP and cGMP, have been discovered by studying growth cone migration [30–32].

4 Collective Cell Migration in Development

Collective cell migration can be defined as the ability of a group of cells to move together, with cells closely affecting each other's behaviour, e.g. through stable or transient cell–cell interactions or intercellular signal transduction. Group sizes can be remarkably variable; for example, only ten border cells migrate collectively during *Drosophila* egg chamber development, but thousands of cells during gastrulation. Cells can even migrate as epithelial sheets, after passing through a partial EMT which leaves adherens and tight junctions and apico-basal polarity mainly intact. These remaining epithelial interactions restrict the motility and velocity of the sheet; however, it is known to occur and cause large-scale tissue deformations during development (e.g. during gastrulation in *Xenopus* or fish) or wound healing [2, 6, 33].

During development, collective cell migration is crucial for gastrulation and neural crest cell migration, vascular sprouting, pronephros development, development of the lateral line in aquatic vertebrates, or tracheal development and border cell migration in *Drosophila*, to name just a few. Though collective migration is deployed during such numerous developmental events with a variety of strategies to effectively achieve it, the core mechanisms required for migration as a group are conserved throughout evolution. Intracellularly, collective cell migration employs the same molecular mechanisms as single-cell migration. However, additional mechanisms are required to enable close collaboration and cooperation between migrating cells. Furthermore, while individually migrating cells can squeeze through the ECM (switching to an amoeboid type of migration), the cell collective has to modify the surrounding ECM if necessary [2, 5, 6, 8].

To some extent the migration of a collective of cells can be indeed compared to the migration of an isolated cell. The collective as a whole is polarised, with the leading edge cell(s) of the migrating cohort forming protrusions to explore the extracellular environment and exerting traction to move the bulk of the cohort forward, equivalent to the leading edge of a single migrating cell (Fig. 4). Front and back of the cluster will have to be linked to achieve successful forward movement of the whole cohort and displacement of the rear of the collective. The trailing edge of single cells is characterised by actomyosin contractility, and a similar organisation can be observed in collectively migrating *Xenopus* neural crest cells. While the cells at the rear of the collective are connected through N-cadherin junctions, a supracellular actomyosin cable forms and spans the entire width of the cell cluster. This allows all rear cells to contract simultaneously [1, 2, 6]. The behaviour of cells within the collective is often orchestrated by a common signal, which requires each individual cell to have access to the same guidance information, as seen in collective migration of neural crest cells or during zebrafish lateral line development. Position of the leader cell is only transient, and cells rapidly exchange positions within the group. Alternatively, the group may be divided more stringently into leader and follower cells, with the leader cells reading and relaying the information to the follower cells through chemical or mechanical signalling [2, 6].

As mentioned earlier, an important difference between collective migration and migration as a single cell are cell–cell interactions. It has been demonstrated that cell–cell adhesions are crucial for the maintenance of mechanical integrity and polarisation within the collective in numerous developmental events [2]. For example, Theveneau et al. [34] showed that N-cadherin-mediated cell–cell adhesion enhances polarisation of collectively migrating *Xenopus* neural crest cells. The cells are attracted towards the chemokine Sdf1 (stromal derived factor 1). Cells within the migratory group exhibit N-cadherin-mediated cell–cell adhesions. If additionally, Sdf1 is present at such a cell–cell contact, cell protrusions and Rac1 activity are inhibited. Only at Sdf1 free N-cadherin cell–cell adhesions are Rac1 activity and cell protrusions promoted; this kind of polarisation and promotion of migration is also known as contact-inhibited locomotion [34]. Apart from mechanical coupling, e.g. through cadherin-dependent cell–cell interactions, cells within a group of collectively migrating cells might be actively attracted to each other or the leading cells, as is the case in the zebrafish lateral line primordium. This cell group is also guided by the chemokine Sdf1, but only leading cells seem to process the signal. The leading cells in the front of the collective secrete FGFs, whereas trailing cells express FGF receptors which in turn enable them to follow the Sdf1-guided cells [6, 35].

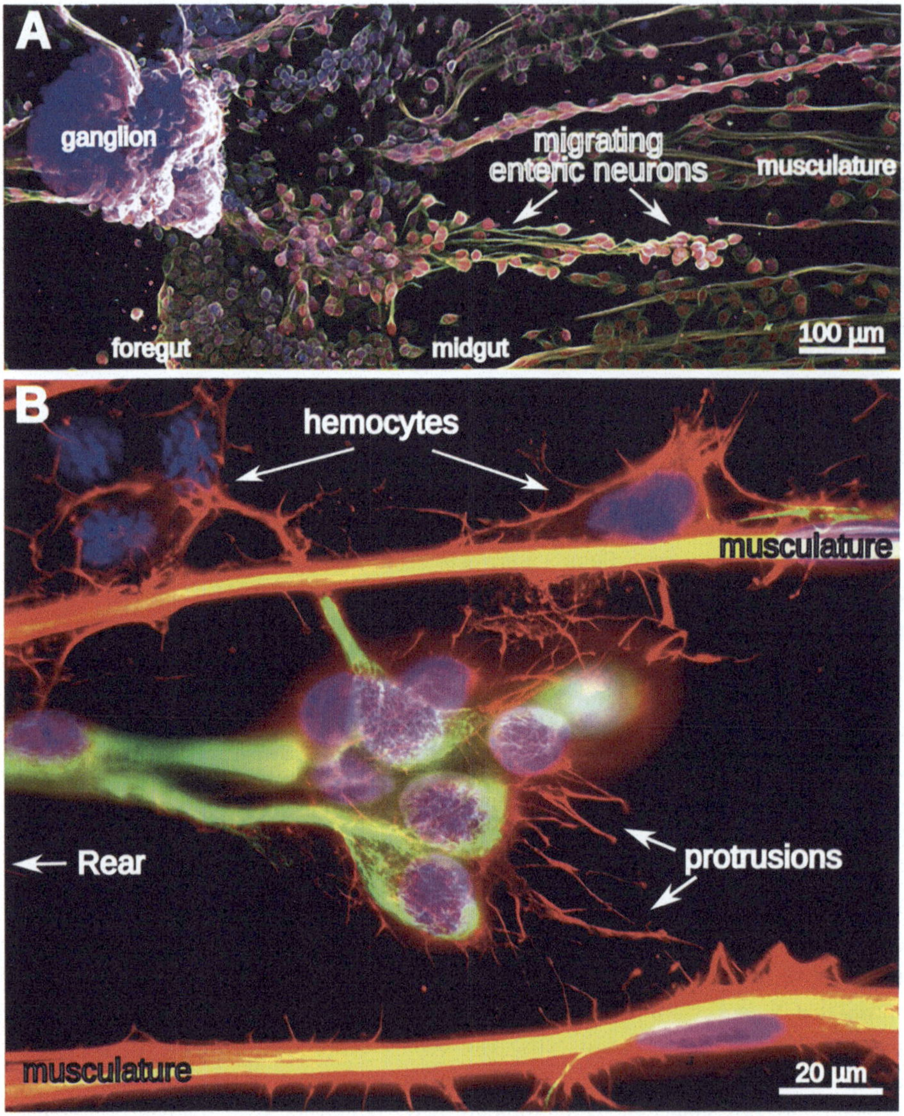

Fig. 4 An example for polarisation of the migrating collective of cells—Development of the enteric nervous system. (**a**) Enteric neurons invade the developing locust embryo midgut as four evenly spaced migratory chains from the anterior foregut–midgut boundary towards the posterior midgut–hindgut boundary. One of these four migratory paths is depicted here. The leading cells leave neurites behind, which are used by follower cells as a guiding scaffold. Maximum projection of a confocal z-stack. Green: actin-immunofluorescence (IF) (AF488), red: alpha-tubulin-IF (Cy3), blue: DAPI. (**b**) High magnification image of the leading cells of the migratory collective. Cells on the leading edge scan the environment with lamellipodia and filopodia, while protrusions are strongly reduced in the follower cells. Epifluorescence image. Green: alpha-tubulin-IF (AF488), red: phalloidin/actin filaments, blue: DAPI

It is suggested that collective migration in general may be more efficient than individual cell migration, since a group of cells is more proficient in finding and interpreting guidance signals for example. A collective cell spans a larger area than a single cell and could therefore detect shallower gradients of guidance molecules. Furthermore, it needs only a few leading cells at the tip of the collective to perceive guidance cues and relay the information to the bulk of the group, whereas individual cells must constantly sample the environment to detect cues, adapting a two-phased migratory behaviour coined "run and tumble". This is seen for example in individually migrating primordial germ cells. These cells frequently exhibit so-called tumbling phases during which they stop migration and explore their close environment. This phase seems to occur cell-autonomously and independently of chemokine signalling, and it might serve to redirect the cells. In collectively migrating *Xenopus* neural crest cells on the other hand, tumbling cells are pulled forward by their neighbouring cells, retaining directional movement at the same speed; even dividing (usually non-motile) cells are pulled forward within the collective of migrating cells [5, 34]. Another advantage for collective migration may be deduced from the behaviour of ants: When confronted with a gap too wide for a single ant to cross, they form a bridge with one another, allowing them to cross the gap as a collective (watch for example this video from National Geographic: https://www.youtube.com/watch?v=4BdjxYUdJS8). Similarly, olfactory neuronal progenitors in the rostral migratory stream in rodents or enteric neuron precursors in insect embryos form migratory chains and slide over each other, using other cells for guidance and support on their journey (Fig. 3) [23, 36, 37].

5 Finding the Way: Cell Guidance During Development

In a developing organism, migrating cells need to find the right way and the appropriate speed to arrive at the right time at their correct destinations. This is facilitated by the capability of migrating cells to act on permissive, attractive, or repellent extracellular guidance cues. The migratory velocity is regulated intracellularly, e.g. inhibition of actin-binding proteins of the Ena/VASP family slows down lamellipodia dynamics and increases cell speed, or extracellularly by cell adhesiveness or impedance of the substrate, e.g. adhesion to the underlying substrate stabilises lamellipodial protrusions, thus promoting migratory speed [5, 8, 11, 25, 33].

A uniformly presented permissive or promigratory signal triggers the basic motility machinery of a cell, resulting in random motility. Though it was originally thought this process always needed an extracellular signal to start, it turned out that embryonic cells are also able to activate and maintain random motility spontaneously and through intracellular self-organising signalling. Howsoever initiated, such a random motility seems to be a prerequisite for initiation of directional migration. For individually migrating cells, random motility allows explorative migration and colonising new territories within the embryo. During zebrafish gastrulation, endodermal cells initially disperse in a "random walk", later switching to a persistently directed migration when cells converge towards the midline in zones with chemotactic cues present. If the migration initiating signal is presented in a concentration gradient, it can even promote directed cell migration along this gradient [1, 5, 24, 25].

Chemotaxis is probably the best-known guidance mechanism for cell migration. It is mediated by the generation of diffusible cues that either attract or repel the migrating cells. If the signal forms a gradient, directed cell migration occurs. Chemoattractants or -repellents comprise a large and diverse group of molecules, including phospholipid metabolites, chemokines, growth factors, or secreted proteins such as Slits, netrins, semaphorins, and ephrins. The second messengers cAMP and cGMP, ATP, and intracellular calcium have been reported to amplify chemotactic signals. Zebrafish posterior lateral line primordium, *Drosophila* border cells, *Xenopus* cephalic neural crest cells, and the life cycle of *Dictyostelium discoideum* are among the most vigorously studied and established in vivo models for chemotaxis. In all these examples, intrinsic random cell motility precedes exposure to a gradient. The subsequent exposure to a chemoattractant gradient activates additionally a receptor-mediated intracellular signalling that shifts the intrinsic random cell motility towards the gradient by directing front–rear cell polarity, protrusion formation, and cell locomotion towards the gradient. We have learned already in Sect. 4 of this chapter how N-cadherin mediated cell–cell adhesion in the presence of Sdf1 influences cell polarisation via contact inhibited locomotion (CIL) and enables chemotactic migration of *Xenopus* neural crest cells. As the cells migrate from the dorsal neural tube to lateral and ventral regions of the embryo, they also follow the chemoattractant Cxcl12 (C-X-C motif chemokine ligand 12) which is produced by epithelial epibranchial placodal precursors. The pre-placodal cells are themselves still motile, and both cell populations coordinate their migration by combining chemotaxis with CIL. Polarised neural crest cells with activated Rac at their leading edge migrate towards the pre-placodal cells following ("chase") the chemoattractant source. Upon cell–cell contact CIL induces cell repolarisation through Rac inhibition and pre-placodal cells move away ("run") from the neural crest cells. This "chase and run" cycle is repeated until such time as the cells reach their destination [11, 24, 34, 38].

In a wide variety of developmental migration events, cells are oriented through contact-mediated guidance along a scaffold. A key example is the migration of neuronal precursors along pre-existing neuronal or glial scaffolds during the development of the nervous system and in the adult brain. Radially migrating neuronal precursors in the cerebral cortex and in the cerebellum wrap their leading processes around radial glia cell, forming a close connection that depends on a cell surface glycoprotein known as astrotactin expressed by the neurons. When the functionality of astrotactin is blocked, e.g. by antibodies masking astrotactin binding sites, or genetic manipulation, neuronal precursor cells are unable to attach to the radial glia cells, and their rate of migration is significantly impaired. In another example, collectively migration olfactory GnRH (gonadotropin-releasing hormone) neurons migrate from the nasal placode to the olfactory bulb following vomeronasal axons. This process is mediated by cell–cell interactions involving integrin-beta1 (expressed in the cell membranes of GnRH neurons) and the GPI-linked semaphorin 7A (expressed on the axon scaffold). Interestingly, the contact with the vomeronasal axons provides only orientation for the GnRH neurons; directional migration depends on a chemotactic gradient of semaphorin 4D. It is important to note that contact-guided cell migration can be regulated by changes in adhesion efficiency and/or changing the intrinsic ability of cells to interpret guidance mechanism on their migratory path. For example, GnRH neurons transform their early directional migration along vomeronasal axons in the final stages by decreasing cell–cell adhesion due to a switch from integrin-beta1 to plexin-mediated adhesion, allowing them to spread away from the axonal scaffold ([23, 24] and citations therein).

Interactions between cells and their migratory surface (be it ECM or other cells) occur most commonly through integrins, which are able to transduce both mechanical and chemical signals from extracellular to intracellular signalling. Interactions of integrins with the substrate can lead on the intracellular side to the activation of Cdc42 and/or Rac, involving GEFs, and intracellular kinases like FAK and Src. This in turn activates signalling pathways (e.g. via downstream effectors of the Scar/WAVE complex) promoting the extension of membrane protrusions and polymerisation of actin filaments at the leading edge. In short, the intracellular machinery we already know from Sect. 3 of this chapter is set in motion. But migrating cells can also change the extracellular matrix composition by the secretion of matrix metallo-proteases (MMPs). For example, FGF-stimulated leading cells in *Drosophila* tracheal development secrete MMP2, and thus remodel ECM fibres to facilitate the movement of the collective. In collective cell migration, follower cells can also be essential for the polarisation of the entire cell group (e.g. through CIL), indirectly influencing polarisation of the leader cells, and thus control their role for pathfinding [2, 33].

Contact inhibition of cell locomotion (CIL) seems to be a driving force for cell dispersion during development. But CIL can also enable cells to switch between collective and individual migration, depending on the balance of attractive and repulsive interactions. The basic principle for CIL is that cell–cell contacts inhibit protrusion formation at the site of collision, thus promoting cell migration in the opposite direction. This repels cells constantly away from the point of highest cell density; therefore, their migration is biased towards unpopulated regions. For example, *Drosophila* hemocytes migrate from the head mesoderm to the ventral midline following a source of PVF1 (PDGF- and VEGF-related factor 1). From this position they migrate laterally to form a linear "three-lined" pattern as a result of CIL. As the hemocytes contact each other, a polarised microtubule bundle, stabilising the leading edge of the cells during migration, collapses, thus enabling cell repulsion, turning, and dispersion. During collective cell migration, CIL rather helps to organise the cell group, than disperse the cells. As CIL inhibits protrusion formation at the site of collision, within a group of migrating cells only those with at least one free side can acquire a mesenchymal character. This limits polarisation to cells on the edges of a collective. For example, collectively migrating neural crest cells express the chemoattractant C3a and the corresponding receptor C3aR on their cell surface, providing attraction to each other, and thus providing cohesion for the group. However, this mutual attraction is balanced out by CIL. If neural crest cells contact each other within the cluster, N-cadherin-mediated cell–cell interactions trigger a repulsive response by activating the planar cell polarity (PCP) pathway and RhoA at the contact site, which leads to cell repolarisation and enhanced cell dispersal. This results in cells within the collective constantly colliding (due to chemoattraction), dispersing (due to CIL), and reassembling of the group, with formation of strongly front-to-back polarised cells at the free edges of the collective [1, 24, 34, 39].

However, for many migration events in development, but also disease, the exact guidance cues, how they are generated, and how migrating cells integrate these cues remain unknown. Future research will hopefully answer many of these long-standing questions.

6 How to Stop Cell Migration

So far, we have seen how migration can be initiated, how changes in the cytoskeleton drive cell polarisation and locomotion, and how migrating cells are guided to their destination. But how do cells know that they have reached their correct destination? How do they terminate migration at the correct position at the right time?

In some cases, termination of migration might just rely on physical impedance based on tissue morphology, or cells find themselves surrounded by repulsive cues. Cells may end migration as they reach regions with high concentrations of a chemoattractant or migrating cells could trigger a downregulation of the chemotactic signal upon reaching their destination. Alternatively, signalling interactions at the destinations may disable cells to respond to guidance cues altogether [5]. For example, primordial germ cells in zebrafish stop their migration upon arrival in the developing gonads where their chemoattractant Sdf1a is highly expressed. Live imaging showed that protrusions are still formed as the cells cease their migration; however, these protrusions are distributed evenly around the (unpolarised) cells, indicating a uniform stimulation of the Sdf1a receptor, and thus further directed migration is prevented [5, 40]. In locust embryos migration of enteric neurons along the developing midgut (see Fig. 4) is driven by extracellular nitric oxide (NO). The enteric neurons are able to respond to this signal due to high intracellular levels of NO-induced cGMP, which activates a signalling cascade towards the cytoskeleton. After reaching their final destinations on the midgut, enteric neurons lose these high cGMP levels and thus are no longer able to respond to the migration permitting nitric oxide signal. However, the cause of the decreasing cGMP levels or if other factors at their destination are involved in the termination of migration is not yet known [36, 41, 42]. Olfactory interneurons migrate through the rostral migratory stream and finish their journey in the olfactory bulb core. Upon arrival at their destination, the cells extend a lateral protrusion from the proximal part of the leading process. The formation of this lateral protrusion is induced by the downregulation of PlexinD1 and local Rac activation. It coincides with microtubule reorganisation within the lateral protrusion, which suppresses further cell body locomotion. The cause for PlexinD1 downregulation is not known; however, it is possible that the timing is intrinsically programmed in each individual cell [43].

The precise mechanisms for the termination of migration are studied to a much lesser extent than other aspects of cellular migration during development [5]. However, termination of migration is an essential part of morphogenesis and uncovering the involved signals and mechanisms is not only important for our understanding of development but could also be of profound interest for clinical applications, e.g. to control cancer cell migration.

Take-Home Message

Numerous complex and versatile modes of cell migration are employed during development of a multicellular organism. However, basic mechanisms and regulating signalling pathways are repeatedly at work in different events and evolutionarily conserved throughout the animal kingdom.

For example, in virtually all migration events proteins of the family of small RhoGTPase are key intracellular regulators of migration. In a simplistic view, Rac is the key regulator of actin polymerisation at the front of the cell, where it inhibits Rho activity. Rho, on the other hand, is the central mediator of actomyosin contractility at the rear of the cell where it in turn blocks Rac activity [1].

Initiation of cell migration, guidance of cells along their route, and termination of migration need to be tightly regulated to enable cells to reach the right position at the right time for a successful migration.

Scientists have studied how these processes work on molecular and mechanistic levels for more than a century, but many riddles remain. Future research in this area will provide exciting insights of great importance not only for developmental cell migration but also for cell migration in disease and regeneration.

References

1. Stock J, Pauli A. Self-organized cell migration across scales—from single cell movement to tissue formation. Development. 2021;148
2. Scarpa E, Mayor R. Collective cell migration in development. J Cell Biol. 2016;212:143–55.
3. Ridley AJ. Rho GTPase signalling in cell migration. Curr Opin Cell Biol. 2015;36:103–12.
4. Bronner ME, LeDouarin NM. Development and evolution of the neural crest: An overview. Dev Biol. 2012;366:2–9.
5. Aman A, Piotrowski T. Cell migration during morphogenesis. Dev Biol. 2010;341:20–33.
6. Weijer CJ. Collective cell migration in development. J Cell Sci. 2009;122:3215–23.
7. Acloque H, Adams MS, Fishwick K, Bronner-Fraser M, Nieto MA. Epithelial-mesenchymal transitions: the importance of changing cell state in development and disease. J Clin Invest. 2009;119:1438–49.

8. Kurosaka S, Kashina A. Cell biology of embryonic migration. Birth Defects Res Part C Embryo Today Rev. 2008;84:102–22.
9. Wolpert L, Jessel T, Lawrence P, Meyerowitz E, Robertson E, Smith J. Principles of development. Oxford: Oxford University Press; 2007.
10. Gilbert SF. Developmental biology. Sunderland, MA: Sinauer Associates; 2000.
11. SenGupta S, Parent CA, Bear JE. The principles of directed cell migration. Nat Rev Mol Cell Biol. 2021;22:529–47.
12. Pollard TD, Earnshaw WC, Lippincot-Schwartz J. Cell biology. Philadelphia, PA: Elsevier; 2004.
13. Ridley AJ, Schwartz MA, Burridge K, Firtel RA, Ginsberg MH, Borisy G, Parsons JT, Horwitz AR. Cell migration: integrating signals from front to back. Science. 2003;302:1704–9.
14. Lee JM, Dedhar S, Kalluri R, Thompson EW. The epithelial–mesenchymal transition: new insights in signaling, development, and disease. J Cell Biol. 2006;172:973–81.
15. Lim J, Thiery JP. Epithelial-mesenchymal transitions: insights from development. Development. 2012;139:3471–86.
16. Kim D, Xing T, Yang Z, Dudek R, Lu Q, Chen Y-H. Epithelial mesenchymal transition in embryonic development, tissue repair and cancer: a comprehensive overview. J Clin Med. 2017;7:1.
17. Lamouille S, Xu J, Derynck R. Molecular mechanisms of epithelial–mesenchymal transition. Nat Rev Mol Cell Biol. 2014;15:178–96.
18. Radisky DC. Epithelial-mesenchymal transition. J Cell Sci. 2005;118:4325–6.
19. Chu Y-S, Eder O, Thomas WA, Simcha I, Pincet F, Ben-Ze'ev A, Perez E, Thiery JP, Dufour S. Prototypical type I E-cadherin and type II cadherin-7 mediate very distinct adhesiveness through their extracellular domains. J Biol Chem. 2006;281:2901–10.
20. Chu Y-S, Thomas WA, Eder O, Pincet F, Perez E, Thiery JP, Dufour S. Force measurements in E-cadherin-mediated cell doublets reveal rapid adhesion strengthened by actin cytoskeleton remodeling through Rac and Cdc42. J Cell Biol. 2004;167:1183–94.
21. Yang J, Weinberg RA. Epithelial-mesenchymal transition: at the crossroads of development and tumor metastasis. Dev Cell. 2008;14:818–29.
22. Kalluri R, Weinberg RA. The basics of epithelial-mesenchymal transition. J Clin Invest. 2009;119:1420–8.
23. Price DJ, Jarman AP, Mason JO, Kind PC. Building brains: an introduction to neural development. Chichester, West Sussex, UK: Wiley; 2011.
24. Reig G, Pulgar E, Concha ML. Cell migration: from tissue culture to embryos. Development. 2014;141:1999–2013.
25. Petrie RJ, Doyle AD, Yamada KM. Random versus directionally persistent cell migration. Nat Rev Mol Cell Biol. 2009;10:538–49.
26. Etienne-Manneville S, Hall A. Cell polarity: Par6, aPKC and cytoskeletal crosstalk. Curr Opin Cell Biol. 2003;15:67–72.
27. Bravo-Cordero JJ, Magalhaes MAO, Eddy RJ, Hodgson L, Condeelis J. Functions of cofilin in cell locomotion and invasion. Nat Rev Mol Cell Biol. 2013;14:405–15.
28. Marin O, Valiente M, Ge X, Tsai L-H. Guiding neuronal cell migrations. Cold Spring Harb Perspect Biol. 2010;2:a001834.
29. Gomez TM, Letourneau PC. Actin dynamics in growth cone motility and navigation. J Neurochem. 2013;129:221–34.
30. Song H-j, Poo M-m. Signal transduction underlying growth cone guidance by diffusible factors. Curr Opin Neurobiol. 1999;9:355–63.
31. Tessier-Lavigne M, Goodman CS. The molecular biology of axon guidance. Science. 1996;274:1123–33.

32. Harrelson AL, Goodman CS. Growth cone guidance in insects: fasciclin II is a member of the immunoglobulin superfamily. Science. 1988;242:700–8.
33. Mayor R, Etienne-Manneville S. The front and rear of collective cell migration. Nat Rev Mol Cell Biol. 2016;17:97–109.
34. Theveneau E, Marchant L, Kuriyama S, Gull M, Moepps B, Parsons M, Mayor R. Collective chemotaxis requires contact-dependent cell polarity. Dev Cell. 2010;19:39–53.
35. Nogare DD, Somers K, Rao S, Matsuda M, Reichman-Fried M, Raz E, Chitnis AB. Leading and trailing cells cooperate in collective migration of the zebrafish posterior lateral line primordium. Development. 2014;141:3188–96.
36. Knipp S, Bicker G. Regulation of enteric neuron migration by the gaseous messenger molecules CO and NO. Development. 2009;136:85–93.
37. Copenhaver PF. How to innervate a simple gut: Familiar themes and unique aspects in the formation of the insect enteric nervous system. Dev Dyn. 2007;236:1841–64.
38. Theveneau E, Steventon B, Scarpa E, Garcia S, Trepat X, Streit A, Mayor R. Chase-and-run between adjacent cell populations promotes directional collective migration. Nat Cell Biol. 2013;15:763–72.
39. Stramer B, Moreira S, Millard T, Evans I, Huang C-Y, Sabet O, Milner M, Dunn G, Martin P, Wood W. Clasp-mediated microtubule bundling regulates persistent motility and contact repulsion in Drosophila macrophages in vivo. J Cell Biol. 2010;189:681–9.
40. Reichman-Fried M, Minina S, Raz E. Autonomous modes of behavior in primordial germ cell migration. Dev Cell. 2004;6:589–96.
41. Knipp S, Bicker G. A developmental study of enteric neuron migration in the grasshopper using immunological probes. Dev Dyn. 2009;238:2837–49.
42. Haase A, Bicker G. Nitric oxide and cyclic nucleotides are regulators of neuronal migration in an insect embryo. Development. 2003;130:3977–87.
43. Sawada M, Ohno N, Kawaguchi M, Huang S-h, Hikita T, Sakurai Y, Nguyen HB, Thai TQ, Ishido Y, Yoshida Y, Nakagawa H, Uemura A, Sawamoto K. PlexinD1 signaling controls morphological changes and migration termination in newborn neurons. EMBO J. 2018;37:e97404.

Cell Migration in Wound Healing

Jessica J. Senior

What You Will Learn in This Chapter

Cell motility in wound healing is critical for tissue regeneration and the restoration of the skin's physical barrier. Cells are able to meticulously organise themselves through orchestrated events during the natural wound healing cascade when succumbed to trauma; however, diseased states such as diabetes are implicated in interrupting the natural progression of wound closure, subsequently resulting in a chronic, non-healing wound that is unable to progress past the inflammation stage. Much has recently been discovered towards a better understanding of the biological mechanisms underpinning complex wound healing pathways, yet much still remains to be realised. To this end, various wound healing assays have been developed in the hope that researchers will shed light on the complex environments within wounds. With this, wound healing models could be used to improve disease treatment and wound management therapies, helping to alleviate the physical, social, and economic burdens associated with wound care worldwide. This chapter first introduces the basic anatomy of skin and its components, referring to the organisation of the epidermis, dermis, and hypodermis. The natural wound healing cascade is then discussed and implications that arise due to trauma or disease during

(continued)

J. J. Senior (✉)
University of Huddersfield, Huddersfield, UK
e-mail: j.j.senior2@hud.ac.uk

A. Brüning-Richardson, S. Knipp (eds.), *Cell Migration in Development, Health and Disease*, Learning Materials in Biosciences,
https://doi.org/10.1007/978-3-031-64532-7_3

this biological process are highlighted in depth with chronic, non-healing wounds being of particular focus. Mechanisms of single and collective cell migration are discussed throughout in both healthy and diseased states, and you will also learn about the standard assays that are used in vitro, in vivo, and *ex vivo* towards the assessment of cell migration in wound healing, with reference to current progress in the field.

1 Introduction

The associations between cells, their extracellular matrix (ECM), and intermediary cell-ECM receptors are the predominant elements in human tissues that allow cell migration. Cell motility is crucial in new tissue formation and homeostasis of the skin and is especially important in wound healing in response to trauma. It has been seen that human epidermal keratinocytes (HEK) and human dermal fibroblasts (amongst other cell types) are recruited and guided to the wound bed site for wound re-fashioning, as directed by a milieu of endogenous growth factors [1]. Although much has been learnt in recent years surrounding the mechanisms behind this, wound healing remains an intricate process with much of the wider underpinning biological crosstalk yet to be understood, including the interplay between cells, cytokines, and surrounding vasculature. In a bid to better comprehend these phenomena, various assays, in vitro and in vivo, have been studied. This research is vital for understanding the processes behind healthy skin homeostasis and the subsequent progression of disease and pathology, and also for clinical translation towards screening of drugs and the generation of relevant skin replacement technologies.

This chapter first introduces the basic anatomy of human skin, followed by the physiological and biochemical changes that are initiated upon wound formation, as described by the wound healing cascade. Key approaches to the assessment of cell migration in wound healing will then be highlighted, and current progress finally outlined.

2 Skin Anatomy and Homeostasis

Native human skin is a hierarchical layered tissue comprised of multiple cell types and ECM components. It is the largest organ of the human body, serving a range of different functions that can become compromised following the establishment of a wound. Predominantly, skin and its cutaneous appendages (that comprise the integumentary system) act as a biological shield to protect the body from environmental factors including physical or chemical damage as well as defence against micro-organisms. In order to maintain this protection and due to the harsh external environments to which epithelia are exposed, skin renewal via cell migration is a stringently regulated process and is critical for human survival [2].

Skin is organised into three major layers, namely the epidermis, dermis, and hypodermis, with the dermis containing papillary and reticular layers and the epidermis being further subcategorised into five parts. The acellular basement membrane lies between epidermis and dermis and consists of the basal lamina and reticular lamina, providing an interface for the passage of macromolecules from the dermis to the epidermis, offers mechanical stability, and aids in the regulation of keratinocyte differentiation [3].

2.1 Epidermis

The epidermis is a compartmentalised tissue which is graded in structure, with the stratum basale being the bottom-most epidermal layer, ranging upwards to the stratum spinosum, stratum granulosum, stratum lucidum (present in glabrous, non-hair-bearing skin), and finally the stratum corneum being the most superficial layer (Fig. 1). The most prevalent cells within the epidermis are keratinocytes (~90%), with the remaining population of cells being melanocytes, Langerhans cells, and Merkel cells. Epidermal epithelia consist of sheets of stratified squamous cells bound by strong intracellular connections within little ECM and rely on dermal vasculature for the provision of oxygen and nutrients as it is an avascular entity [4].

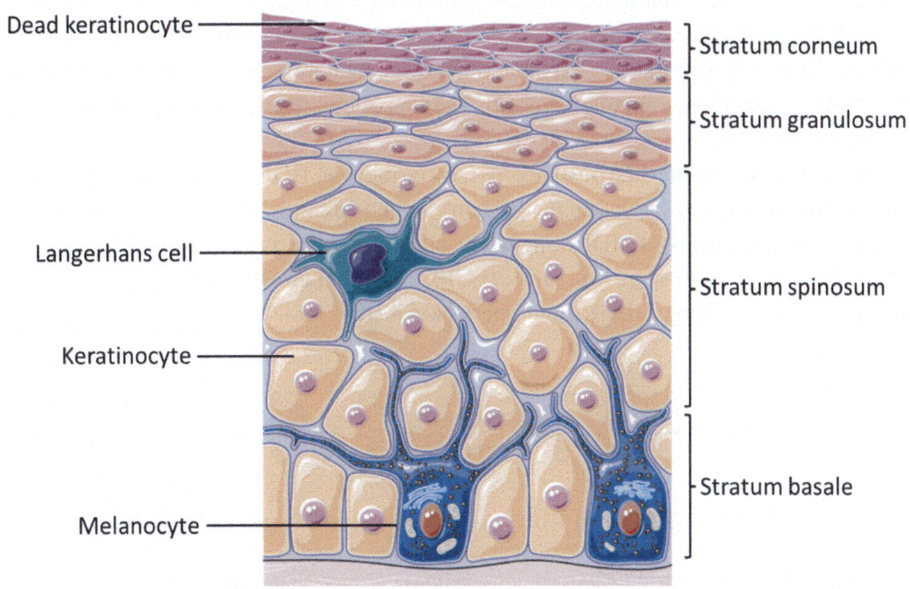

Fig. 1 Schematic illustration of the epidermal substrata and its associated cells

The stratum basale is the base epidermal layer and is made of keratinocyte-producing stem cells. These cells migrate upwards whilst differentiating to maturation for replenishment of cornified keratinocytes at the surface. This process, on average, typically lasts between 45 and 75 days depending on epidermal thickness and age amongst other factors yet can take as little as 3–7 days in conditions such as psoriasis meaning that keratinocytes do not reach full maturation before arriving at the surface [5]. Cells are able to migrate as a result of orchestrated cytoskeletal-mediated events, including protrusion and polarisation, formation of adhesion points, and subsequent cell contraction and tail-end retraction in repeated succession. The ECM and associated growth factors, such as keratinocyte growth factor (KGF), support this movement by providing focal adhesions and pro-migratory signals to keratinocytes. KGF is a potent paracrine signal that is produced by mesenchymal cells but acts on epithelial cells. It has been seen that other growth factors, including epidermal growth factor (EGF) and insulin-like growth factor 1 (IGF-1), are responsible for cell membrane protrusion, spreading, and contraction, respectively [6].

2.2 Dermis

Dermal connective tissue has a much higher abundance of ECM than the epidermis and comprises the main bulk of skin. The predominant cell type of these layers are human dermal fibroblasts (HDF), with endothelial cells, smooth muscle cells, and mast cells also studded within these matrices. The papillary layer provides mechanical anchorage in the form of a dermo-epidermal junction and is loosely packed with fibrous collagen bundles, as opposed to the underlying reticular layer that exhibits much denser fibres that owe to its elastic and flexible properties. HDF are therefore much more prevalent in the papillary layer than they are in the reticular layer [7]. The location of the dermis within human skin is depicted in Fig. 2.

Type I collagen in the first instance is responsible for HDF motility and migration. Growth factors that are released in the surrounding area are what govern the directionality and chemotaxis of HDFs, including platelet-derived growth factor BB and AA (PDGF-BB/AA), fibroblast growth factor (FGF), epidermal growth factor (EGF), vascular endothelial growth factor (VEGF), and interleukin 4 (IL-4) [8].

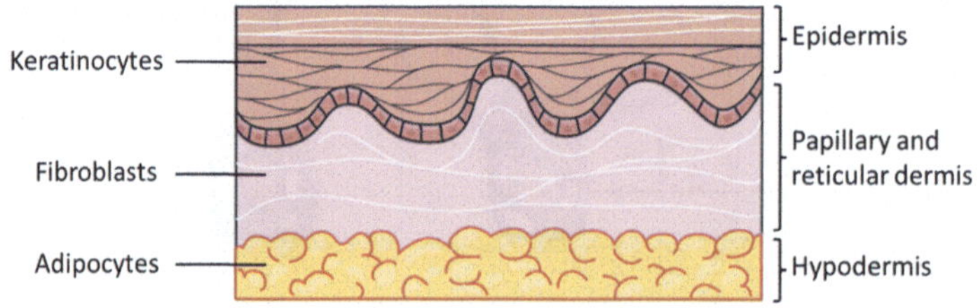

Fig. 2 Human skin comprised of epidermis, dermis, and hypodermis and their associated predominant cell type

2.3 Hypodermis

The hypodermis (also termed subcutis or subcutaneous tissue) is located between the dermis and the fascia which separates underlying muscle and organs. It is classed as a loose connective tissue, inhabited by adipocytes and mesenchymal stem cells (MSC), and behaves as a protective gliding plane, acts as storage for fat, and also has roles in secreting various hormones and cytokines. MSCs within the hypodermis have a remarkable influence on the overall skin microenvironment, coordinating keratinocyte and fibroblast proliferation, regulating angiogenesis, and contributing to inflammatory response [9].

Research has shown that adipocytes remain relatively local to their original site whilst nondifferentiated preadipocytes are much more motile, migrating further, and faster [10]. Upon mechanical stimulation, adipogenesis is accelerated whereby round lipid-filled adipocytes are formed from elongated adipocyte precursors. The resulting phenotype presents less organised cytoskeletal arrangements and therefore impacts on cell migration behaviour. Preadipocytes are also much more polarised, with a distinct leading edge and a retracting tail compared to spherical, lipid-filled adipocytes when cultured in 2D. Locomotion and chemokinesis in preadipocytes appear to be promoted by plasminogen activator inhibitor-1 (PAI-1), macrophage inflammatory protein (MIP)-related protein-2 (MRP-2), and extracellular nucleotides (such as ATP).

Following wound formation induced by trauma, skin tissue microenvironments experience a shift in growth factor and chemokine release, initialising migration of cells towards the wound bed for procedure through the typical wound healing stages.

3 Wounds and Wound Healing

Wounds are injuries to the body that typically manifest as a disruption of tissue continuity. The term 'wound' is often used as a descriptor for impaired skin having experienced laceration or contusion; however, other underlying tissues such as muscle and tendon may also be compromised [11]. Following tissue trauma and wound formation, the human body initiates a natural physiological response to repair injured tissue, known as the wound healing cascade. This phenomenon is governed by a series of delicate and complex signalling events and involves orchestrated crosstalk between several different cell types and mediatory signals. The natural progression of wound healing routinely occurs across four major stages, namely haemostasis, inflammation, proliferation, and remodelling.

3.1 Wound Healing Cascade

The first official stage of the wound healing cascade is known as haemostasis (Fig. 3). The first critical steps during this period are to prevent further blood loss from severed

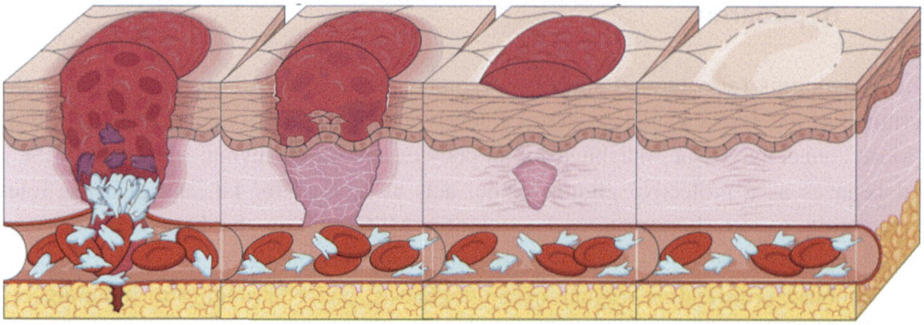

Fig. 3 The four stages of the wound healing cascade: haemostasis, inflammation, proliferation, and remodelling

blood vessels and to restore the physical barrier that has been compromised by establishing a clot. Here, thrombin cleaves fibrinogen to form a fibrin mesh which captures platelets and enables attachment to capillary walls. Associated coagulation factors are subsequently upregulated, and at full formation, the clot comprises platelets, red blood cells, and various ECM components. This chain of events typically lasts around 3 h, during which the rapid onset of platelet-derived growth factor (PDGF), vascular endothelial growth factor (VEGF), and transforming growth factor beta (TGF-β) release occurs which are crucial in later stages of wound healing [12].

Following haemostasis, the wound enters the inflammatory stage, ascribed by the influx of white blood cells named macrophages and phagocytic neutrophils to the wound location. The area is cleared of any foreign bodies, bacteria, and necrotic cell components, and fibroblast migration to the wound site is stimulated. During the proliferation stage, newly recruited fibroblasts lay down temporary ECM scaffolding, primarily composed of collagen that is commonly known as granulation tissue. Endothelial cells respond to elevated levels of VEGF and TGF-β that cause them to release matrix metalloproteinases (MMPs) for digestion of the basement membrane. Endothelial cells are liberated from parent vessels and new capillaries sprout from pre-existing vasculature. Fibroblasts then proliferate and differentiate into myofibroblasts that assist in wound contraction and closure, allowing keratinocytes to migrate over the wound surface for re-epithelialisation. Finally, during the fourth wound healing stage that is remodelling, granulation tissue transitions into acellular, collagen-rich scar tissue and previous blood-vessel densities are resumed.

The way in which cells migrate during wound healing is known as collective cell migration and it is this hallmark that defines much of the success of skin regeneration.

3.2 Collective Cell Migration in Wound Healing

Collective cell migration entails the movement of a group of cells en masse as opposed to migrating as single entities. During this mode of migration, cells move transiently in an interconnected fashion, bound by strong cadherin-based junctions, which is the fundamental mechanism in cell motility during embryonic morphogenesis, tissue regeneration, and cancer invasion [13].

During the reparative process of skin, cells proliferate and migrate into the centre of the defect during wound closure. Most of the cell recruitment occurs during the proliferative phase, driving migration of a milieu of cells with the vast majority being keratinocytes, fibroblasts, and endothelial cells. All of these cell types tend to traverse the wound bed in a collective manner, yet each of them has distinct biological mechanisms in which they do so.

As keratinocytes inhabit the epithelial layer that is the epidermis, these colonies migrate initially as monolayer sheets and transition into multi-layered stratified epidermal tissue de novo after a period of hours/days. Mechanically stimulated keratinocytes at the wound edge go through partial epithelial-mesenchymal transition (EMT), assuming an invasive and migratory phenotype. During this transformation, top-to-bottom polarity is replaced by front-to-rear polarity and cells mediate changes in their adherens junctions to a state more conducive to migrating as epithelial sheets [14]. These primary cell rows utilise $\alpha2\beta1$, $\alpha5\beta1$, and $\alpha v\beta3$ integrins to create focal contacts and traction within the collagen and fibrin-rich ECM and follower cells employ $\alpha6$ integrins for movement along the new basement membrane as it is being synthesised by leader cells [15]. Front row polarity is promoted by EGF-EGFR signalling and by reactive oxygen species (ROS). E-cadherin, desmoglein 1, desmoglein 3, and desmosomes mediate intercellular contacts during migration and are stabilised by the intracellular cortical actin cytoskeleton in the protruding lamellipodia, dependent on Rho GTPase. Keratinocytes also recognise signals from fibroblasts within the surrounding stroma such as FGF, keratinocyte growth factor, and TGF-β, which transduces intracellular mitogen-activated protein kinase (MAPK) signalling and subsequently propagates in a wave-like manner from cell to cell in a rearward direction [16]. Both keratinocytes and fibroblasts in unison are essential in the constitution of tissue-specific ECM and jointly deposit laminin 1, laminin 5, collagen IV, and nidogens for the formation of the basement membrane [17] (Fig. 4a).

During neo-angiogenesis and vascular sprouting, however, endothelial cells collectively migrate in strands by penetration of the provisional fibronectin and fibrin-rich matrix previously laid down by recruited fibroblasts in the proliferation stage. A single header cell with protruding actin-rich filopodia engages $\alpha v\beta3$ integrin with membrane type 1 matrix metalloproteinase (MT1MMP; also known as MMP14) to proteolytically remodel the ECM. This tip cell leads a stalk of trailing endothelial cells, held together by vE-cadherin that, together with stromal pericytes, deposits a basement membrane [18]. During this process, FGFR and VEGFR maintain tip cell differentiation, leading to expression of Notch ligands Delta-like 4 (DLL4) and Jagged 1 and downstream Notch signalling

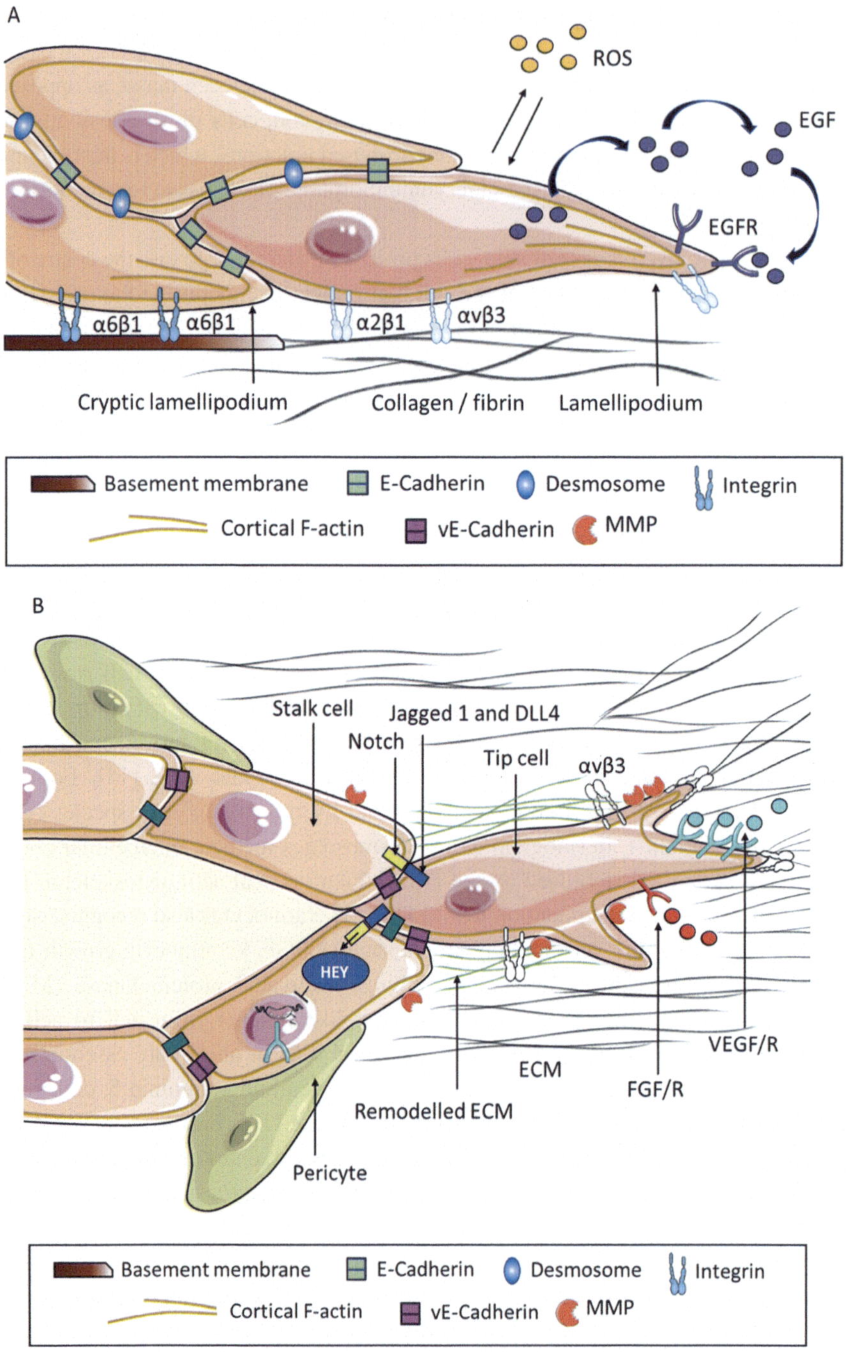

Fig. 4 Mechanisms of collective cell migration during (**a**) epidermal regeneration and (**b**) neoangiogenesis in wound healing

being directed at trailing stalk cells. Notch then communicates via transcriptional repressor protein HEY towards VEGFR silencing and maintenance of stalk cell differentiation [13] (Fig. 4b).

Depending on wound location and severity, skin appearance and function may not always be regained despite the intricate mechanisms that cells have adapted to successfully migrate during wound repair, and in some more severe cases, chronic wounds may develop.

3.3 Chronic Wounds

Disruption to the natural healing process can lead to a chemical and physiological imbalance and the progression of a chronic wound, characterised by prolonged inflammation, a lack of re-epithelialisation, and a lack of sufficient vasculature (Fig. 5) that typically exhibits >12 weeks of non-healing. In cases such as these, the inherent lack of cell migration associated with chronic wounds is often responsible for inadequate tissue regeneration. These pathological extremes of abnormal healing processes are classified as exhibiting both disease and trauma elements, with cases often manifesting in the forms of diabetic foot ulcers, venous leg ulcers, pressure ulcers, and arterial insufficiency ulcers. Although they often only cover a small area of the skin, they are a major healthcare problem, are associated with poor prognosis, and require a great deal of time and costs during specialised treatment. Chronic wound care often involves debridement of dead tissue, slough, biofilm, debris, and exudate from the wound bed, administration of therapeutic agents and antimicrobials, and compression bandaging of the wound to prevent infection and encourage natural wound healing [19]. In even more extreme cases, patients may undergo surgery for skin excision and replacement with partial-thickness, full

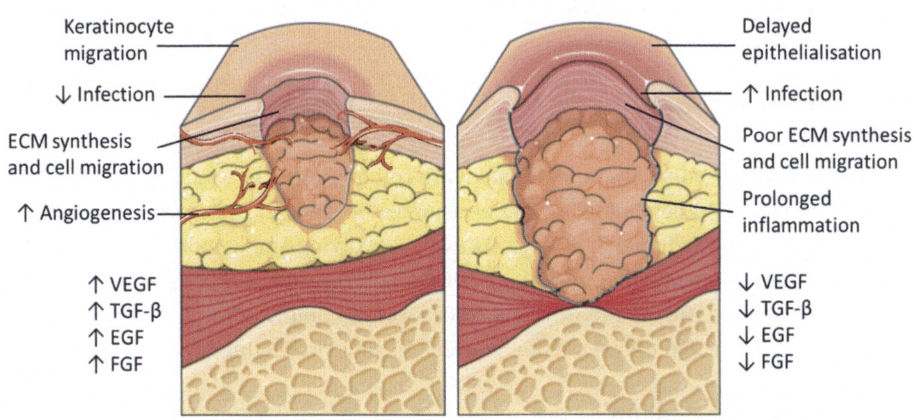

Fig. 5 Key differences between acute (healing) and chronic (non-healing) wound microenvironments

thickness, or artificial skin to address time-consuming nursing care, cost-effectiveness, and psychological distresses associated with the prolonged treatment of chronic wounds.

A major implication of aged and diabetic wound pathology is the senescent progression of cells induced by sustained hyperglycaemia. Senescent, non-mitotic cells assume a non-proliferative yet hypersecretory state, establishing a secretome that is abundant in pro-inflammatory cytokines and tissue-degrading proteases, contrary to natural wound healing progression. Chronicity is thus perpetuated through the sustained destruction of wound tissue. Epithelial remodelling is stunted due to the lack of dermal matrix that keratinocytes use as scaffolding during wound closure. Dermal fibroblasts also display senescence and impaired ECM deposition. Furthermore, keratinocytes at the wound perimeter display abnormal levels of nuclear β-catenin and elevated c-myc, which has strong associations with delayed migration [14]. The healthy microbiome in healing wounds is also overthrown by pathogenic bacteria in chronic wound counterparts, which consequently leads to infection, biofilm formation, and a physical barrier that prevents cell migration.

4 Models for Cell Migration in Wound Healing

Wound healing assays are employed to study the biological mechanisms underpinning wound repair and for the exploration into potential drug candidates towards wound treatment. An array of different wound healing assays now exists, each with slight technical differences, yet all aim to address the same rudimentary questions. These assays can be undertaken in 2D or 3D, in vitro, or in vivo and with varying set-up approaches based on protocols followed. Selection of the appropriate assay is often dependent on the research outcomes one wants to address amongst other factors such as assay complexity, cost, and availability of equipment.

4.1 In Vitro Models

The most commonly adopted assays for studying cell migration in wound healing in vitro include the 2D wound healing assay, 3D invasion assay, and 3D organ explant, with this chapter focusing on the first two assays. 2D assays involving single monolayer sheets are a suitably relevant model when mimicking keratinocyte migration during wound closure. 3D models, however, are able to offer a greater level of complexity that better resembles the 3D nature of skin and the interaction between its interconnecting layers.

4.1.1 2D Wound Healing Assay

Most 2D wound healing protocols consist of three basic principles which are wounding of a confluent cell monolayer, monitoring cell movement over time, and the acquisition of data for analysis. The most typical way in which the 2D wound healing assay is performed

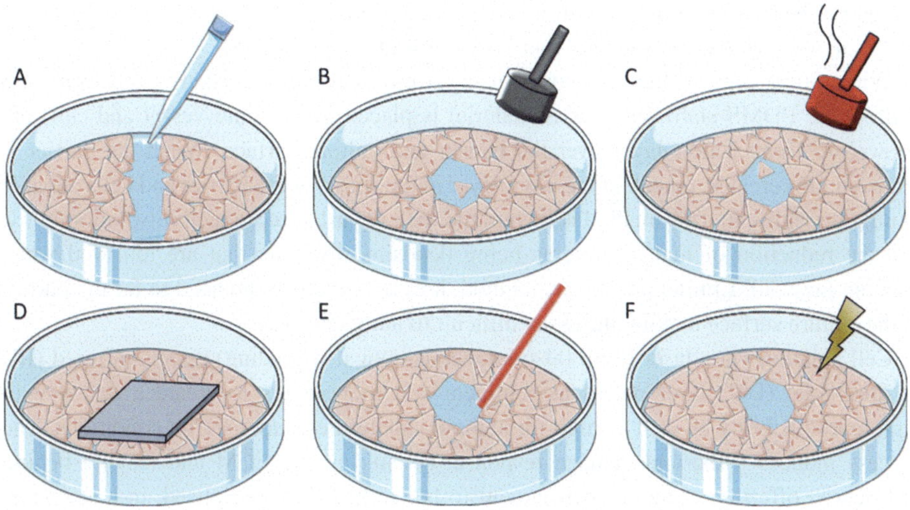

Fig. 6 2D wound healing observation via (**a**) scratch assay, (**b**) stamp assay, (**c**) thermal ablation, (**d**) physical barrier, (**e**) laser ablation, and (**f**) electrical wounding

is via the classical scratch assay; however, methods of wounding by stamping, thermal or laser ablation, a physical barrier, and electrical injury have also been developed [20] (Fig. 6).

Scratch assays are a simple and common technique used to monitor cell migration and wound closure. Injury is induced by performing a scratch within a confluent cell monolayer using a sharp object such as a pipette tip or cell scraper. Once the defect has been prepared, cell migration and wound healing can be assessed as a function of gap closure in real time. Using advanced microscopy techniques and a range of readily available analysis software, it is possible to quantify the functional behaviour of cells, including migration velocity and directionality, as well as qualitatively evaluating cell polarity. Using this set-up, it is also possible to introduce potential therapeutic drugs or migratory agents towards clinical research. Although this technique is well established, users often experience irregular scratches leading to lack of reproducibility, and therefore, it is often considered the first 'port of call' for collection of preliminary results. The act of scratching can also impair the coating or matrix on which the cells are bound and anti-mitotic agents such as mitomycin c are often required to prevent proliferating cells from skewing data [21].

By using a stamp for the creation of a wound, it is possible to preserve the underlying matrix coating from being impaired. A stamp of preferred shape (usually made from rubber or PDMS) can either be used manually or using robotics to remove cells using a pressing force and various patterns can be engraved on the stamp for more intricate systems. Again, this method is associated with inconsistent cell removal (especially with manual stamping) yet is conducive to studying the influence of cell debris on migration. Incorporation of heat can add a thermal ablation element to the stamping process which may be useful if one wishes to assess the effects of heat transfer to cells at the periwound.

When preparing a simulated wound using a physical barrier (often in the form of a hydrophilic PDMS slab), the barrier material is placed in a culture vessel and cells are allowed to form a monolayer over the barrier and culture surface. When the barrier is removed, this creates a break in the monolayer in the shape of the barrier and cells are able to migrate within it. This method is favourable with regard to reproducibility, tailorability, and the reduction of matrix proteins being deposited by cells that are removed when creating the void. Limitations of this method, however, include the need to fix the barrier to the culture surface making the assay difficult to automate [22].

Cell ablation via a laser beam (also known as optical wounding) directs infrared (IR) or ultraviolet (UV) laser pulses at confluent cell monolayers to create a defect with high reproducibility as well as high-throughput capabilities under sterile conditions. The process induces a thermal damage response, denaturing the ECM and leaving behind cell debris, offering a unique environment to be studied, yet specialised, in-house laser technology is required.

One other method for cell removal is electrical wounding whereby a local current is applied to cells via gold-coated electrodes embedded in the culture vessel causing electroporation and subsequent cell death. These electrodes can measure the impedance in electron flow caused by cells in the culture vessel and can thus be used as a measure of cell migration, removing errors or bias due to human intervention. Like laser ablation, this method requires expensive and specialised equipment as well as keeping culture environments under strict conditions to prevent temperature and pH influencing impedance measurements [20].

4.1.2 3D Invasion Assay

3D invasion assays allow the assessment of single cell migration or collectively migrating cell strands within 3D ECM, modelled by cell seeding on 3D scaffolds, 3D bioprinting cell-laden polymers, or observation/implantation of cell spheroids within 3D culture (Fig. 7).

Similar to simulating wound formation in 2D, lasers can be utilised to create local hyperthermia for selective cell ablation in epithelial spheroids, stimulating collective cell migration in 3D structures [23]. This system allowed analysis using live cell actin staining during the wound healing response with associated roles of ROCK and Notch1-Dll4 signalling in regulating 3D epithelial wound healing being implicated. 3D bioprinting also provides great opportunities to develop skin models in which cell migration can be studied following implantation into a simulated wound model [24].

To examine capillary sprouting in vitro, early models of 3D collagen-based angiogenesis were developed [25]. Endothelial cell spheroids were subject to tractional forces transduced through type I collagen ECM that were sufficient to guide directional outgrowth of endothelial cells. Organotypic co-culture systems containing fibroblasts and squamous cell carcinoma (SCC) cells embedded in collagen and Matrigel have been used to assess

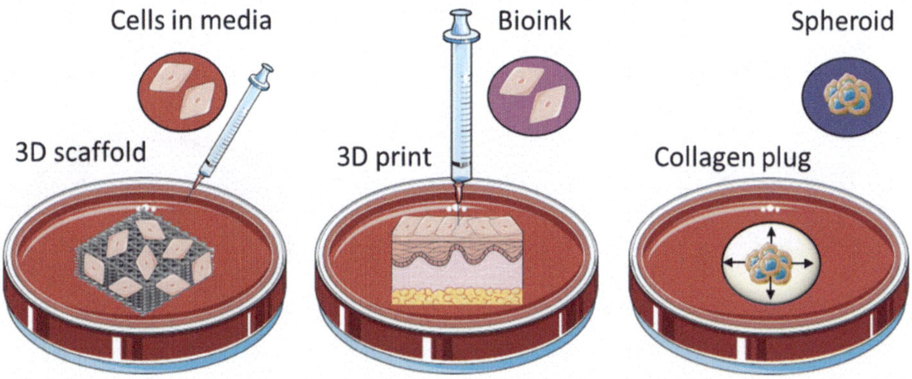

Fig. 7 Methods for 3D invasion assays including cell seeding of a 3D porous scaffold (left), 3D bioprinting bioinks consisting of various polymers and cells (middle), and implantation of spheroids within a 3D matrix (right)

EMT in cancer invasion [26]. Researchers found that stromal fibroblasts have a clear role in leading collective cell migration, forming tracks within the ECM along which SCC cells would follow. Mechanisms of protease- and force-mediated matrix remodelling in both fibroblasts and SCC cells were noted yet via alternative pathways, making the 3D invasion assay a suitable observable model.

4.2 In Vivo and *Ex Vivo* Models

In vivo modelling involves the use of animals to examine various biological mechanisms, and in the context of cell migration in wound healing, murine (mouse and rat) and porcine (pig) models often apply (Fig. 8a). Although both in vitro and in vivo models are necessary to understand biological pathways and pharmacokinetic profiles, in vivo studies are often thought as more 'reliable' or relevant due to the complexity of the 3D in vivo model system greater matching that of human bodily systems. *Ex vivo* modelling, however, involves the assessment of tissue taken from the body yet kept in optimal conditions to mimic natural living conditions (Fig. 8b). These systems address the issues associated with potential ethics and overcome the limited utility of in vitro tests and animal models of wound healing and often employ *ex vivo* human tissue as a more bio-relevant model.

4.2.1 In Vivo Wound Healing

In vivo modelling allows the potential for many areas of wound healing to be investigated including the application of herbal medicines or drugs, incorporation of hydrogel dressings, cell therapies, or scaffolds towards accelerated wound healing, or simply studying the mechanisms underpinning the natural wound healing cascade and disease development. In murine models, punch biopsies are often taken from a distinct area of the animal for

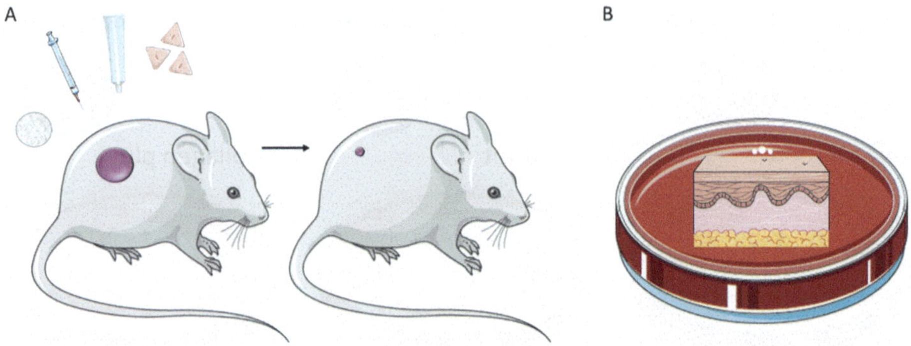

Fig. 8 (**a**) In vivo testing in mice models to assess the potential effects of scaffolds, drugs, creams, or cell therapy on wound healing. (**b**) *Ex vivo* testing of human skin explants in culture

the formation of a wound, whereas larger animals such as pigs require skin removal via scalpel. Researchers looking at the role of voltage-gated calcium channels (Cavs) and their β subunits on wound healing utilised a 2 mm punch biopsy method in mouse models and found that the loss of Cavβ3 protein enhanced Ca^{2+} release [27]. This was causative of accelerated cell migration and enhanced collagen secretion and contributed a better understanding towards natural wound healing. Towards accelerating cutaneous wound healing, lecithin-based deferoxamine nanoparticles were topically applied in diabetic rats and assessed over 11 days [28]. Compared with control groups who received pluronic gel only, mice that received deferoxamine nanoparticles incorporated in pluronic gel exhibited accelerated wound closure, enhanced neovascularisation, and collagen deposition. Nanoparticles vs standard formulation of deferoxamine also had a greater efficacy in enhancing wound healing. HIF-1α, VEGF, SDF-1α, TGF-β_1, and IL-10 protein levels were significantly higher in the treated group with lower levels of TNF-α. Taken together, these results suggested a potential therapy in the treatment of diabetic wounds. Using a porcine wound model in vivo, a study to create a clinically relevant delayed wound healing model was established [29]. Half of the wounds were fitted with silicon blocks to simulate delayed wound healing, whereas the other half of the wounds received no further intervention and analysis was undertaken following 3 weeks. The model successfully simulated wound ischemia, necrosis, and persistent inflammation ascribed by the elevated levels of IL-1 and TNF-α in the treated cases and a marked reduction in closure following removal of the silicon blocks.

4.2.2 *Ex Vivo* Wound Healing

Ex vivo wound healing utilises skin explants (often obtained from humans during routine elective cosmetic surgery) for the examination of the wound healing process and relevant applicable therapies. Photodynamic therapy is one such treatment that has been assessed in *ex vivo* cultures [30]. Models were first optimised by modulating wound geometry, support system, growth media, and cellular/matrix biomarkers, with the consensus that explants embedded in collagen and supplemented with DMEM were most successful. Following this, the effect of photodynamic therapy on the wound healing organ cultures was assessed, revealing that re-epithelialisation, neo-epidermis differentiation, proliferation, and neo-collagen deposition were significantly enhanced versus untreated controls. Similar *ex vivo* models of corneal culture have also been applied for the quantification of myofibroblast development and scarring characterisation [31]. Porcine eyes with wounded corneas were mounted on a collagen/agar base, cultured for 2 weeks, and stabilised with vitamin C to prevent cell proliferation and ECM deposition by resident fibroblasts. The model is reported to be conducive to testing of putative healing agents and administration of siRNAs into the wound, as well as holding potential in assaying viral transduction. With all models, limitations were highlighted in that organ cultures lack functional limbal vasculature, tears, and aqueous humour and that investigators must therefore assess applicability in their studies.

Take-Home Message

- Wound healing often describes a biological process where compromised human skin is regenerated and restored through a series of cell migratory events.
- Chronic wounds are caused by an interruption in normal healing processes and the reasons for this remain difficult to understand.
- Collective cell migration, whereby cells move as collective sheets, is known to be a key mechanism underpinning the natural wound healing cascade and is currently a major area of research in re-epithelialisation and neovascularisation of skin.
- Such models involved in assessing cell migration in wound healing are in vitro, in vitro, or *ex vivo* in nature, each with their advantages and disadvantages, and are used as tools towards a better understanding of healthy and diseased states of wound closure.

Acknowledgements Some of the materials in Figs. 1, 2, 3, 4, 5, 6, 7, and 8 were obtained from Servier Medical Art (http://smart.servier.com/), licensed under a Creative Common Attribution 3.0 Generic License (https://creativecommons.org/licenses/by/3.0/).

References

1. Li W, Fan J, Chen M, Woodley D. Mechanisms of human skin cell motility. Histol Histopathol. 2004;19(4):1311–24.
2. Tortora G, Derrickson B. Principles of anatomy & physiology. 14th ed. Hoboken, NJ: Wiley; 2014.

3. Standring S. Gray's anatomy. 42nd ed. Amsterdam: Elsevier; 2020.
4. Abdo J, Sopko N, Milner S. The applied anatomy of human skin: A model for regeneration. Wound Med. 2020;28:100179.
5. Rajguru J, Maya D, Suri P, Bhardwaj S, Patel N, Kumar D. Update on psoriasis: A review. J Family Med Prim Care. 2020;9(1):20.
6. Haase I, Evans R, Pofahl R, Watt F. Regulation of keratinocyte shape, migration and wound epithelialization by IGF-1- and EGF-dependent signalling pathways. J Cell Sci. 2003;116(15):3227–38.
7. Falanga V, Faria K, Bollenbach T. Bioengineered skin constructs. In: Lanza R, Langer R, Vacanti J, editors. Principles of tissue engineering. San Diego: Academic; 2014. p. 1619–43.
8. Li W, Fan J, Chen M, Guan S, Sawcer D, Bokoch G, et al. Mechanism of human dermal fibroblast migration driven by type I collagen and platelet-derived growth factor-BB. Mol Biol Cell. 2004;15(1):294–309.
9. Diegel K, Danilenko D, Wojcinski Z. The integumentary system. In: Wallig MA, Haschek WM, Rousseaux CG, Bolon B, editors. Fundamentals of toxicologic pathology. Cambridge, MA: Academic; 2018. p. 791–822.
10. Lustig M, Zadka Y, Levitsky I, Gefen A, Benayahu D. Adipocytes migration is altered through differentiation. Microsc Microanal. 2019;25(5):1195–200. Cambridge University Press
11. Leaper D, Harding K. Wounds: biology and management. Oxford: Oxford University Press; 1998.
12. Wallace H, Basehore B, Zito P. Wound healing phases [Internet]. Treasure Island, FL: StatPearls; 2022 [cited 9 May 2022]. Available from https://www.ncbi.nlm.nih.gov/books/NBK470443/
13. Friedl P, Gilmour D. Collective cell migration in morphogenesis, regeneration and cancer. Nat Rev Mol Cell Biol. 2009;10(7):445–57.
14. Wilkinson H, Hardman M. Wound healing: cellular mechanisms and pathological outcomes. Open Biol. 2020;10(9):200223.
15. Grose R, Hutter C, Bloch W, Thorey I, Watt F, Fässler R, et al. A crucial role of β1 integrins for keratinocyte migration in vitro and during cutaneous wound repair. Development. 2002;129(9):2303–15.
16. Nikolić D, Boettiger A, Bar-Sagi D, Carbeck J, Shvartsman S. Role of boundary conditions in an experimental model of epithelial wound healing. Am J Physiol Cell Physiol. 2006;291(1):C68–75.
17. Smola H, Stark H, Thiekötter G, Mirancea N, Krieg T, Fusenig N. Dynamics of basement membrane formation by keratinocyte–fibroblast interactions in organotypic skin culture. Exp Cell Res. 1998;239(2):399–410.
18. Li L, Jiang J, He Y, Zhao M. Collective cell migration: Implications for wound healing and cancer invasion. Burns Trauma. 2013;1(1):21.
19. Frykberg R, Banks J. Challenges in the treatment of chronic wounds. Adv Wound Care. 2015;4(9):560–82.
20. Stamm A, Reimers K, Strauß S, Vogt P, Scheper T, Pepelanova I. In vitro wound healing assays—state of the art. BioNanoMaterials. 2016;17(1–2)
21. Pijuan J, Barceló C, Moreno D, Maiques O, Sisó P, Marti R, et al. In vitro cell migration, invasion, and adhesion assays: from cell imaging to data analysis. Front Cell Dev Biol. 2019;7
22. Fong E, Tzlil S, Tirrell D. Boundary crossing in epithelial wound healing. Proc Natl Acad Sci USA. 2010;107(45):19302–7.
23. Xiao Y, Riahi R, Torab P, Zhang D, Wong P. Collective cell migration in 3D epithelial wound healing. ACS Nano. 2019;13(2):1204–12.
24. Moakes R, Senior J, Robinson T, Chipara M, Atansov A, Naylor A, et al. A suspended layer additive manufacturing approach to the bioprinting of tri-layered skin equivalents. APL Bioeng. 2021;5(4):046103.

25. Korff T, Augustin H. Tensional forces in fibrillar extracellular matrices control directional capillary sprouting. J Cell Sci. 1999;112(19):3249–58.
26. Gaggioli C, Hooper S, Hidalgo-Carcedo C, Grosse R, Marshall J, Harrington K, et al. Fibroblast-led collective invasion of carcinoma cells with differing roles for RhoGTPases in leading and following cells. Nat Cell Biol. 2007;9(12):1392–400.
27. Belkacemi A, Hui X, Wardas B, Laschke M, Wissenbach U, Menger M, et al. IP3 receptor-dependent cytoplasmic Ca2+ signals are tightly controlled by Cavβ3. Cell Rep. 2018;22(5):1339–49.
28. Qayoom A, Aneesha V, Anagha S, Dar J, Kumar P, Kumar D. Lecithin-based deferoxamine nanoparticles accelerated cutaneous wound healing in diabetic rats. Eur J Pharmacol. 2019;858:172478.
29. Jung Y, Son D, Kwon S, Kim J, Han K. Experimental pig model of clinically relevant wound healing delay by intrinsic factors. Int Wound J. 2012;10(3):295–305.
30. Mendoza-Garcia J, Sebastian A, Alonso-Rasgado T, Bayat A. Optimization of an ex vivo wound healing model in the adult human skin: Functional evaluation using photodynamic therapy. Wound Repair Regen. 2015;23(5):685–702.
31. Castro N, Gillespie S, Bernstein A. Ex vivo corneal organ culture model for wound healing studies. J Vis Exp. 2019;144

Cell Migration in Immune Responses

Giuliana Clemente

What You Will Learn in This Chapter

Cell migration is a fundamental process intimately underlying the development of multicellular organisms, and its dysregulation is associated with a variety of diseases. The ability to confidently navigate a complex extracellular space is at the core of an effective immune response. Indeed, motility represents a key aspect of immune cell behavior, fundamental to eliminating invading pathogens, scanning tissues for damage, and detecting developing tumors. The convenience of live imaging and the genetic tractability of certain in vivo model organisms (i.e., flies and zebrafish) offer a unique opportunity to better understand mechanisms of immune cell motility in an in vivo context and to gain novel insights into how these cells fulfill their numerous functions by promptly modifying their migratory behavior in response to a wide variety of extrinsic cues.

In this chapter, we explore how the use of in vivo model organisms has greatly advanced the understanding of the delicate interplay between immune cells and their surrounding tissues, in different physiological and pathological scenarios. Moreover, we discuss how multidisciplinary approaches, bringing together the acquisition of experimental data and mathematical modeling, can better describe the

(continued)

G. Clemente (✉)
University of Bristol, Bristol, UK
e-mail: giuliana.clemente@bristol.ac.uk

A. Brüning-Richardson, S. Knipp (eds.), *Cell Migration in Development, Health and Disease*, Learning Materials in Biosciences,
https://doi.org/10.1007/978-3-031-64532-7_4

spatiotemporal regulation of complex biological systems and guide the development of new therapeutic treatments in the clinic.

1 Introduction

Our body is under constant attack from threatening foreign invaders, such as bacteria and viruses, and it is constantly exposed to noxious substances, such as toxins and chemicals. To maintain overall health, our body is therefore equipped with powerful defensive mechanisms that all fall under the umbrella term of "immunity." Anatomical and physiological barriers provide the first line of defense, blocking pathogens from entering the body (mechanical barrier) or creating a hostile environment difficult for the microbes to endure (chemical barrier). Beyond these protective strategies, the immune system also comprises a complex and diverse collection of cells that act as "cellular sentinels" and constantly patrol the extracellular space searching for dangerous intruders or dysfunctional cells, such as tumorigenic or dying cells. In addition to the physical barrier provided by the skin and mucous membranes, a subset of cells named leukocytes, as neutrophils and macrophages, constitute a branch of the immune system known as innate immunity. This is an antigen-independent, fast, non-specific type of defense, unleashed by the host shortly after the encounter with the antigen [1]. Another subgroup of cells, B- and T-cells or lymphocytes, show an antigen-dependent and antigen-specific response known as adaptive immunity. Cells of the adaptive immunity retain a "molecular memory" of their first encounter with the pathogens so that, upon subsequent encounters, they can mount a quicker and more robust response tailored against the specific antigens [1].

These two lines of defense are not mutually exclusive, but they rather work synergistically to provide full protection, and defects in either result in an inadequate immune response and increased host vulnerability.

Motility is the primary characteristic that supports every aspect of an efficient immune surveillance response, from the long-range patrolling throughout the body in search of invading pathogens to the local recruitment of immune cells to target tissues, scanning for the presence of injuries or developing tumors.

In their journey throughout the body, cells of the immune system experience diverse environments. Yet they successfully navigate the different tissues by integrating various extracellular cues and translating such diverse information into changes in the cytoskeleton dynamics.

Understanding how cells modify their migratory pattern depending on the various physical, chemical, and mechanical cues is still a big challenge that, if addressed, will help to design more sophisticated therapeutic approaches for the treatment of migration-related dysfunctions. In this regard, much progress has been done by performing in vitro assays. However, despite being informative, in vitro experiments can scarcely reproduce the complexity and variability of an in vivo setting [2]. In recent years, the use of model organisms has helped to build a more holistic understanding of the multi-factorial nature of immune cell motility and its relevance for immune cell function in vivo. Moreover, the opportunity to fluorescently label specific immune cell subpopulations as well as the surrounding tissues has allowed us to draw a more comprehensive picture of the plethora of functions carried out by immune cells in different physiological and pathological scenarios and to watch live their interaction with the environment.

Given their optical translucency, genetic tractability, and tremendous translational potential, *Drosophila* and zebrafish have been extensively used in recent years to study mechanisms of immunity. Thanks to the pioneering work of Lemaitre et al. [3, 4] which uncovered evolutionary conserved pathways and mechanisms of immunity, *Drosophila* has emerged as a powerful model to study innate immunity and its parallelism to the vertebrate counterpart. Flies possess a rudimentary immune system that lacks adaptive immunity. The major class of immune cells, colloquially called hemocytes, are equivalent to vertebrate macrophages and serve as a case model not only to investigate macrophage specification and activation by local cues but also to address their numerous functions not directly linked to their phagocytic activity. Indeed, if for decades macrophages have been mainly considered "big eaters" that eliminate invaders and aberrant cells, today these cells have been implicated in the maintenance of tissue homeostasis by other means, including deposition of extracellular matrix components and maintenance of stem cell niches, during development and in the adults [5].

The zebrafish has been widely used as a model to study vertebrate hematopoiesis and is emerging as a powerful system to dissect mechanisms of innate immunity [6]. Indeed, the crucial role played by innate immunity in orchestrating a robust immune response can be easily studied during the first weeks of zebrafish larval development. During this developmental window, zebrafish larvae grow in the absence of an adaptive immune system, which will become mature and reactive only 4 weeks post-fertilization [7]. This poses an advantage compared to higher vertebrate models, where the unique contribution of the innate immune response can be addressed only upon pharmacological inhibition of the adaptive immunity.

In this chapter, we review current studies in *Drosophila* and zebrafish that have deepened our knowledge of the complexity of innate immune response and its relevance to host health, during both homeostasis and disease. Moreover, we discuss how mathematical modeling has greatly helped in describing the immune cell response in different clinical settings (i.e., tissue repair and cancer), contributing to the design of novel therapeutic strategies.

2 Mechanism(s) of Immune Cell Adhesion and Motility

Cell migration is a complex and heterogeneous process that is essential for development, proper immune response, and tissue repair, and its deregulation often results in pathologies including cancer metastasis, birth defects, and chronic non-healing wounds [8].

In general terms, cell migration can be classified into two main modes: amoeboid and mesenchymal migration [9, 10]. The mesenchymal mode of migration involves the formation of strong focal attachment to the extracellular matrix (ECM), proteolytic degradation of the ECM, elongated spindle-like cell shape characterized by long membrane protrusion, and slow cell movement. Conversely, the amoeboid mode is characterized by the acquisition of a round, ellipsoid cell shape with a highly contractile rear, known as the uropod. Amoeboid movement is generally fast and does not involve the formation of strong adhesion sites to the substrate.

The highly diverse environments experienced by the cells of the immune system significantly influence their mode of migration. Immune cells do not commit to either the mesenchymal or amoeboid mode of migration but rather use the two interchangeably based on their activation status and the properties of the surrounding environment [11–13].

In this paragraph, we will broadly discuss the differential mode of migration adopted by the cells of the innate immunity (neutrophils and macrophages), with a particular focus on the modulation and interplay of the different cytoskeletal components.

2.1 Mode of Immune Cell Migration and Cytoskeletal Dynamics

Innate immune cells are motile from the time they are born. The first migratory task they face is the migration along several defined routes to disperse throughout the embryo and populate the various developing tissues. Following this initial developmental dispersal, immune cells either remain stationary, patrolling the tissue where they reside, or they follow the bloodstream, traveling within the vasculature. Leukocyte migration in vivo is largely guided by chemoattractants which, by signaling through heterotrimeric G-protein-coupled receptors (GPCRs, Fig. 1), induce the acquisition of a front-to-back polarity [14].

The diversified repertoire of GPC receptors expressed at the leading edge allows these cells to sense and integrate multiple extracellular signals, ultimately dictating the mode and direction of migration.

Neutrophils primarily engage in the fast, adhesion-free amoeboid movement [15, 16]. In these cells, early polarization is induced by the formation of mutually exclusive gradients of the Rho-family monomeric G-protein, Rac, Cdc42, and RhoA. At the front, a phosphoinositol-3-kinase (PI3K)-dependent control of Rac and Cdc42 modulates the constant and dynamic rearrangement of the actin network necessary to probe the environment and power cell migration forward (Fig. 1).

The mid-region and the back of the cell (uropod) accumulate the phosphate and tensin homolog (PTEN) to counteract the activity of PI3K and prevent the formation of

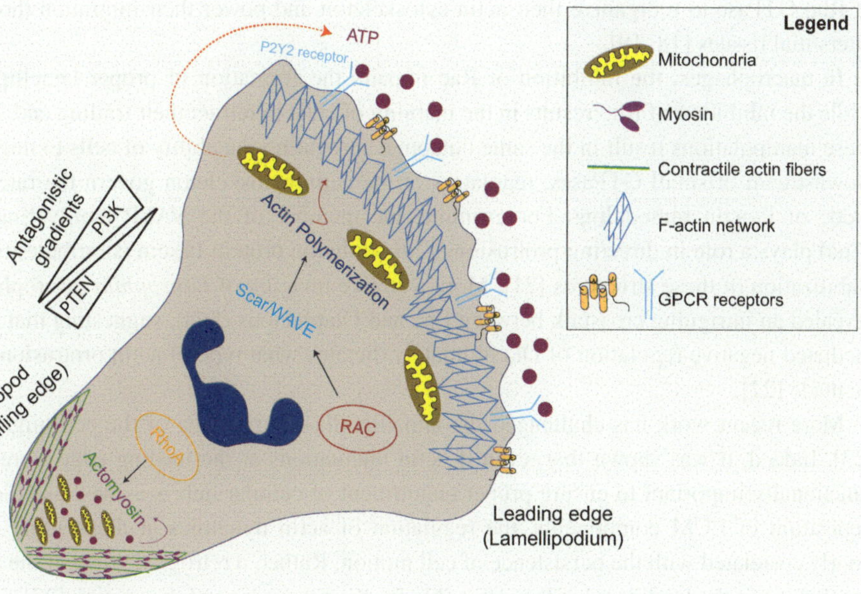

Fig. 1 Neutrophil polarization during cell migration. At the front of a migrating cell, Rac- and Cdc42-dependent polymerization of F-actin contributes to the formation of a stable membrane sheet, known as lamellipodium, which pushes the cell forward. At the back (uropod), the contraction of actomyosin fibers allows the detachment of the trailing edge from the substrate. The uropod is also characterized by the accumulation of mitochondria to provide sufficient ATP for the phosphorylation of myosin light chain and therefore myosin contraction. A small subset of mitochondria localizes at the leading edge where they release ATP extracellularly (red circles). The autocrine binding of ATP to P2Y2 receptors expressed at the leading edge reinforces the polarization dictated by pre-existing chemotactic cues. A mutually exclusive gradient of PI3K and PTEN contributes to the maintenance of the established front/back polarity

ectopic protrusion, to maintain a front-to-back polarity. Moreover, by regulating Myosin II contractility, PTEN not only allows for rear contraction, propelling the cell forward, but also creates a refractory zone insensitive to receptor signaling, reinforcing cell polarity. Therefore, the intracellular antagonist gradient of PI3K/PTEN activation establishes an asymmetric response to GPC receptors and, consequently, an asymmetric activation of Rho GTPases, with the highest levels of Cdc42 defining the direction of cell movement. At the back, high levels of RhoA and myosin activity create a cell rear unresponsive to receptor inputs, maintaining cell directionality. However, by modulating localization and activation of Myosin II, it is possible to tune the plasticity of the cell rear and the ability of cells to change the direction of migration. Cell polarization is therefore somewhat fluid to allow the system to change the direction of movement when needed [17].

Contrary to neutrophils, macrophages mostly adopt a mesenchymal-like, adhesion-based mode of migration [16]. Like neutrophils, macrophages also rely on the activity of Rho GTPase to reorganize their actin cytoskeleton and power their migration through interstitial tissues [18, 19].

In macrophages, the inhibition of Rac impairs the formation of proper lamellipodia while the inhibition of Rho results in the inability of cells to retract their trailing end. Both these manipulations result in the same outcome, abrogating the ability of cells to migrate. Downstream of small GTPases, regulators of the actin cytoskeleton govern the rate and mode of F-actin remodeling. For example, the member of the VASP family Enabled (Ena) plays a role in directing protrusions [20] while the protein Fascin contributes to the stabilization of these structures [21]. Moreover, live imaging of *Drosophila* macrophages revealed an intriguing crosstalk between Ena and Diaphanous (Dia), suggesting that Ena-mediated negative regulation of Dia ultimately dictates what type of actin protrusion will be made [22].

More recent work has challenged the lamellipodia-centric view of the cell migration [23]. Indeed, it was shown that, even if actin fluctuations at the leading edge were still functionally important to ensure proper engulfment of cellular debris as well as efficient deposition of ECM components, the regulation of actin dynamics at the leading edge poorly correlated with the persistence of cell motion. Rather, a retrograde flow of the actin network from the leading edge is responsible for the persistence of the motion [23].

While the remodeling of the actin network is the engine that powers cell migration, the microtubule cytoskeleton influences cell motility via various mechanisms, including transports of cargoes to the cell edge, remodeling of cell adhesions, and regulation of the actin network. Neutrophils show a strong polarization of the microtubules toward the rear of the cells [16, 24, 25]. This peculiar polarization allows these cells to concentrate most of their mitochondria at the rear. Mitochondria engage with the motor protein dynein through the adaptor Miro [26] and travel along the microtubule shaft toward the cell rear to provide enough ATP to power the contraction of the actomyosin cytoskeleton [27], fueling repeated cycles of actomyosin contractility (Fig. 1). Cell migration, and especially contraction of the actomyosin network, is indeed an energetically consuming process. A small subset of mitochondria localizes at the front of the cell. ATP released at the front acts autocrinally on purinergic P2Y2 receptors to promote dynamic actin polymerization, reinforcing neutrophil response to chemotactic gradients (Fig. 1) [28].

In macrophages, microtubules show a less obvious polarization toward the back, and they are more organized into bundles that extend in the actin-based lamellipodia [16, 29]. In *Drosophila* macrophages, these microtubule bundles are required to control the directionality of the movement. It was shown that microtubules work as a compass that helps cells orient their movement, actively dictating the direction of lamella polarization. This was even more evident in response to a polarizing chemotactic cue as the one released from the edge of an epithelial wound. Upon tissue injury, macrophages in the immediate proximity quickly reorganized their microtubule network assembling an arm in the direction of the wound, before polarizing the actin cytoskeleton accordingly. Moreover, microtubules also regulate cell–cell interaction during contact inhibition of locomotion (CIL) [29, 30]. Despite chemotaxis being the driving signal that directs cell migration during development or in response to tissue damage, contact with neighboring cells can provide spatial and directional cues during migration. Mathematical modeling of *Drosophila* macrophage dispersal during embryonic development showed how cell repulsion in vivo is sufficient to guide cell distribution within the embryos [30], despite soluble factors such as the platelet-derived and vascular endothelial growth factors (PDGF and VEGF, Pvf ligands in flies) still act as important dispersal cues [31]. In recent years, numerous studies have suggested that it is not a stochastic event but acts as an instructive migratory cue that requires the synchronized reorganization of cytoskeletal components between colliding cells.

High-resolution imaging of these cells has contributed to clarifying the importance of cytoskeleton dynamics in the regulation of CIL. Microtubules also control in vivo cell–cell repulsing behavior upon collision. Colliding cells indeed align their microtubule arms before repulsion occurs and cells with a compromised microtubule cytoskeleton lose the ability to repolarize their movement and therefore remain clumped together [29]. Collision also forces a rearrangement of the actin cytoskeleton. Colliding cells indeed organize a transient focal adhesion that leads to the formation of stress fiber and the consequent release of tension that drives repulsion [32].

3 Innate Immunity and Development

In all vertebrates, macrophages are the first class of leukocytes to be specified during embryonic development [33]. This lineage of "primitive macrophages" arises from monocyte precursors in the yolk sac or the aortic endothelium. Following their specification, they either differentiate locally or infiltrate the fetal liver, contributing to the initiation of definitive hematopoiesis. Once liver hematopoiesis is established the number of phagocytes has increased exponentially, and they finally become the most abundant cell type in the embryo, constituting up to 15% of the total cells of each developing tissue. During the early phases of embryonic development, macrophages are most abundant in those tissues undergoing active remodeling. Tissue remodeling by apoptosis is a very well-documented mechanism that shapes organs and defines their correct size [34]. Macrophages crucially support this process by scavenging apoptotic corpses and cellular debris, hence facilitating tissue reorganization. Not surprisingly, the position of macrophages during embryonic development closely overlaps with areas of intense apoptosis. For instance, in the fly embryo, macrophages cover the entire length of the developing ventral nerve cord (VNC) to actively remove dying neurons. In this scenario, altering macrophage dispersal by genetic means greatly compromises neuronal development [35–37]. Similarly, macrophages populate the interdigital tissue of a developing mouse limb, helping to complete the process of digit separation within a few hours [38]. In most cases, macrophages recognize a dying cell with high levels of specificity thanks to the presence of various "eat-me" signals and act accordingly, eventually eliminating the dysfunctional cells. However, there are circumstances in which macrophages source the killing signal causing the death of their target cells through the release of pro-apoptotic mediators. Analysis of macrophage behavior in the developing rodent eyes showed that these scavengers can elicit the activation of the apoptotic program in adjacent vascular endothelial cells, setting the correct timing for capillary regression [39]. Endothelial cell apoptosis is induced through the release of WNT7b and the activation of the canonical Wnt pathway [40].

Since their discovery more than a century ago by Elie Metchnikoff [41], the "big eaters" have drawn a lot of attention as their developmental function is not merely linked to phagocytosis. Fetal macrophages are for example important during branching morphogenesis. Mice null for transcription factors (TFs) involved in macrophage differentiation for example show clear defects in the development of organs, such as the lung, kidneys, and mammary glands, where branching morphogenesis is crucial.

In these scenarios, local activity of macrophages supports branching by deposition of extracellular matrix components, as collagen [42]. Similar function has also been described in the developing *Drosophila* embryos, where macrophages support correct kidney migration and positioning within the body by de novo deposition of Collagen IV [43].

Finally, during development macrophages contribute to vessel sprouting and vasculature remodeling. Despite not being directly involved in the initial phase of blood vessel formation (vasculogenesis), macrophages regulate vessel anastomosis by acting as "cellular chaperones". In fact, they can simultaneously interact with multiple endothelial tip cells, bringing them in sufficient proximity for fusion [44].

Moreover, they contribute to the remodeling of the vessel network by inducing the regression of superfluous vessels as well as by secreting soluble anti-angiogenetic factors which act by sequestering the vascular endothelial growth factor A (VEGF-A), ultimately inhibiting the main angiogenesis-inducing signal [45]. After vessels formation, this immune cell population is essential for the maintenance of vascular homeostasis, through the secretion of a variety of pro-angiogenetic growth factors as well as the recruitment of perivascular cells (or pericytes), which provide structural support to the vessel by reinforcing adhesion between endothelial cells [46].

4 Innate Immunity and Infection

The development of antimicrobial resistance (AMR) is considered a worldwide emergency that is estimated to cause the death of ten million people per year by 2050 [47]. AMR occurs when microbes, such as bacteria and fungi, become over time unresponsive to the treatments with drugs already available on the market, making it much harder to control the infection and posing a serious threat to public health [48]. Therefore, there is a pressing need to develop novel strategies to treat such infections, to ultimately override the problem of pathogen persistence.

Mouse models have been an invaluable resource to model the pathogenesis of infectious diseases and to test new drugs and vaccines for their effectiveness and toxicity before proceeding to human clinical trials [49]. In addition, the use of domesticated livestock has allowed studying the zoonotic transmission of certain infectious diseases, successfully contributing to the development of new vaccines [50]. However, these models do not easily lend themselves to describe with accuracy the cellular response to the infection. The use of genetically tractable and visually transparent organisms, such as zebrafish, has allowed us to gain useful molecular and cellular insights into the activation of the host inflammatory response to a wide variety of clinically relevant pathogens while happening in a living organism. Compared to higher vertebrate models, zebrafish offers the unique opportunity to study the dynamics of the inflammatory response in isolation from the adaptive immunity given that in the first 4 weeks of larva development the innate immunity is the only line of defense of the developing organisms.

Moreover, the possibility of injecting fluorescently labeled microbes combined with the use of transgenic lines to visualize specific immune cell populations has allowed us to describe with unprecedented cellular resolution the immune cell response to an infection and to decipher what strategies the pathogens adapt to evade the host immunity and disseminate [51].

Over the years, zebrafish have also become an extremely valuable model organism to investigate the molecular mechanism behind the acquisition of antibiotic resistance which underlies the recurrence of infection. For example, zebrafish have contributed to gaining a better understanding of a phenomenon known as the L-form switching, whereby *E. coli* strains treated with antibiotics targeting the cell-wall switch into a cell-wall-deficient state (the so-called L-form), becoming unresponsive to the treatment [52]. The optical translucency and the suitability for high-throughput drug screening make the zebrafish an excellent model organism to gain more mechanistic details behind the molecular nature of phenomena such as the L-form switching and to tackle the problem of antimicrobial resistance. In this paragraph, we will describe how studies in zebrafish have contributed to identifying the signaling pathways responsible for directing innate immune cells toward infected tissues and to describe the dynamic interaction of innate immune cells with invading pathogens.

4.1 Directional Migration of Leukocytes to Sites of Local Infection

Neutrophils and macrophages are the main effectors of the initial host immune response against invading pathogens and therefore play a crucial role in their recognition and subsequent activation of a robust immune response. At the site of infection, these cells are involved in the active removal of the pathogens from the extracellular space and in the release of antimicrobial effectors. Additionally, they are also responsible for the release of cytokines which enhance the recruitment and activation of other immune cell types.

Neutrophils are the most abundant leukocytes in both humans and zebrafish. Their crucial role in controlling the spread of infection is well documented by the numerous reports of the increased susceptibility to a wide variety of pathogens, typical of patients affected by warts, hypogammaglobulinemia, infections, and myelokathexis (WHIM) syndrome, a genetic condition characterized by severe neutropenia despite the abnormal accumulation of mature neutrophils in the bone marrow [53]. This syndrome is caused by a mutation in the gene *CXCR4*, which encodes for a G-protein-coupled receptor. CXCR4 regulates leukocyte mobilization from the bone marrow and their trafficking to the peripheral blood. Studies in zebrafish support this function and suggest that the retention of neutrophils to the caudal hematopoietic tissue (CHT, equivalent to the fetal liver in mammals) is mediated by the interaction between CXC receptor 4 (CXCR4) and a chemokine, known as stromal-derived growth factor 1 (SDF-1). This signaling axis is impaired in a zebrafish model of the autosomal dominant disorder WHIM and, as in humans, the fish suffer from severe neutropenia due to the abnormal retention of neutrophils in the bone marrow [54].

The rapid mobilization of neutrophils from the bloodstream is key for the successful resolution of infection. In healthy individuals, neutrophils are quickly recruited to infected tissues where they actively phagocytose bacteria, helping to reduce the microbial burden. Elegant work in zebrafish larvae monitored the fast accumulation of these phagocytes at the site of infection shortly after bacteria injection, until neutrophil number peaks 6 h post-inoculation [55]. The nature of the signals that mediate the long-range recruitment of neutrophils from the bloodstream still needs careful characterization. Hydrogen peroxide (H_2O_2), a known chemoattractant signal responsible for the recruitment of leukocytes to sites of tissue damage, has been regarded as a potential candidate. However, elegant work in zebrafish larvae revealed that H_2O_2 is indispensable for the recruitment of neutrophils to sites of localized infections. Further work has identified CXCL8, also known as interleukin-8 (IL-8), as a potent chemoattractant that, signaling through the receptor CXCR2, allows the infection-induced mobilization of neutrophils and their directional migration to the target tissue through a complex 3D environment [56, 57].

Increased mobilization of neutrophils from hematopoietic tissue and increased levels of neutrophils in circulation (neutrophilia) are not always indicative of an effective and successful response to bacterial infection. Indeed, once mobilized, neutrophils must be able to infiltrate the infected tissue. Not surprisingly, patients affected by Leukocyte Adhesion Deficiency (LAD) syndrome are characterized by increased peripheral distribution of neutrophils and much reduced infiltration to sites of infection. Mutation of the Rho GTPase Rac2 recapitulates the neutrophilia phenotype of LAD patients, suggesting that polarized neutrophil migration is required for infiltration of the infected tissues. By introducing a human inhibitory Rac2^{D57N} mutation in zebrafish neutrophils, Deng and colleagues show that Rac2 signaling is required in neutrophils for efficient migration in 3D environments, by allowing the accumulation of PI3(K) phospholipid products and regulating actin protrusion dynamics [58]. The work in zebrafish, therefore, provides a mechanistic explanation for the neutrophilia observed in human patients affected by LAD or carrying the same inhibitory RAC mutation and provides a powerful platform for future investigations.

Rac2 also regulates the motility of macrophages and enhances their migration toward sites of infections. It is not surprising therefore that *rac2* null mutants are extremely susceptible to infections and that re-expression of *rac2* in either neutrophils or macrophages enhances the host's resistance to bacterial infection [59]. However, *rac2*$^{-/-}$ macrophage migration defects are less severe than those of *rac2*$^{-/-}$ neutrophils, suggesting a differential requirement of this Rho GTPase between the two cell types or might simply reflect the differential mode of migration adopted by these two different classes of leukocytes. As for the neutrophils, also the recruitment of macrophages to localized infections is mediated by chemokines mediators. Mobilization of macrophages from the bone marrow and efficient migration to sites of infection are mediated by the CCL2–CCR2 signaling axis. Macrophages indeed constitutively express the CCR2 receptor, which is well conserved between fish and human [60, 61]. Another key chemokine receptor, which stimulates macrophage recruitment and regulates several other aspects of their biology, is CXCR3 which signals through the ligands CXCL9, CXCL10, and CXCL11 [62]. Given the importance of CCR2 and CCR3 in mediating leukocyte recruitment, it is not surprising that these receptors have been the targets of several pharmacological treatments aiming at reducing further infiltration of these cells to sites of severe inflammation [63].

4.2 Mechanisms of Pathogen Recognition and Modes of Elimination

The powerful defense provided by innate immune cells against invading pathogens crucially relies on their effective recognition, leading to the activation of a robust immune response. The process of recognition depends on the expression of pattern recognition receptors (PRRs). The existence of PRRs was firstly postulated in 1989 by Charles Janeway, which suggested that a collection of receptors, expressed on the surface of immune cells, mediated the recognition of evolutionarily conserved structures on the pathogens, known as pathogen-associated molecular patterns (PAMPs) [64]. Since then, intense investigations have been carried out to clarify how the activation of such receptors initiates intracellular signaling pathways that culminate with the activation of a proper pro-inflammatory response against the dangerous invaders, including the production of chemokines and cytokines.

So far, different classes of PRRs have been identified. Of these, one of the best characterized is the family of Toll-like receptors (TLRs), which were originally identified in *Drosophila* as crucial in mediating the anti-fungal response [65]. TLRs are transmembrane glycoproteins that oligomerize upon binding with their ligands, either directly or through intermediate molecules known as PAMP-binding molecules. TLRs recognize bacterial products, lipopolysaccharide, and lipoteichoic acids. While it is commonly accepted that TLR4 recognizes gram-negative bacteria while TLR2 is involved in the recognition of gram-positive, most bacteria can simultaneously activate multiple TLRs via multiple

PAMPs. Most TLRs are expressed on the cell membrane; however, some of them localize in intracellular compartments, such as lysosomes and endosomes, and are involved in the recognition of nucleic acids. In addition to these intracellular TRLs, there is a large class of cytosolic PRRs, generally categorized as retinoic acid-inducible gene I (RIG-I)-like receptors (RLRs) and nucleotide-binding oligomerization domain (NOD)-like receptors (NLRs), which have been characterized as mediators of the anti-viral response.

Activation of TRLs stimulates a robust pro-inflammatory response which includes the production of cytokines, vasodilation to recruit more leukocytes to the site of infection, and production of antimicrobial components, including reactive oxygen and nitrogen species (ROS and RNS, respectively) [66]. ROS and RNS are potent microbicidal weapons produced in high amount at the phagosomes of both neutrophils and macrophages in the so-called oxidative burst.

Neutrophils can also clear dangerous pathogens by other means than phagocytosis, including the release of a variety of toxic molecules. Most of these compounds are stored within intracellular granules that either fuse with the phagolysosomes, to help the digestion of internalized pathogens, or get released in the extracellular space (degranulation). The granules contain antimicrobial peptides and proteases, such as serine proteases and matrix metalloproteinases (MMPs). These enzymes are utilized not only to kill bacteria but also to digest the basement membrane and the extracellular matrix, which facilitate further recruitment of neutrophils from the circulation. Neutrophils have also perfected another valuable host defense strategy to fight infections, which is known as suicidal NET or NETosis [67, 68]. NETosis is a specialized form of cellular suicide, which is stimulated by IL-8. This extremely beneficial mechanism of defense consists in the release of structures of decondensed chromatin, proteases, and granular proteins, known as Neutrophil Extracellular Traps (NETs), which adhere to the invading bacteria with the final goal to trap them and prevent their further dissemination. In more recent years, a non-suicidal form of NETosis, named vital NETosis, has been identified. Vital NETosis is triggered by microbe recognition by specific host pattern recognition receptors and consists in the release of NETs without cell lysis, leaving neutrophils intact and able to migrate in response to chemoattractant gradient and phagocytose pathogens. While neutrophils can adopt this varied array of strategies to keep infections under control, the production of ROS (such as superoxide) has been shown to be crucial when the site of infection is not readily accessible, as in the notochord infection model presented by Phan et al. [69]. In this scenario, neutrophils achieve long-range killing of bacteria by swarming near the site of infection and releasing a high amount of superoxide in a Nox-dependent manner. This strategy could be adopted not only to kill bacteria that invade tissues that leukocytes access with difficulty but also to fight those pathogens that try to evade immune surveillance by escaping phagocytosis [69].

As for the neutrophils, the first macrophage response to a pathogen is its phagocytosis and killing. However, another pathway that controls the spreading of those microbes that escape the phagosome is autophagy, which relies on the expression of the autophagy marker Lc3 and the receptor p62.

Another alternative strategy that allows macrophages to gain control over a bacterial infection consists in the activation of a specialized form of cellular suicide, known as pyroptosis. Activation of cytosolic PRRs triggers the assembly of a large inflammatory complex, called the inflammasome, which ultimately activates Caspase-1. This response culminates in the activation of inflammatory cytokines, including IL-1β and IL-18, and finally in cell death by pyroptosis. At the molecular level, Gasdermin D (GSDMD) is the crucial mediator of pyroptosis. GSDMD is cleaved by Caspase-1, and this active form oligomerizes to form pores in the plasma membrane, disrupting cell integrity and eventually killing the leukocyte. Activated GSDMD also shows affinity for cardiolipin present on the bacteria plasma membrane to which it binds leading to membrane pore formation and consequent bacteria killing. Macrophage death by pyroptosis induces the release of cleaved GSDMD, which will target those bacteria present in the extracellular space, and of the pro-inflammatory mediators IL-1β and IL-18, which contribute to amplify the inflammatory state induced by pyroptosis with consequent recruitment of more leukocytes to the area of interest [70–72].

Finally, another strategy of host defense performed by macrophages consists in the formation of multicellular structures, known as granulomas. The formation of these structures is triggered by the encounter with the pathogen *Mycobacterium tuberculosis*, the agent causing tuberculosis in humans. This disease is difficult to cure due to several reasons, including the development of anti-drug resistance and the ability of the pathogen to enter a state of dormancy, followed by reactivation and disease recurrence. For a long time, the formation of granuloma lesions has been considered a beneficial protective strategy specialized in trapping the pathogens and blocking their replication in the extracellular space. However, additional studies in fish larvae have revolutionized the initial paradigm whereby macrophage-induced granulomas were considered solely protective structures. It has been shown that the mycobacteria can stimulate granuloma formation by promoting further infiltration of macrophages to the site of infection through the release of MMP-9, the secretion of which has been targeted therapeutically as a means to fight the infection. Once phagocytosed by macrophages, these bacteria can escape phagosomal degradation and gain control of the cytosol, where they replicate. Moreover, they stimulate the upregulation of efflux pumps which promote the development of drug resistance. Eventually, this drug-resistant population of bacteria kills the host cell and infects other macrophages, supporting its own replication. Overall, in recent years zebrafish has been used to watch live granuloma formation, offering the unprecedented opportunity to get a better understanding of how pathogens evade this system and providing novel venues for the design of more innovative strategies to control bacteria dissemination.

5 Leukocyte Migration to Sites of Tissue Damage

Migration of leukocytes to the site of tissue damage is crucial for the proper healing of tissues upon wounding. This phenomenon was reported for the first time by Ilya Mechnikov, who observed small motile "elements" being rapidly drawn to sites of local injury in the starfish larvae [73]. Since the pioneering studies of Mechnikov, a lot of work has been done to understand how inflammatory cells drive the healing process. Nowadays, the use of translucent animal models, the possibility to fluorescently label different immune cell populations, and the tremendous improvement of imaging techniques allow us to observe live the process of leukocyte infiltration to sites of tissue damage with an unprecedented temporal and spatial resolution (Fig. 2).

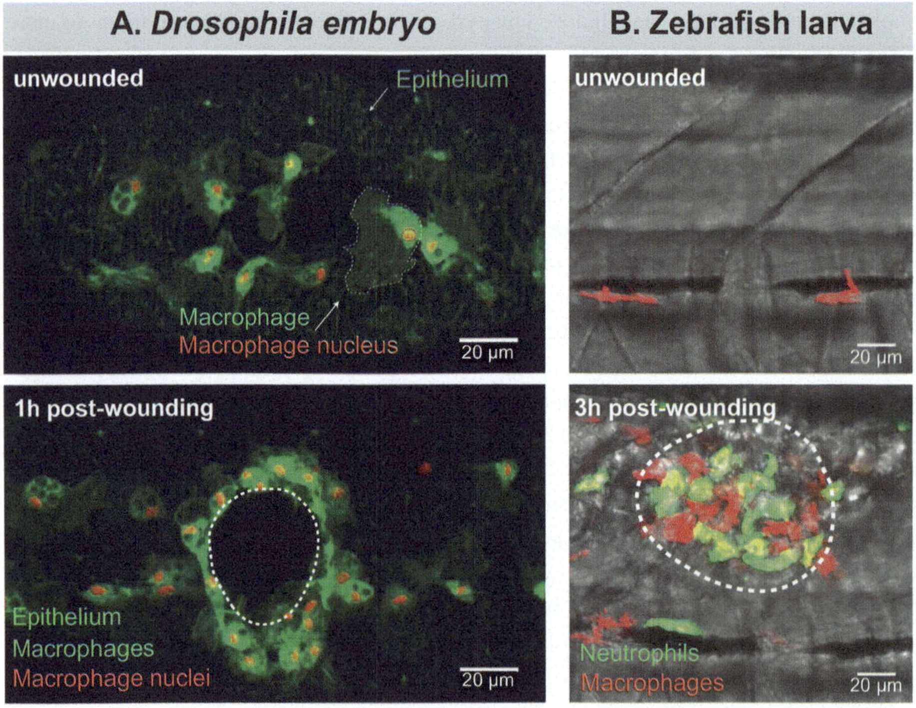

Fig. 2 Recruitment of immune cells to sites of tissue damage. A laser-induced wound in the ventral epithelium of a *Drosophila* embryo (**a**) or 5-day post-fertilization (dpf) zebrafish larva (**b**) triggers the quick recruitment of cells of the innate immunity from the surroundings. In (**a**), the epithelium is labeled in green, and fly macrophages (or hemocytes) are co-labeled in green and red. In the unwounded panel, a representative macrophage is outlined by a dotted white line. In (**b**), a double transgenic zebrafish has neutrophils labeled in green and macrophages in red. In both post-wounding panels, the site of damage is marked by a white dotted line. Fly image courtesy of Dr. Giuliana Clemente, University of Bristol, and fish image courtesy of Dr. David Gurevich, University of Bath

In this session, we will discuss what are the signals that help immune cells confidently navigate the three-dimensional (3D) environment toward sites of tissue damage, how inflammatory cells recognize these signals, and finally the resolution of the inflammatory response.

5.1 Signals Guiding Leukocytes Recruitment to Sites of Tissue Damage

Leukocyte recruitment to sites of damage is a chemotactic guided process controlled through the passive release from the injured tissue of small endogenous molecules, named alarmins or damage-associated molecular patterns (DAMPs) (Fig. 3).

To be able to act as damage signal molecules, most DAMPs are localized in intracellular compartments, hidden from the inflammatory cells. Only upon damage, they are released

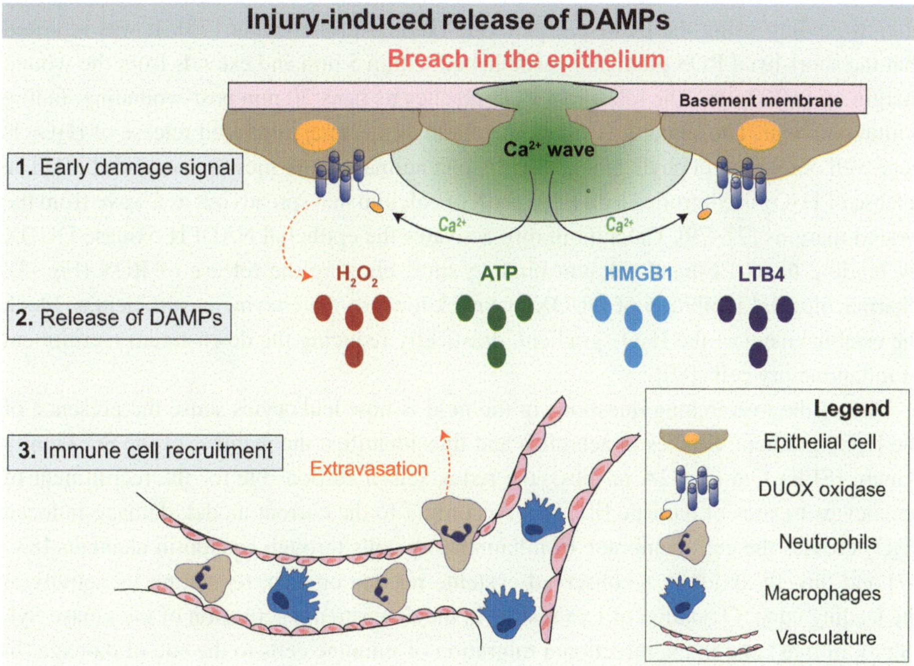

Fig. 3 Damage-associated molecular patterns (DAMPs) are released after tissue injury. Following damage, calcium spreads like a wave (in green) from the wound margin. Calcium binds to the calcium-sensitive EF-domains of the epithelial NADPH oxidase DUOX (in blue, EF-hand domain in orange), resulting in the release of hydrogen peroxide (H_2O_2, in red) as one of the earliest injury-induced DAMPs. The gradient of DAMPs is sensed by the cells of the innate immunity through G-protein-coupled receptors (GPCRs), inducing their polarization and migration toward the site of injury

in the extracellular space and bind to specialized receptors on the surface of the immune cells, inducing cell polarization and migration toward the site of damage. Strategically, immune cells themselves can actively release some alarmins to enhance the inflammatory state and further sustain leukocyte infiltration to the repairing tissue.

Since Polly Matzinger proposed the "danger model" in 1994 [74], several molecules have been classified as alarmins. In fact, today it is known that tissue injury results in the release of a complex blend of inflammatory and chemoattractant signals that must be hierarchically interpreted to guide the timely recruitment of leukocytes to sites of injury (Fig. 3).

5.2 Reactive Oxygen Species

Reactive oxygen species (ROS) are among the earliest damage signals responsible for the coordinated recruitment of leukocytes at the location of injury [75, 76].

Niethammer and colleagues described the establishment of a gradient of ROS shortly after wounding using the hydrogen peroxide (H_2O_2) probe Hyper [75]. It was reported that the short-lived ROS gradient is established within 5 min and extends from the wound margin up to 200 μm. The release of ROS reaches its peak 30 min post-wounding, fading within one hour. The signaling cascade regulating the damage-induced release of H_2O_2 is very well conserved in phylogenetically distinct animals, from flies to mammals [77]. The release of H_2O_2 is controlled by a rapid flash of calcium that spreads out as a wave from the wound margins [77–79]. Calcium in turn activates the epithelial NADPH oxidase DUOX by binding to its EF-hand calcium binding sites, eliciting the release of ROS (Fig. 3). Pharmacological inhibition of DUOX or knockdown of the enzyme is sufficient to block the establishment of the H_2O_2 gradient, drastically reducing the downstream recruitment of inflammatory cells [80].

One of the overarching questions in the field is how leukocytes sense the presence of the H_2O_2 gradient. Studies in zebrafish and flies identified the member of the Src Family Kinase (SFK) Lyn (src42A in flies), the redox sensor responsible for the recruitment of leukocytes to sites of damage [81, 82]. According to the current model, damage-induced H_2O_2 crosses the cell membrane of inflammatory cells through aquaporin channels [83–87] and directly oxidizes a conserved cysteine residue on Lyn, regulating its activity at the leading edge. Oxidation of Lyn results in the downstream activation of the kinase Syk (Shark in flies) and in the directional migration of immune cells to the site of damage. In fact, mammalian Syk is required for the polarized accumulation of phosphoinositide 3-kinase (PI3K) at the leading edge of leukocytes, where PI3K is required for the regulation of lamellipodia dynamics and directional migration [88]. Interestingly, activation of PI3K is also essential in flies for efficient macrophage chemotaxis toward epithelial wounds [31], suggesting a conserved Syk-dependent mechanism of PI3K polarization.

Wound-induced ROS do not serve exclusively as a chemoattractant, but they also act on immune cells by enhancing the expression and cortical localization of damage receptors.

For example, *Drosophila* macrophages increase the levels of Draper, an apoptotic cell clearance receptor of the CED-1 family, and this molecular change primes immune cells for the robust detection and response to subsequent tissue damage [82, 89]. Moreover, the increased levels of intramolecular ROS upon injury activate the Toll receptor and stimulate the production of cytokines. This wound-induced activation of the inflammatory response trains the immune cells to quickly respond to infection, even before an encounter with dangerous pathogens that, if unchallenged, could ultimately cause life-threatening sepsis [87].

The beneficial effects of wound-induced ROS on tissue repair go beyond the modulation of the innate immune response. Basal levels of ROS support tissue repair also by promoting keratinocyte migration and proliferation [90] as well as fibroblast proliferation and matrix deposition, which favors the formation of a barrier that prevents further passage of microbes in the host [91, 92].

ROS is an umbrella term that refers to various oxygen-containing molecules, characterized by several unpaired electrons that make them particularly prone to interact and damage primary cellular constituents, such as DNA, proteins, and lipids. As uncontrolled release of ROS can therefore cause irreversible and devastating damage to otherwise healthy tissues, wound repair requires a finely tuned balance between positive and noxious effects of ROS. For this reason, the cells in close proximity to a wound upregulate a complex cytoprotective network to withstand the bystander damage caused by these highly reactive compounds [93, 94]. Not surprisingly, chronic non-healing wounds in the elderly are often characterized by levels of ROS which exceed the low concentration of 100–250 μM characteristic of normal wounds [95] as well as an age-related decline in ROS scavenging strategies [93]. Given the multiple important healing roles of ROS at sites of damage, the ability to manipulate ROS levels can represent an exciting route for improving wound healing responses. Wound dressings represent a particularly advantageous strategy to treat wounds and burns and are extensively used in the clinic as an effective approach to prevent infections as well as to keep a moist environment to sustain the healing process [96, 97].

Indeed, wound dressings are often combined with anti-bacterial material, as zinc oxide nanoparticles, or with growth factors and cytokine that, if delivered in a functional, active state and at the appropriate concentration, can help the regeneration of the tissue. For example, hyaluronic acid-based hydrogels have been successfully used to locally deliver appropriate levels of vascular endothelial growth factor (VEGF) to promote angiogenesis at the wound site [98].

Wound dressings could potentially also be used to control levels of ROS at the site of damage. Recently, wound dressings made of oxidized regenerated cellulose (ORC) or a mixture of collagen and ORC have shown greater ROS quenching capacity than collagen dressings on their own [99]. Another strategy to alleviate the overload of ROS characteristic of non-healing wounds might consist in the local delivery of ROS detoxification enzymes, such as superoxide dismutase (SOD) or catalase [100, 101]. The use of polymerizable ROS-processing enzymes has given promising results in alleviating inflammatory liver injury caused by ischemia and reperfusion of the tissue as well as in minimizing inflammatory damage associated with transplantation procedure.

5.3 High Mobility Group Box-1 (HMGB-1)

Identified in 1973, HMGB-1 is a non-histone DNA binding protein that facilitates the binding of transcription factors to the chromatin [102]. Upon damage, this nuclear protein is released in the extracellular space and stimulates inflammatory cells through the activation of Toll-like receptors (TLRs) 2 and 4 and receptor of advanced glycation end products (RAGE) [103, 104].

As a cytokine, HMGB-1 promotes tissue repair through multiple mechanisms. Firstly, HMGB-1 has been shown to promote the recruitment of leukocytes [105], especially neutrophils, to sites of tissue damage. In fact, HMGB-1 facilitates neutrophil infiltration by increasing the expression of vascular cell-adhesion molecule 1 (VCAM-1), intercellular adhesion molecule 1 (ICAM-1), and endothelial cell selectin (E-selectin) on endothelial cells, and therefore increases the adhesion of leukocytes to the vessel walls, helping their translocation across the epithelial barrier [106]. Moreover, HMGB-1 exerts a direct effect on neutrophils by stimulating the release of NETs, enhancing their antimicrobial activity.

In addition to its role as a leukocyte chemoattractant, HMGB-1 promotes wound healing by stimulating the migration of stem cells to sites of injury as well as their proliferation. Exposure to extracellular HMGB-1 enhances tissue regeneration by stimulating the transition of stem cells from G_0 to "alert" G_{alert} phase. Transition into G_{alert}, therefore, primes stem cells for cell cycle entry, enabling the tissue to rapidly respond to injury and stress conditions [107]. Injection of purified HMGB-1 into the peri-infarcted myocardium ameliorates tissue repair by increasing migration and proliferation of c-kit cardiac stem cells at the site of injury, greatly improving tissue function compared to control [108]. In some contexts, pharmacological pre-administration of HMGB-1 confers protection against hepatic, lung, and kidney reperfusion injury, suggesting that low-level exposure to molecular mediators of injury as HMGB-1 can precondition tissues, providing tolerance against deleterious stresses [109–111].

Finally, HMGB-1 also acts as a potent angiogenetic factor. As such, it stimulates the RAGE receptors present on endothelial cells, significantly increasing the sprouting and de novo growth of vessels [112]. In addition, it also supports angiogenesis by stimulating the macrophage-dependent release of pro-angiogenetic cytokines, such as TNF-α and IL-8 [113].

5.4 Purinergic Signaling

ATP signaling has been involved in the regulation of the inflammatory response upon injury and in the healing process. In epithelial cells, a 10^6-fold difference in the levels of intracellular (10 mM) and extracellular (10 nM) levels of ATP has been recorded under basal conditions [114]. Therefore, the rupture of cells upon injury induces the release of high levels of ATP and elicits a response in the surrounding viable cells through the binding to P2Ys and P2Xs purinergic receptors.

Purinergic signaling positively regulates the initial steps of the healing process for example by regulating platelet aggregation. The formation of this platelet "plug" prevents further blood loss and regulates the release and diffusion of cytokines with mitogenic and chemotactic properties. ATP has also been implicated in the targeting of neutrophils to the site of damage. Treatment with apyrase or inhibitors of those receptors drastically reduces the recruitment of neutrophils to the site of damage, reinforcing the knowledge of ATP as a chemotactic cue.

Migrating leukocytes stimulated by chemoattractant molecules, such as the N-formylated tripeptide fMet-Leu-Phe (fMLP), release mitochondria-produced ATP from the leading edge, generating an auto-induced ATP gradient that reinforces the pre-existing chemoattractant signal [115]. More recent work shows that ATP favors neutrophils recruitment not acting as a chemoattractant molecule per se but stimulating the macrophage-dependent release of pro-inflammatory mediators, eventually targeting neutrophils to location of injury [116].

5.5 Lipid Mediators

Most of the chemoattractant molecules discussed above are short-lived signals. For example, ROS production upon damage is transient and its intensity declines within one hour. Therefore, more persistent, long-lived signals might be needed to sustain prolonged, long-range leukocyte infiltration. Studies in *Drosophila*, which combine mathematical simulation of the inflammatory response with live imaging of the tissue damage response, suggest that early DAMPs, as ROS or ATP, contribute to the generation of a favorable environment that sustains immune cells migration toward the site of injury [117]. In other words, these molecules amplify a pre-existing signal generated by a more persistent, less diffusive attractant released at the wound site [117]. However, the exact nature of this chemoattractant is still elusive. Recent work looking at the transmigration of *Drosophila* pupal macrophages through the vessel walls (extravasation) toward sites of epithelial damage identifies the fatty-acid sensing GPCR Tre-1 as important for the process [118]. During embryogenesis, Tre-1 works in the same signaling pathway as the phosphate phosphatase Wunen to promote germ cell migration [119]. The requirement of Tre-1 for macrophage transmigration upon wounding raises the intriguing possibility that a gradient of phospholipids generated from the wound edge might be the primary wound-released signal orchestrating leukocyte chemotaxis toward an epithelial wound. The concept of a gradient of lipid mediators acting as a long-range chemotactic signal is a venue of research that is also currently being investigated in vertebrate models. In fact, in mouse and zebrafish models of wound healing, lipid mediators such as leukotriene

B4 (LTB4) have been already implicated in the process [120, 121]. LTB4 is synthesized from arachidonic acid and acts as a chemoattractant through the G-coupled receptor BLT-1. Upon wounding, LTB4 is required to drive the initial recruitment of neutrophils to the site of injury, but it also acts at later stages to sustain neutrophil infiltration. During the early stages of neutrophil recruitment, neutrophil-secreted LTB4 reinforces chemotaxis acting autocrinally and stabilizing polarized migration [122]. Moreover, LTB4 promotes early wound infiltration, favoring intravasation. Indeed, LTB4 stimulates the upregulation of integrins and therefore adhesion of neutrophils to endothelial cells [121]. Long-term gradients of LTB4 are crucial for the exponentially amplified recruitment of neutrophils, known as swarming. Indeed, murine neutrophils lacking the LTB4 receptor BLT-1 were able to reach a laser-induced injury only if present in its close proximity, while more distal cells failed [121]. Recent work from Poplimont and colleagues investigates the role of LTB4 in the long-range recruitment of neutrophils and aims to address the intricate interplay between different alarm signals in supporting neutrophil swarming [123]. The authors suggest that those neutrophils that are recruited first to the site of damage form a cluster of tightly associated cells that communicate through connexin-43 (Cx43) hemichannels. The early damage signal ATP opens the hemichannels allowing for a calcium wave to travel across the neutrophil cluster. This close interplay between ATP and calcium works as a signal-relay mechanism that amplifies the original "calcium alarm" signal and culminates with the local secretion of the chemoattractant LTB4, promoting further recruitment of leukocytes to the wound [123]. This relay mechanism favors the formation of a dense "plug" of neutrophils that tightly seal the breach in the tissue, providing an effective barrier against dangerous invaders. Inhibition of the connexin-mediated cell–cell communication strongly impairs swarming, leaving the host more vulnerable to the presence of opportunist pathogens.

5.6 Differential Contributions of Neutrophils and Macrophages to the Healing of Wounded Tissue

Accumulation of inflammatory cells to the site of damage is crucial for a successful outcome of the healing process. Neutrophils are the first and most abundant responders at the wound site, where they congregate within minutes from the injury (Fig. 4). At the site of damage the main function of this tight cluster is to limit the access of opportunist pathogens through the breach in the epithelium, preventing any life-threatening septic event, through either phagocytosis or the release of NETs ([124], Fig. 4).

In addition to their main role in killing invading microbes, neutrophils are also involved in the regulation of the inflammatory response through the release of mediators, such as LTB4, prostaglandins, and interleukin-8 (IL-8), which stimulate the recruitment of additional inflammatory cells [125].

The pro-inflammatory environment at the site of damage results in the quick activation of tissue-resident macrophages as well as an influx of monocytes that leak through the

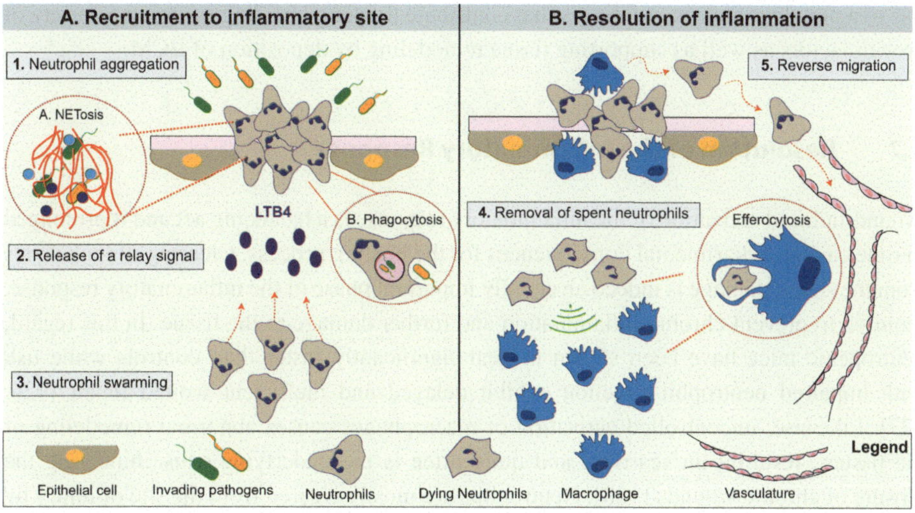

Fig. 4 Differential role of neutrophils and macrophages at the site of damage. Neutrophils are the first immune cells to be recruited to sites of injury (panel **a**). Here, neutrophils seal the breach in the epithelium to minimize the number of pathogens accessing the host tissue. Moreover, they actively fight infections by either trapping the invading pathogens by the extracellular release of chromatin and antimicrobial molecules (NETosis, A) or alternatively by internalizing and digesting them at the phagosomes (phagocytosis, B). The early recruitment of neutrophils is further amplified by a relay signal (LTB4) which elicits neutrophil swarming from the circulation. Dying neutrophils stimulate the recruitment of macrophages to the site of injury (panel **b**). The recruited macrophages phagocytose spent neutrophils (efferocytosis) and initiate the resolution of the inflammatory phase by inducing neutrophil reverse migration

disrupted vessel network to the wound bed. These early monocytes differentiate into M1 pro-inflammatory macrophages which express iNOS, IL-6, and IL-1b and join neutrophils in the battle against invading microbes. Moreover, they also voraciously phagocytose spent neutrophils (efferocytosis) and matrix debris, whose accumulation would eventually impair proper regeneration (Fig. 4).

Macrophages at the site of damage aggregate in stable, persistent clusters. The formation of these clusters implies that macrophages turn off the repelling behavior typical of CIL. How macrophages fine-tune CIL is still an area of intense investigation. A hypothesis is that macrophage clustering could be facilitated by the transient upregulation of E-cadherin and the establishment of homophilic cell–cell interactions.

At later stages, the healing process is driven by M2 anti-inflammatory macrophages which contribute to the re-vascularization of the healing tissue as well as to its remodeling through the de novo deposition of extracellular matrix. CCR2-expressing monocytes will differentiate in pro-angiogenetic macrophages which express high levels of VEGF and initiate de novo vessel sprouting and growth. CX3CR1-expressing monocytes will give rise to a population of macrophages that contribute to wound healing sourcing a variety of growth factors as well as supporting tissue remodeling by deposition of ECM.

5.7 Resolution of the Inflammatory Response

Wound-induced activation of the inflammatory response is a balancing act and if prolonged in time can have detrimental consequences for the healing process. Clearing of leukocytes from the site of damage is indeed an equally important phase of the inflammatory response, required to prevent chronic inflammation and further damage to the tissue. In this regard, neutropenic mice have been shown to heal significantly faster than controls while fish with impaired neutrophil function exhibit delayed and inefficient wound repair [126, 127]. Likewise, uncontrolled activation of macrophages causes abnormal remodeling of the tissues, resulting in scarring, and quite often is the underlying cause hindering the closure of chronic wounds [128]. Neutrophils are mostly cleared from the site of injury by macrophage efferocytosis (Fig. 4). This is recognized as an extremely important step in the resolution of the inflammatory response as it stimulates the macrophage-dependent release of anti-inflammatory mediators, such as TGF-β and IL-10. Macrophages also release interleukin-1 receptor antagonist (IL-1Ra), which counteracts the effects of IL-1alpha and IL-1 beta, overall dampening the inflammatory response. The release of IL-1Ra indeed reduces the infiltration of neutrophils at the site of damage and induces their apoptosis (IL-1 is a survival signal for neutrophils).

Additionally, neutrophils can also migrate away from the site of injury back to the vasculature, in a process known as "reverse migration" (Fig. 4) [129, 130]. Even if the molecular details of this phenomenon are unknown, macrophages are thought to be involved. Macrophages recruited at the wound site have been shown to physically interact with neutrophils and display neutrophil-repelling behavior. Consequently, neutrophils would be forced to migrate away from the injury back into the vasculature, initiating the resolution of inflammation. In support of this working model, depletion of macrophages reduces the number of neutrophils leaving the wound site [131]. Resolution of neutrophils from the wound area also supports the formation of a permissive angiogenetic area and the recruitment of anti-inflammatory macrophages, expressing high levels of VEGF-A which dynamically interact with endothelial tip cells, favoring vessel sprouting [132]. Moreover, these M2 macrophages take upon the challenge of supporting vessel regression and remodeling to restore a functional vasculature and they do so by phagocytosing dying endothelial cells [132].

Less clear is the mechanism driving the resolution of macrophages from sites of injury. Reactivation of CIL, and therefore of the cell–cell repulsive behavior, might provide a potential mechanistic explanation for the resolution of macrophage clusters. Understanding how macrophages regulate CIL might therefore shed light on some aspects of the resolution of the inflammatory response. However, this aspect of the healing process still needs further investigation.

5.8 Mathematical Modeling of Wound Healing

The successful outcome of the process of skin repair requires the complex and coordinated interplay between multiple cell types and the hierarchical and timely integration of a diversified cocktail of chemical mediators. When the process of skin repair and regeneration fails, it often results in the development of chronic non-healing wounds or excessive scarring [133]. Chronic wounds are defined as wounds that take longer than 6 weeks to heal. Moreover, chronic wounds are often associated with the development of severe and persistent infections and microbial films, which contribute to a persistent inflammatory state at the wound site. Scarring instead results from the dysregulated deposition of ECM components and uncontrolled proliferation of keratinocytes at the site of injury. Therefore, chronic wounds and scarring represent the opposite side of the same coin, and they both severely compromise patients' quality of life, posing a huge economic challenge to the health system worldwide [133]. Given the pivotal role played by the immune system during the various phases of tissue repair, it is not surprising that a dysregulated immune response, and particularly a prolonged inflammatory phase, is the underlying cause of a poor healing process. Hence, research in the field has been focusing on deepening our understanding of the molecular pathways through which the immune system influences the healing process, and on developing novel immune-based therapeutic strategies to prevent the formation of both chronic wounds and scars.

Nowadays, the use of animal models and the tremendous advancement of imaging techniques allow the collection of a great amount of biological data with an unprecedented spatio-temporal resolution. Therefore, there has been a substantial need to develop cross-interdisciplinary collaborations to provide a deeper insight into aspects of the healing process that would otherwise remain unexplored or that would require much more difficult and costly approaches to be investigated. Moreover, mathematical modeling is helping to understand the complex interaction between the different factors (or species) involved in the healing process, and it is becoming an invaluable tool to describe quantitatively and qualitatively the nature of these interactions. Sharrett and Murray developed the first mathematical model of epidermal wound healing describing how a chemical mediator can activate epidermal cells, stimulating their division and driving therefore tissue repair [134]. This early contribution from 1990 has been later used as a reference to mathematically describe processes such as cell migration, division, and death.

An important aspect underlying the healing process is the timely activation and resolution of the inflammatory response. The partial pressure of oxygen at the wound site greatly influences the outcome of this very first step of the tissue repair cascade. Neutrophils accumulating at the wound bed indeed consume a high amount of oxygen in the respiratory burst, a process whereby the production of ROS leads to the elimination of the phagocytosed pathogens. Recently, Saiko and colleagues put forward a 2D multi-agent stochastic model of the healing response looking at the interaction between neutrophils and bacteria in the presence of local hypoxia [135]. This model predicts that the local pressure of oxygen influences the number of recruited neutrophils and the duration of the inflammatory phase. Specifically, severe hypoxia often correlates with a significantly higher number of immune cells at the wound and a much longer inflammatory phase, independently from the presence of an infection. Interestingly, a number of trial studies suggest that increased tissue oxygenation can greatly reduce the likelihood of infection in postoperative wounds [136]. Particularly, it was found that peri-operative supplementation of oxygen reduced the likelihood of surgical site infection (SSI) up to 85%. Local tissue oxygen tension also impacts the process of wound angiogenesis, an extremely important process required to supply sufficient oxygen and nutrients to the repairing tissue. As described, macrophages help orchestrate the development of the new vessel network by secreting pro-angiogenetic growth factors. A model that considers local levels of oxygen, capillary density, and macrophage-derived growth factors predicts the establishment of a feedback loop whereby decreased levels of oxygen stimulate recruitment of macrophages and increased release of pro-angiogenetic growth factors which ultimately results in higher capillary density. Therefore, the model anticipates that the local oxygen tension will dictate the number of macrophages actively secreting growth factors and therefore vessel growth [137].

Mathematical simulations of how local oxygen tension influences angiogenesis allow predicting the efficacy of hyperbaric oxygen therapy (HBOT) in facilitating the healing of normal and chronic cutaneous wounds. The model predicts that, while an extremely hypoxic environment impairs angiogenesis, hyperoxia promotes de novo vessel formation and predicts a range of oxygen concentrations beyond which the positive effects of hyperoxia on the healing tissues are reversed. Based on this model, intermittent oxygen treatments could be particularly beneficial for the cure of infected wounds and could represent a promising therapeutic approach to stimulate wound angiogenesis in the healing tissue [138].

Other computational studies have been modeling the impact of the exogenous addition of growth factors on the rate of wound closure. Molecules such as the epidermal growth factor (EGF) have been shown to stimulate the proliferation and migration of epithelial cells during the healing process. Mathematical modeling of the healing response upon exogenous supplementation of EGF not only recapitulates the experimental data but also helps in predicting the optimal therapeutic levels of EGF administration necessary to improve tissue repair [139, 140]. Recently, a 6-species model focusing on the role of the transforming growth factor β (TGF-β) on wound healing and its relation to scar formation has been developed [141, 142]. In embryonic tissues, wound healing occurs without scarring, and this has been correlated to low levels of TGF-β. A long-standing research goal is to recapitulate the embryonic environment in the context of the adult wound healing process to promote scar-free tissue repair [143]. TGF-β influences cell motility, proliferation, and collagen deposition. However, these effects were not alone sufficient to explain the free-scarring property of TGF-β. This property seemed to depend on the ability of TGF-β to modulate the interaction between fibroblasts and collagen fibers, ultimately changing the orientation of the latter.

Overall, the intense collaborative work between biologists and mathematicians has led to a better and more detailed understanding of each stage of wound healing, helping to clarify the contribution of each individual element to the process as well as the consequences of their interaction on a tissue scale level.

6 Cancer and Immunity

Cancer progression initiates when a single mutated cell proliferates abnormally within otherwise healthy tissue. Since the early 1800, the anatomical progression of this malignancy has been described as a step-wise process, starting with the accumulation of mutations that confer a proliferative advantage to the transformed clone over the wild-type neighboring cells, and culminating with the dissemination of transformed cells from the primary tumor to the more distal organs [144]. In more recent years, it has been well established that dissemination of transformed cells from the primary tumor occurs very early on, even well before the primary tumor is diagnosed [145, 146]. These disseminated cancer cells (DCCs) persist in a dormant state for a long period of time before their reactivation and overgrowth into a metastasis, causing tumor relapse [147]. For tumor cells to grow and metastasize, they must necessarily evade the host immune response. The role of the immune system in protecting the host from tumor growth has been commonly accepted and investigated from as early as 1900, and it is now a fundamental pillar of cancer immunotherapy. Indeed, mice lacking various immune cell subpopulations show a greater predisposition to spontaneously develop cancer than their immune-competent counterparts, reinforcing the idea that neoplastic lesions are under immunosurveillance [148–150]. Interestingly, the interaction of the transformed cells with the immune system not only sets the rate of tumor growth but also shapes the ability of cancer cells to induce a robust immune response, a characteristic known as immunogenicity. Tumors derived from immunocompetent mice tend to grow at a faster rate than tumors recovered from immunocompromised animals when transplanted into immunocompetent animals of the same genetic background [151]. This evidence suggests that over time cancer cells exposed to an activated immune system develop a set of new features that will help them hide from the host immune defense with the ultimate goal of evading it.

Overall, these experimental data reveal that the interplay between cancer cells and the host immune system is quite intricate, with tumor-infiltrating immune cells playing opposing roles in the early stages of cancer progression, including cancer cell dissemination. Indeed, if it is true that immune cells display the most powerful cancer-fighting weapons to achieve tumor control (anti-tumor immunity), tumors over time adopt various strategies to achieve immune tolerance and even hijack immune cells to positively influence their own growth and progression [152]. This provides a potential explanation of why immunotherapy is often ineffective against some very aggressive forms of so-called cold tumors, such as prostate or breast cancer, which are characterized by highly immunosuppressive tumor microenvironments (TMEs).

The early phases of tumor initiation are difficult to image in the opaque tissues of mammals, such as mice and humans, compromising the possibility of gaining a better mechanistic understanding of the initial interaction between the tumor and the immune cells. In this session, we will discuss how the use of translucent and genetically tractable animals, such as zebrafish and flies, has greatly contributed to the study of the complex crosstalk between pre-neoplastic lesions and the host immune system while occurring in vivo. Furthermore, we will describe how cancer cells can ultimately take advantage of this interaction for their own benefit (i.e., to sustain their growth and the invasion of distant tissues). Finally, we will address how the mathematical modeling of tumor-immune cell interaction can better inform the system and support the development of more effective cancer treatments.

6.1 Detection of Pre-neoplastic Lesions: The Early Chemoattractant(s)

To infiltrate the tumor microenvironment (TME) and interact with the malignant clones, immune cells must be able to sense the presence of the early pre-neoplastic lesions. To gain a better understanding of how immune cells detect small tumors, even when they still exist as single cells [80], we will need to revisit what we learned about the immune detection of epithelial wounds.

"Tumors are wounds that never heal" [153]. This famous quote by Harold Dvorak well summarizes his observations that tumors subvert the wound healing response of the host tissue to be able to grow and expand. Dvorak drew striking parallelisms, at both the cellular and molecular levels, between actively healing wounds and cancer, concluding that epithelial tumors remodel their stroma by uncontrollably activating the wound healing response of the host to promote proliferation, infiltration, and eventually metastasis [154].

The tumor microenvironment is characterized by the abundance of inflammatory cells. Immune cell infiltration of the TME occurs with mechanisms similar to the ones adopted during the recruitment of leukocytes to sites of epithelial damage. Recognition of the pre-neoplastic lesion occurs by sensing a gradient of oxygen peroxide that, as previously discussed, is a very well-known chemoattractant drawing leukocytes to locations of tissue damage (Fig. 5) [80, 155]. ROS production is therefore recognized as a universal mechanism that signals and allows the detection of the disruption of tissue homeostasis. Inhibition of epithelial DUOX greatly reduces the number of immune cells reaching the transformed clones and intriguingly results in fewer transformed cells, hinting at the possibility that immune cells might perform trophic and supportive roles during the early phases of cancer progression. Other alarm signals that would normally direct immune cells to epithelial wounds could also attract immune cells to the aberrant cells growing within otherwise healthy tissues. For example, in a mouse model of late-stage melanoma, HMGB-1 is released by irradiated cancer cells, suggesting that this potent wound attractant might mediate the detection of cancer cells in certain circumstances [156].

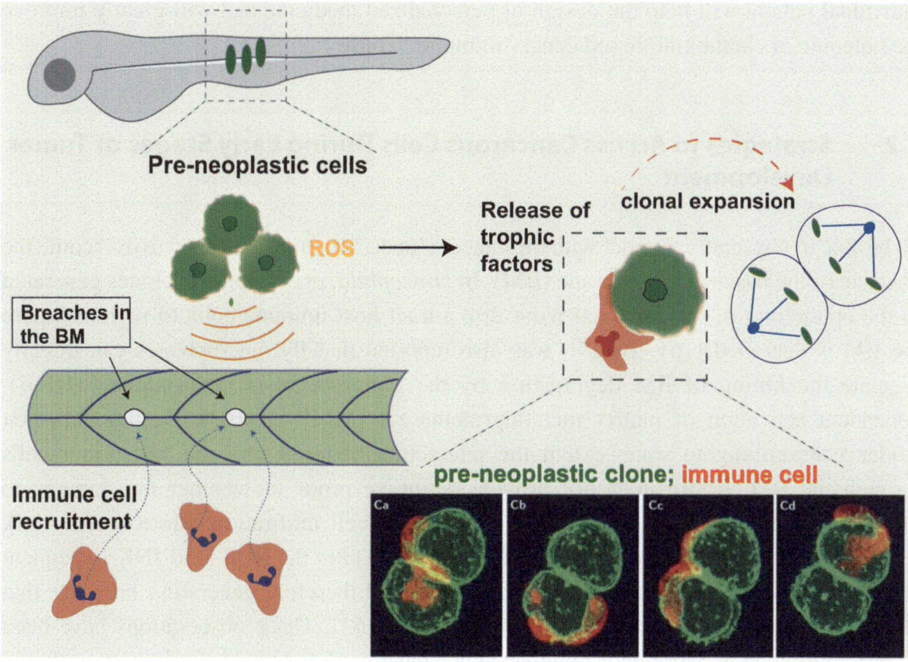

Fig. 5 The intricate interaction between immune cells and pre-neoplastic lesions. The release of hydrogen peroxide (ROS) draws immune cells to early pre-neoplastic lesions, squeezing through pre-existing breaches in the basement membrane (BM). Once recruited, immune cells closely interact with tumor cells and release nurturing factors (such as PGE_2), which stimulate tumor growth. In the schematic, pre-neoplastic cells are in green and immune cells in red. Micrographs are courtesy of Professor Paul Martin, University of Bristol [157]

Chemokines are also important mediators that direct immune cell migration toward growing tumors and help the infiltration of the TME [158]. For example, studies in zebrafish larva show that CXCL8-CXCR2 and CXCL8-CXCR1 signaling axes guide infiltration of leukocytes to the tumor microenvironment. Pharmacological or genetic inhibition of these chemokine-receptor pairs dramatically reduces neutrophil infiltration to early cancer lesions [159, 160], suggesting that chemokine signaling can be potentially exploited as an anti-cancer therapy. Analysis of various pre-clinical models had led to the identification of multiple chemokine receptors as promising candidates that steer immune cell migration to the TME, and since then, these chemokine axes have been therapeutically targeted to reduce immune cell infiltration. However, the results on the effectiveness of these treatments are difficult to interpret as the expression of certain chemokine receptors is not exclusive to one specific immune cell type. Therefore, treatments need to be administered after careful consideration of potential pleiotropic effects that might result in directly opposing outcomes on the tumor. In this sense, a profiling of the specific levels and combination of chemokine-receptor pair present on tumor and immune cells of each individual patient will help the design of personalized medicine and will greatly improve the outcome of chemokine-based cancer immunotherapy.

6.2 Strategies to Access Cancerous Cells During Early Stages of Tumor Development

To be able to physically interact with the cancer clones, immune cells need to overcome the obstacle of the basement membrane (BM). In *Drosophila*, pre-neoplastic clones generated in the epithelium of the imaginal wing disc attract host immune cells to regions where the BM is degraded [161, 162]. It was also reported that fly macrophages can directly regulate the timing of BM degradation by the tumor necrosis factor-alpha (TNF-α)-dependent activation of matrix metalloproteinase-1 (MMP-1) [163]. Despite this first evidence describing to some extent the interaction between immune and tumor cells through the BM, it still does not provide definitive proof of whether the damage to the BM is the cause or a consequence of immune cell infiltration. More recent work shows how changes in the tumor microenvironment lead to the ROS- and JNK-dependent activation of MMP-2 causing damage to the BM and therefore generating breaches that allow for immune cell recruitment and infiltration [164]. These observations have been further reinforced by work on a zebrafish skin cancer model. Animals harboring epithelial HRasG12V mutant cells have been used to study how immune cells circumvent the BM to gain access to the pre-neoplastic cells, shedding further light into this initial step of immune cell infiltration in the TME [165]. It was shown that active proteolytic degradation of the BM does not seem necessary in the early phases of immune cell infiltration. Indeed, leukocytes gain access to the epidermis by using accessible "gates" already present in the BM, which would normally be used for immunosurveillance (Fig. 5). Therefore, cancer cells forming in close vicinity of these "weak spots" in the BM have better chances to interact with the infiltrating immune cells and benefit from their pro-tumorigenic role (see the next paragraph for more details). Similarly, in a chordoma zebrafish model, severe notochord sheet wounds are used by leukocytes as sites of preferential access to a pre-neoplastic lesion, supporting the already proposed model of early immune cell infiltration [166]. Similar breaching of the BM has been observed during development as well as during immune cell diapedesis and extravasation, suggesting that immune cells repurpose well-oiled developmental strategies to infiltrate the tumor microenvironment [167, 168].

6.3 Establishment of Early Leukocyte–Cancer Cell Connections and Trophic Role

High-resolution live imaging of the immune cell/neoplastic cell interaction in the translucent zebrafish larvae revealed the formation of transient, connecting structures between the two cell types (Fig. 5). These cell–cell connections are similar to those observed in vitro between interacting T-cells, named nanotubes, transient structures that mediate signaling and transfer of viruses between cells [169]. The formation of these tethers is thought to strengthen and prolong the interaction between tumor and immune cells, by holding the latter at the TME for longer [80]. Often the establishment of these cell–cell connections is followed by phagocytic events, with neutrophils engulfing small portions of the cell membrane and macrophages engulfing the whole transformed cell [80]. Pre-neoplastic cells show some level of heterogeneity in the ability to attract and form cell–cell contact with the immune cells. It is plausible to hypothesize that some transformed clones might exploit their vicinity to pre-neoplastic cells tethered to immune cells to indirectly benefit from the nurturing effect of immune cells on the tumor, without incurring the risk of being phagocytosed and eliminated. Indeed, the initial interaction with immune cells has largely beneficial effects on tumor growth.

In this regard, studies in zebrafish larvae showed the importance of the COX-PGE$_2$ pathway in supporting cancer progression (Fig. 5) [170]. Interestingly, this pathway has been found to be upregulated in the tumor microenvironment and the use of non-steroidal anti-inflammatory drugs (NSAIDs) like aspirin has been proven successful in alleviating morbidity and mortality associated with many cancers [171, 172]. More recent reports even suggest that regular intake of low doses of these drugs reduces the risk of developing tumors [173]. A study in zebrafish larvae showed that prostaglandin E$_2$ released by inflammatory cells is the signal that nurtures transformed cells, supporting their expansion. Indeed, knockdown of leukocyte PGE$_2$ production, as well as pharmacological inhibition of COX2 enzyme, reduced the number of transformed epithelial cells [170]. These elegant experiments by Feng and colleagues are backed up by work on APC$^{min/+}$ mice. Lifetime administration of aspirin reduced the development of tumors in this mouse model of intestinal cancer. Therefore, anti-inflammatory drugs were proven to be most effective in cancer prevention suggesting again prostaglandin production crucially supports the earliest stages of cancer initiation [174]. Interestingly, prostaglandin not only acts on cancer cells by reducing their proliferation rate, but it appears to act autocrinally on inflammatory cells, specifically on macrophages. In particular, inhibition of the COX-PGE$_2$ pathways promotes the acquisition of an anti-tumoral, pro-inflammatory phenotype, enhancing the engagement of macrophages with the neoplastic cells and the increased clearance of the latter [170, 175].

Another mechanism by which the tumor supports its own expansion beyond 1 mm in diameter is the initiation of the growth of novel vessels to supply nutrients and oxygen. To drive vessel sprouting, tumors exploit the immune cells in the tumor microenvironment. One of the most abundant populations of inflammatory cells at the TME is represented by the pro-angiogenetic TIE2-expressing macrophages. When stimulated by tumor-derived angiopoietin-2 (ANG-2), these cells unleash their pro-angiogenetic potential, by increasing the expression of proteases (such as MMP9) and growth factors (such as VEGF-A) [176]. This observation implies that the use of anti-angiogenetic drugs targeting specific cancer-immune cell signaling axis such as the ANG-TEM axis harbors potentially beneficial therapeutic consequences [177].

In some contexts, macrophages also take part in the process of vascular mimicry (VM), whereby aggressive cancer cells stimulate vessel formation in the absence of endothelial cells [178]. VM is often associated with very high cancer recurrence and poor prognosis. The contribution that the immune cells make to the process of vascular mimicry opens the way to the development of novel VM-targeting therapies based on modulation of immune cell functions [179, 180].

6.4 A Role for Immune Cells in Cancer Cell Spreading and Metastasis

Metastasis, i.e., the ability to develop secondary tumors far from the primary site, is a hallmark of cancer and causes the death of more than 90% of cancer patients worldwide [181]. To grow away from the primary tumor, cancer cells must detach, migrate to the bloodstream while escaping the host immunosurveillance, and finally nest into a new organ. Cells of the immune system influence the different steps of this metastatic cascade, and indeed the immune cell composition of the TME can be a good predictor of the metastatic potential of the disseminating cells. For example, in a study analyzing samples of colorectal cancer from 800 individuals, it was observed that a reduced number of cytotoxic T-cells, combined with a reduced density of lymphatic vessels at the invasive margins of the primary tumor, favor the formation of metastasis by reducing the chances of an anti-tumorigenic immune response [182].

One important step of the metastatic cascade is the intravasation, which is the colonization of the vasculature lumen to endure the very challenging journey to secondary organs. Immune cells closely interact with circulating tumor cells (CTCs) to help them reach the nearest vessels as well as to protect them while in circulation. Intravital imaging of cancer cells in the mouse mammary fat pad reveals a clear role for macrophages in guiding the directional movement of cancer cells toward nearby vessels, favoring their dissemination to secondary sites (Fig. 6) [183].

The close interaction between macrophages and cancer cells favors the highly directional and fast movement of cancer cells along fibers of a dense mash of collagen matrix toward nearby blood vessels. This migration is a result of chemotaxis. Macrophages associated with blood vessels indeed secrete epidermal growth factor (EGF) that acts paracrinally on cancer cells expressing the EGF receptor (EGFR). Concomitantly, cancer cells produce and secrete CSF-1 which attracts macrophages and stimulates them to produce even higher levels of EGF. This signaling crosstalk ensures a close association between macrophages and cancer cells guiding them toward the nearest blood vessel [184]. This paracrine loop also helps intravasation; indeed pharmacological inhibition of EGFR signaling, as well as experimental ablation of macrophages, drastically impair this step of the metastatic cascade [185, 186].

In addition, immune cells also favor metastasis formation by protecting cancer cells in circulation. Indeed, while some circulating tumor cells (CTCs) travel as single entities, most of them tend to aggregate into multicellular clusters, whose presence is often

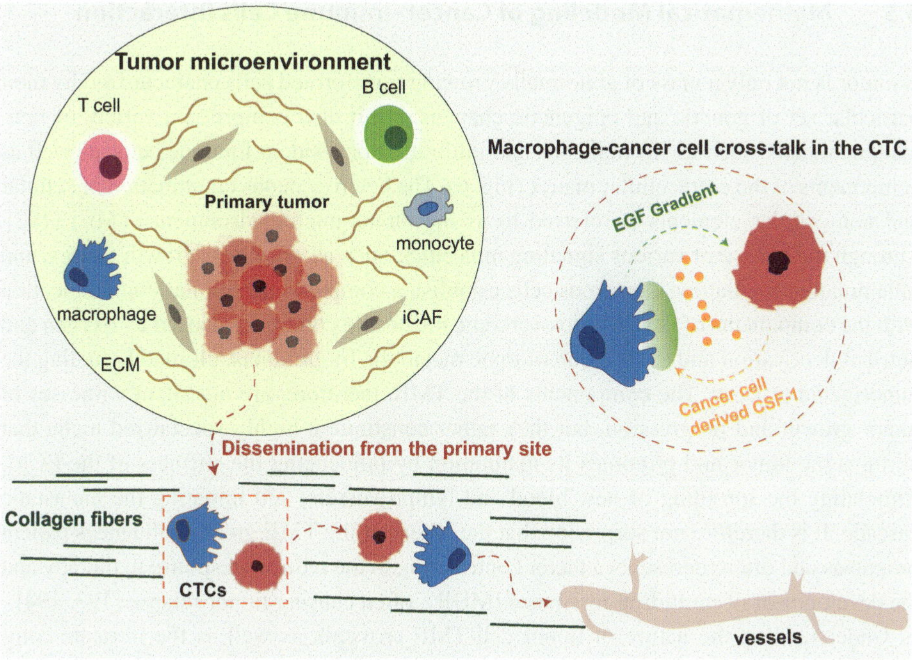

Fig. 6 The close association between macrophages and cancer cells ensures the directional migration of CTCs toward the vasculature. Transformed clones (in red) migrate away from the primary tumor and navigate a dense mesh of highly aligned collagen fibers (in green) toward the vasculature, to disseminate and seed more distal sites. The migration of the circulating tumor clusters (CTCs, red dotted square) toward nearby vessels is facilitated by their close association with macrophages (in blue), and it is maintained through the establishment of a paracrine EGF-CSF1 signaling loop

associated with very poor clinical outcomes. These clusters have generally better chances of survival and show heterogeneity in composition, with stromal and immune cells contributing to their formation [187]. The interaction between cancer cells and white cells within these clusters enhances the ability of CTCs to metastasize. Neutrophils for instance support cluster formation by suppressing leukocyte activation and therefore enhancing the survival of the cluster [188]. Moreover, neutrophils also sustain cell cycle progression of the cluster while in circulation and facilitate metastasis seeding. Indeed, mice injected with neutrophil-CTC clusters grow a much higher number of metastases, while suppression of neutrophils in animals bearing breast tumors significantly decreases the formation of neutrophil-CTC clusters [189].

Like neutrophils, platelets also associate closely with the clusters, providing structural support to successfully withstand the physical stress of the circulation but also hiding the presence of circulating cancer cells from the host immune system [190, 191]. Further characterization of the adhesive mechanisms at the core of the formation of these malignant clusters will provide new invaluable therapeutic tools to limit the formation of metastasis and improve patient prognosis.

6.5 Mathematical Modeling of Cancer–Immune Cells Interaction

A tumor is not only a mass of abnormally growing transformed cells characterized by their particular set of genetic and epigenetic changes, but it also comprises a variety of non-malignant cells, such as stromal cells and infiltrating or resident immune cells, as well as components of the extracellular matrix (Fig. 6). The heterogeneous combination of cellular and non-cellular elements is referred to as the tumor microenvironment (TME) [192]. Through the release of various signaling molecules, such as cytokines, growth factors, and inflammatory mediators, cancerous cells establish a complex intercellular communication with the components of the TME to overcome the hostile conditions caused by oxygen and nutrient deprivation and the buildup of toxic metabolic by-products, ultimately fueling the tumorigenic process. The components of the TME, therefore, are not silent witnesses of tumor growth and progression, but they rather constitute a highly specialized niche that nurtures the tumor and promotes its maturation by influencing the stiffness of the ECM, stimulating the sprouting of new blood and lymph vessels, and initiating the metastatic cascade. It is therefore not surprising that the nature of the TME greatly influences patient prognosis and often represents a factor contributing to the reduced response to therapy and the development of multidrug resistance (MDR), often causing tumor relapse [193, 194].

Understanding the nature of tumor cell-TME crosstalk as well as the intricate communication between the components of the TME can benefit the clinic and help the design of new and more targeted treatments, depending on the stages of tumor progression [195]. Often the outcome of these interactions is quite difficult to interpret with the sole contribution of experimental data. In this respect, mathematical modeling can be a powerful ally in defining the characteristics of such complex regulatory networks. In this paragraph, we will discuss current knowledge of immune-mediated response to tumor growth with a particular focus on T-cell and macrophage-mediated response. Moreover, we will provide insights on how mathematical modeling can provide a powerful framework for studying the dynamics of cell–cell interactions within the tumor microenvironment and helping to predict more effective anti-cancer therapies.

6.6 Cytotoxic T-Lymphocytes

Immune cells represent a big constituent of the TME. Components of both innate and adaptive immunity infiltrate the TME and, depending on the context and the nature of the tumor, can either antagonize or promote tumor growth (Fig. 6). B- and T-lymphocytes represent the adaptive mediators of the anti-tumor response. Of all the cells of the adaptive branch of immunity, cytotoxic $CD8^{+}$ T-lymphocytes (CTLs) are considered the most effective anti-cancer weapon and are the focus of most cancer immunotherapies [196].

Tumors characterized by a high infiltration of CTLs are classified as immune-inflamed or “hot” tumors. CTLs are known for their potent pro-apoptotic activity toward cancer cells. Mathematical modeling of CTLs response to tumors predicts that this branch of adaptive immunity is particularly efficient at controlling the growth of small tumors [197, 198]. The numerical simulation predicts that the number of CTLs rapidly grows at the early stages of tumor initiation and that the number of cancer cells decreases proportionally to CTLs expansion and activation. Interestingly, if the size of the tumor is inferior to 2 mm (<6×10^5 transformed cells), the tumor is not eliminated completely but rather remains small and stable. This model explains the recurrence of that phenomenon known as "cancer without disease," whereby people live with microscopic lesions called in situ tumors without ever developing any symptoms [199]. The same models suggest that CTLs fail at controlling the growth of larger tumors (> 1.2×10^8 transformed cells). The significance of this last prediction is that the presence of CTLs in the TME is not always a synonym of a successful and effective anti-tumor response. In fact, experimental data suggest that accumulation of $CD8^+$ T-cells can often lead to the local activation of a negative regulatory mechanism which might result in a lack of immune response and ultimately favor tumor escape, despite the initial inflamed phenotype [200]. An arm of this negative mechanism is mediated by cancer-associated fibroblasts (CAFs) which are the main producer of the potent immunosuppressive cytokine, TGF-β. TGF-β stimulates the differentiation of naïve CD4+ T-lymphocytes into regulatory T-cells (T-reg) which, through the release of interleukin-2 (IL-2) and interferon-γ (IFN-γ), suppress the activity of CTLs, curbing the anti-tumor response [201]. Recently, pre-treatment of a murine model of melanoma with chemo agents such as cyclophosphamide followed by an anti-$CD4^+$ regimen greatly enhances the anti-tumorigenic activity of $CD8^+$ CTLs, greatly improving the host survival [202].

The role of $CD4^+$ T-lymphocytes in supporting tumor growth is quite controversial. Indeed, if regulatory T-cells act by suppressing CTLs response, $CD4^+$ conventional T-cells (T-conv) have diametrically opposed roles, promoting anti-tumor immunity [203]. Recently, mathematical modeling of immune cell–tumor cell interaction in the presence or not of chemotherapy regimen reveals the potential benefit of CD4+ T-cell targeted therapies. This model not only suggests that CD4+ T-cell therapy could be used in combination with chemotherapy and have the additional benefit of reducing the associated toxicity and side effects but, depending on the initial size of the tumor, could even completely replace the use of chemotherapeutic agents [204].

Another aspect contributing to the suppression of CTLs anti-tumor activity relates to the expression of programmed death ligand-1 (PD-L1). PD-L1 is normally expressed on host cells and represents an important immune checkpoint against autoimmunity. The receptor PD-1 is normally expressed on the surface of T-lymphocytes, and the interaction with its cognate ligand results in the suppression of T-cell activity. Expression of PD-L1 is often increased in some types of tumors which hijack the PD-L1/PD-1 signaling axis to evade immunosurveillance. Overall, because of a pre-existing anti-tumor T-cell response, hot tumors can be more successfully treated in the clinic with the use of therapeutic strategies aimed at reactivating exhausted or dysfunctional $CD8^+$ T-cells. In particular, the use of immune checkpoint inhibitors (ICIs), singularly or in combination, has been proven effective and with durable results in several clinical trials [205].

Tumors classified as immune-excluded or immune-deserted are extremely resistant to ICI-based therapies [206]. In these tumors, generally referred to as "cold" tumors, cytotoxic $CD8^+$ T-lymphocytes localize at the periphery and fail to infiltrate the mass, or alternatively, they are completely absent from the TME. In this case, a better understanding of the mechanisms that allow T-cell homing to the TME as well as of the multiple strategies adopted by the tumor to achieve T-cell exclusion could potentially "fire up" unresponsive tumors and improve the effectiveness of more classical cancer treatment [207]. Over the past 50 years, the use of nanoparticles (NPs) has been introduced in the clinic as an alternative drug delivery approach to chemotherapy aiming at reverting the TME of cold tumors into a more immunoreactive one [208]. Mathematical modeling has been a powerful tool to better understand the interaction of nanocarriers with biomaterial aiming to optimize the targeted delivery of highly toxic or insoluble drugs and improve the effectiveness of the treatment [209].

6.7 Macrophages

Tumor growth is often associated with a certain degree of inflammation, and unsurprisingly, it has been found that the tumor microenvironment is infiltrated with components of innate immunity. Macrophages at the TME are referred to as tumor-associated macrophages (TAMs) and belong to quite different phenotypic classes. M1 macrophages are classified as pro-inflammatory and inhibit tumor growth through the release of cytotoxic compounds, such as nitric oxide (NO) and reactive oxygen species (ROS). Moreover, M1 macrophages modulate the activity of T-helper lymphocytes and show robust phagocytic activity. A second class of TAMs is classified as anti-inflammatory M2 macrophages, which on the contrary promote tumor progression [210]. M2 macrophages inhibit CD8+ T-lymphocytes, facilitate tumor cell proliferation through the release of TGF-β, and promote angiogenesis by releasing VEGF-A. Finally, a third class of recently described macrophages is the so-called Tie-expressing cells (TEMs). TEMs are thought to promote tumor progression by activating an angiogenetic program, which stimulates vessel sprouting and promotes the development of a vascular network around the tumor.

Macrophages are particularly abundant at the TME; they can even become the main immune population of the TME, constituting up to 50% of the tumor mass.

From a clinical perspective, a higher proportion of macrophages (especially M2-like) at the TME is considered unfavorable to the patients and correlates with poor prognosis [211, 212]. Indeed, macrophages support tumor growth and metastasis by favoring tumor cell migration and facilitating their intravasation into nearby vessels. Moreover, macrophages sustain the survival of metastatic lesions which otherwise would regress and eventually disappear. However, mathematical modeling of triple-negative breast cancer and colorectal cancer conceptually shows that the heavy infiltration of macrophages might instead promote tumor regression [213, 214]. Indeed, if infiltration of a small number of macrophages positively influences tumor progression as described, infiltration of immune cells at a higher number might result in an overcrowded environment, inhibiting tumor growth.

Given the multifaceted nature and roles of macrophages in tumor growth, drawing a more comprehensive picture of the dynamic interplay between the different macrophage populations and their impact on tumor progression could better inform the development of highly targeted and effective therapies.

Therapeutic interventions that inhibit monocyte and macrophage infiltration to the TME might not represent the most powerful approach in the fight against tumor growth as they would compromise the possibility to harness the tumoricidal activity of M1 macrophages. Forcing macrophages to switch phenotype might be a more fruitful approach. In other words, forcing macrophages to revert their phenotype from an M2-like to an M1-like state might accelerate the normalization of the tumor microenvironment and be more beneficial in the clinic.

Macrophage polarization is quite fluid; the system indeed shows some degree of plasticity and can revert between the different phenotypes.

Cancer cells actively interact with TAMs and influence their repolarization, with more epithelial-like cancer cells favoring macrophage polarization toward the M1 phenotype, while mesenchymal-like cancer cells induce M2 polarization. A recent mathematical model predicts that a therapeutic regimen aiming at increasing the transition from a mesenchymal-to-epithelial phenotype (MET) and at suppressing cancer cell growth could favor the conversion to an M1-dominated tumor microenvironment and greatly reduce tumor aggressiveness. However, the same model predicts that despite the ability of macrophages to repolarize, once the system (TME) has reached its steady state, its "normalization" (meaning reverting it from a mesenchymal-M2 state to an epithelial-M1 state) is much harder to achieve and any pharmaceutical manipulation aimed at stimulating the M2-M1 phenotype interconversion might be unsuccessful [215]. This conclusion is in line with the knowledge that the rate at which the M2-M1 repolarization occurs impacts the speed at which the mass grows and consequently its size [216].

Another aspect to consider is the degree of vascularization of the tumor and the role of Tie2-expressing macrophages (TEMs) in the tumor microenvironment. Recently, it has been predicted that the rate of tumor growth is largely dependent on the proportion of M2 macrophages present at the TME regardless of the presence of TEMs and the extent of vascular development [217]. Drug treatments that inhibit TME activity might therefore not help in inhibiting or even limiting cancer progression, especially when M2-like cells represent the most abundant class of macrophages populating the TME. Counterintuitively, patients may benefit from promoting the effects of TMEs on vasculature formation, especially when this happens in combination with a chemotherapeutic regimen. Indeed, there are cases in which the inhibition of de novo vessel formation around the primary tumor would not stop cancer progression but rather result in more malignant and aggressive disease.

Moreover, promoting the development of a well-perfused vessel network might facilitate cancer treatment by favoring a more homogeneous delivery of chemotherapeutic drugs into the tissue. In this regard, the mathematical prediction suggests that a higher percentage of tumor regression could be achieved in the clinic if initial cycles of immunotherapy (to increase the proportion of TMEs and therefore vessel growth) would be alternated with cycles of chemotherapy, to drastically reduce the pro-tumorigenic influence of both TMEs and M2 macrophages [217]. An alternative strategy to repress tumor growth relies on harnessing the tremendous phagocytic potential of macrophages to use these cells as a vehicle to deliver active drugs into the transformed tissue [218]. In more recent years, macrophages have been engineered thanks to the application of nanotechnology in the field of cancer therapy. Nanoparticles (NPs) are organic or inorganic structures with a dimension in the range of nanometers. Several methods have been tested to efficiently generate NP-loaded macrophages (NPL-Ms), including electroporation or in vitro coincubation methods. More recently, NPs encapsulated by apoptotic bodies (i.e., artificially damaged red blood cells) have been injected intravenously to allow the generation of NPL-Ms in vivo, simply exploiting the phagocytic activity of macrophages in circulation. These engineered macrophages are still able to navigate their way toward the tumor to eventually deliver the drug-loaded NPs, achieving successful anti-cancer activity [219]. Nanotechnology has also been successfully utilized to induce macrophage repolarization toward an M1-like state. In particular, mannose-coated NPs have been shown to be preferentially phagocytosed by M2 macrophages and induce conversion into M1 phenotype when loaded with polyinosinic-polycytidylic acid (poly IC), enhancing anti-tumor activity [220]. NPs can also be utilized to enhance macrophage infiltration to the TME. This can be achieved by loading macrophages with magnetic NPs and the subsequent application of a magnetic field to the lesion [221, 222]. Recently, the effect of combining conventional cyclophosphamide-based therapy with the use of engineered macrophages has been explored through mathematical modeling [223]. Macrophages were genetically modified to express hypoxia-inducible cytochrome P450 and were loaded with magnetic nanoparticles. The application of a magnetic field to the tumor lesion increases the influx of macrophages to heavily hypoxic regions of the tumor and stimulates the local activation of the drug cyclophosphamide, achieving tumor killing. This mathematical model predicts that the simultaneous use of conventional and macrophage-based strategies shows much more enhanced effects on limiting tumor growth than the additive effects achieved by the sequential administration of each therapy.

Overall, the simultaneous combination of multiple therapeutic approaches can enhance the activity of this versatile population of immune cells to unleash its full anti-tumor potential.

Take-Home Message

While in vitro studies have deepened our knowledge of the numerous functions performed by leukocytes, in vivo model organisms have further pushed the boundaries, clarifying what processes are happening in a living organism and allowing us to describe the intricate interaction of the innate immune cells with the environment and their contribution during both normal development and pathologies. *Drosophila* and zebrafish are excellent model organisms to dissect the complexity of the innate immune response. We explored how these models allow us to image, at high spatial and temporal resolution, immune cell behavior, while they patrol their native environment and accurately quantify their effort in maintaining tissue homeostasis when facing challenging conditions, such as infections by foreign pathogens, tissue repair upon wounding, or cancer. We also outlined how mathematical modeling and computational studies inform our understanding of such complex biological systems. For decades, mathematical modeling has provided insightful details on the intimate connection between immune cells and the environment and has contributed to address how the disruption of such interactions results in the development of pathologies. A blended approach, which brings together the acquisition of in vivo biological data and computational and mathematical modeling, will pinpoint the most critical steps in the process and will assist the clinic in the design of more efficient and personalized therapeutic regimens.

References

1. Marshall JS, et al. An introduction to immunology and immunopathology. Allergy Asthma Clin Immunol. 2018;14(Suppl 2):49.
2. Friedl P, et al. New dimensions in cell migration. Nat Rev Mol Cell Biol. 2012;13(11):743–7.
3. Lemaitre B, et al. A recessive mutation, immune deficiency (imd), defines two distinct control pathways in the Drosophila host defense. Proc Natl Acad Sci U S A. 1995;92(21):9465–9.
4. Lemaitre B, et al. Functional analysis and regulation of nuclear import of dorsal during the immune response in Drosophila. EMBO J. 1995;14(3):536–45.
5. Mase A, Augsburger J, Bruckner K. Macrophages and their organ locations shape each other in development and homeostasis—a Drosophila perspective. Front Cell Dev Biol. 2021;9:630272.
6. Gore AV, et al. The zebrafish: A fintastic model for hematopoietic development and disease. Wiley Interdiscip Rev Dev Biol. 2018;7(3):e312.
7. Lam SH, et al. Development and maturation of the immune system in zebrafish, Danio rerio: a gene expression profiling, in situ hybridization and immunological study. Dev Comp Immunol. 2004;28(1):9–28.
8. Franz CM, Jones GE, Ridley AJ. Cell migration in development and disease. Dev Cell. 2002;2(2):153–8.
9. Pankova K, et al. The molecular mechanisms of transition between mesenchymal and amoeboid invasiveness in tumor cells. Cell Mol Life Sci. 2010;67(1):63–71.

10. Sanz-Moreno V, et al. Rac activation and inactivation control plasticity of tumor cell movement. Cell. 2008;135(3):510–23.
11. Vesperini D, et al. Characterization of immune cell migration using microfabrication. Biophys Rev. 2021;13(2):185–202.
12. Moreau HD, et al. Integrating physical and molecular insights on immune cell migration. Trends Immunol. 2018;39(8):632–43.
13. Renkawitz J, Donnadieu E, Moreau HD. Editorial: Immune cell migration in health and disease. Front Immunol. 2022;13:897626.
14. Janetopoulos C, Jin T, Devreotes P. Receptor-mediated activation of heterotrimeric G-proteins in living cells. Science. 2001;291(5512):2408–11.
15. Lam PY, Huttenlocher A. Interstitial leukocyte migration in vivo. Curr Opin Cell Biol. 2013;25(5):650–8.
16. Barros-Becker F, et al. Live imaging reveals distinct modes of neutrophil and macrophage migration within interstitial tissues. J Cell Sci. 2017;130(22):3801–8.
17. Hadjitheodorou A, et al. Directional reorientation of migrating neutrophils is limited by suppression of receptor input signaling at the cell rear through myosin II activity. Nat Commun. 2021;12(1):6619.
18. Stramer B, et al. Live imaging of wound inflammation in Drosophila embryos reveals key roles for small GTPases during in vivo cell migration. J Cell Biol. 2005;168(4):567–73.
19. Paladi M, Tepass U. Function of Rho GTPases in embryonic blood cell migration in Drosophila. J Cell Sci. 2004;117(Pt 26):6313–26.
20. Tucker PK, Evans IR, Wood W. Ena drives invasive macrophage migration in Drosophila embryos. Dis Model Mech. 2011;4(1):126–34.
21. Zanet J, et al. Fascin is required for blood cell migration during Drosophila embryogenesis. Development. 2009;136(15):2557–65.
22. Bilancia CG, et al. Enabled negatively regulates diaphanous-driven actin dynamics in vitro and in vivo. Dev Cell. 2014;28(4):394–408.
23. Yolland L, et al. Persistent and polarized global actin flow is essential for directionality during cell migration. Nat Cell Biol. 2019;21(11):1370–81.
24. Yoo SK, et al. The role of microtubules in neutrophil polarity and migration in live zebrafish. J Cell Sci. 2012;125(Pt 23):5702–10.
25. Eddy RJ, Pierini LM, Maxfield FR. Microtubule asymmetry during neutrophil polarization and migration. Mol Biol Cell. 2002;13(12):4470–83.
26. Morlino G, et al. Miro-1 links mitochondria and microtubule Dynein motors to control lymphocyte migration and polarity. Mol Cell Biol. 2014;34(8):1412–26.
27. Campello S, et al. Orchestration of lymphocyte chemotaxis by mitochondrial dynamics. J Exp Med. 2006;203(13):2879–86.
28. Bao Y, et al. mTOR and differential activation of mitochondria orchestrate neutrophil chemotaxis. J Cell Biol. 2015;210(7):1153–64.
29. Stramer B, et al. Clasp-mediated microtubule bundling regulates persistent motility and contact repulsion in Drosophila macrophages in vivo. J Cell Biol. 2010;189(4):681–9.
30. Davis JR, et al. Emergence of embryonic pattern through contact inhibition of locomotion. Development. 2012;139(24):4555–60.
31. Wood W, Faria C, Jacinto A. Distinct mechanisms regulate hemocyte chemotaxis during development and wound healing in Drosophila melanogaster. J Cell Biol. 2006;173(3):405–16.
32. Davis JR, et al. Inter-cellular forces orchestrate contact inhibition of locomotion. Cell. 2015;161(2):361–73.

33. Lichanska AM, Hume DA. Origins and functions of phagocytes in the embryo. Exp Hematol. 2000;28(6):601–11.
34. Suzanne M, Steller H. Shaping organisms with apoptosis. Cell Death Differ. 2013;20(5):669–75.
35. Defaye A, et al. Genetic ablation of Drosophila phagocytes reveals their contribution to both development and resistance to bacterial infection. J Innate Immun. 2009;1(4):322–34.
36. Martinek N, et al. Haemocyte-derived SPARC is required for collagen-IV-dependent stability of basal laminae in Drosophila embryos. J Cell Sci. 2008;121(Pt 10):1671–80.
37. Evans IR, et al. Interdependence of macrophage migration and ventral nerve cord development in Drosophila embryos. Development. 2010;137(10):1625–33.
38. Wood W, et al. Mesenchymal cells engulf and clear apoptotic footplate cells in macrophageless PU.1 null mouse embryos. Development. 2000;127(24):5245–52.
39. Diez-Roux G, Lang RA. Macrophages induce apoptosis in normal cells in vivo. Development. 1997;124(18):3633–8.
40. Lobov IB, et al. WNT7b mediates macrophage-induced programmed cell death in patterning of the vasculature. Nature. 2005;437(7057):417–21.
41. Cavaillon JM. The historical milestones in the understanding of leukocyte biology initiated by Elie Metchnikoff. J Leukoc Biol. 2011;90(3):413–24.
42. Ingman WV, et al. Macrophages promote collagen fibrillogenesis around terminal end buds of the developing mammary gland. Dev Dyn. 2006;235(12):3222–9.
43. Bunt S, et al. Hemocyte-secreted type IV collagen enhances BMP signaling to guide renal tubule morphogenesis in Drosophila. Dev Cell. 2010;19(2):296–306.
44. Fantin A, et al. Tissue macrophages act as cellular chaperones for vascular anastomosis downstream of VEGF-mediated endothelial tip cell induction. Blood. 2010;116(5):829–40.
45. Stefater JA 3rd, et al. Regulation of angiogenesis by a non-canonical Wnt-Flt1 pathway in myeloid cells. Nature. 2011;474(7352):511–5.
46. Sweeney M, Foldes G. It takes two: endothelial-perivascular cell cross-talk in vascular development and disease. Front Cardiovasc Med. 2018;5:154.
47. Antimicrobial Resistance C. Global burden of bacterial antimicrobial resistance in 2019: a systematic analysis. Lancet. 2022;399(10325):629–55.
48. Prestinaci F, Pezzotti P, Pantosti A. Antimicrobial resistance: a global multifaceted phenomenon. Pathog Glob Health. 2015;109(7):309–18.
49. Sarkar S, Heise MT. Mouse models as resources for studying infectious diseases. Clin Ther. 2019;41(10):1912–22.
50. Buddle BM, Vordermeier HM, Hewinson RG. Experimental infection models of tuberculosis in domestic livestock. Microbiol Spectr. 2016;4(4)
51. Torraca V, Mostowy S. Zebrafish infection: from pathogenesis to cell biology. Trends Cell Biol. 2018;28(2):143–56.
52. Mickiewicz KM, et al. Possible role of L-form switching in recurrent urinary tract infection. Nat Commun. 2019;10(1):4379.
53. Badolato R, Donadieu J, WHIM Research Group. How I treat warts, hypogammaglobulinemia, infections, and myelokathexis syndrome. Blood. 2017;130(23):2491–8.
54. Summers C, et al. Neutrophil kinetics in health and disease. Trends Immunol. 2010;31(8):318–24.
55. Deng Q, Harvie EA, Huttenlocher A. Distinct signalling mechanisms mediate neutrophil attraction to bacterial infection and tissue injury. Cell Microbiol. 2012;14(4):517–28.
56. Deng Q, et al. Localized bacterial infection induces systemic activation of neutrophils through Cxcr2 signaling in zebrafish. J Leukoc Biol. 2013;93(5):761–9.

57. de Oliveira S, et al. Cxcl8 (IL-8) mediates neutrophil recruitment and behavior in the zebrafish inflammatory response. J Immunol. 2013;190(8):4349–59.
58. Deng Q, et al. Dual roles for Rac2 in neutrophil motility and active retention in zebrafish hematopoietic tissue. Dev Cell. 2011;21(4):735–45.
59. Rosowski EE, et al. Rac2 functions in both neutrophils and macrophages to mediate motility and host defense in larval zebrafish. J Immunol. 2016;197(12):4780–90.
60. Cambier CJ, et al. Mycobacteria manipulate macrophage recruitment through coordinated use of membrane lipids. Nature. 2014;505(7482):218–22.
61. Cambier CJ, et al. Phenolic glycolipid facilitates mycobacterial escape from microbicidal tissue-resident macrophages. Immunity. 2017;47(3):552–65 e4.
62. Zakrzewska A, et al. Macrophage-specific gene functions in Spi1-directed innate immunity. Blood. 2010;116(3):e1–11.
63. Sommer F, et al. Inhibition of macrophage migration in zebrafish larvae demonstrates in vivo efficacy of human CCR2 inhibitors. Dev Comp Immunol. 2021;116:103932.
64. Janeway CA Jr. Approaching the asymptote? Evolution and revolution in immunology. Cold Spring Harb Symp Quant Biol. 1989;54(Pt 1):1–13.
65. Lemaitre B, et al. The dorsoventral regulatory gene cassette spatzle/Toll/cactus controls the potent antifungal response in Drosophila adults. Cell. 1996;86(6):973–83.
66. Li Y, et al. Roles of toll-like receptors in nitroxidative stress in mammals. Cells. 2019;8(6)
67. Tadie JM, et al. HMGB1 promotes neutrophil extracellular trap formation through interactions with Toll-like receptor 4. Am J Physiol Lung Cell Mol Physiol. 2013;304(5):L342–9.
68. Castanheira FVS, Kubes P. Neutrophils and NETs in modulating acute and chronic inflammation. Blood. 2019;133(20):2178–85.
69. Phan QT, et al. Neutrophils use superoxide to control bacterial infection at a distance. PLoS Pathog. 2018;14(7):e1007157.
70. Bergsbaken T, Fink SL, Cookson BT. Pyroptosis: host cell death and inflammation. Nat Rev Microbiol. 2009;7(2):99–109.
71. Liu X, Lieberman J. A mechanistic understanding of pyroptosis: the fiery death triggered by invasive infection. Adv Immunol. 2017;135:81–117.
72. Vincent WJ, et al. Macrophages mediate flagellin induced inflammasome activation and host defense in zebrafish. Cell Microbiol. 2016;18(4):591–604.
73. Gordon S. Elie metchnikoff, the man and the myth. J Innate Immun. 2016;8(3):223–7.
74. Matzinger P. Tolerance, danger, and the extended family. Annu Rev Immunol. 1994;12:991–1045.
75. Niethammer P, et al. A tissue-scale gradient of hydrogen peroxide mediates rapid wound detection in zebrafish. Nature. 2009;459(7249):996–9.
76. Moreira S, et al. Prioritization of competing damage and developmental signals by migrating macrophages in the Drosophila embryo. Curr Biol. 2010;20(5):464–70.
77. Razzell W, et al. Calcium flashes orchestrate the wound inflammatory response through DUOX activation and hydrogen peroxide release. Curr Biol. 2013;23(5):424–9.
78. Yoo SK, et al. Early redox, Src family kinase, and calcium signaling integrate wound responses and tissue regeneration in zebrafish. J Cell Biol. 2012;199(2):225–34.
79. Antunes M, et al. Coordinated waves of actomyosin flow and apical cell constriction immediately after wounding. J Cell Biol. 2013;202(2):365–79.
80. Feng Y, et al. Live imaging of innate immune cell sensing of transformed cells in zebrafish larvae: parallels between tumor initiation and wound inflammation. PLoS Biol. 2010;8(12):e1000562.
81. Yoo SK, et al. Lyn is a redox sensor that mediates leukocyte wound attraction in vivo. Nature. 2011;480(7375):109–12.

82. Evans IR, et al. Draper/CED-1 mediates an ancient damage response to control inflammatory blood cell migration in vivo. Curr Biol. 2015;25(12):1606–12.
83. Poole LB, Nelson KJ. Discovering mechanisms of signaling-mediated cysteine oxidation. Curr Opin Chem Biol. 2008;12(1):18–24.
84. Paulsen CE, Carroll KS. Orchestrating redox signaling networks through regulatory cysteine switches. ACS Chem Biol. 2010;5(1):47–62.
85. Bienert GP, Schjoerring JK, Jahn TP. Membrane transport of hydrogen peroxide. Biochim Biophys Acta. 2006;1758(8):994–1003.
86. Miller EW, Dickinson BC, Chang CJ. Aquaporin-3 mediates hydrogen peroxide uptake to regulate downstream intracellular signaling. Proc Natl Acad Sci U S A. 2010;107(36):15681–6.
87. Chakrabarti S, Visweswariah SS. Intramacrophage ROS primes the innate immune system via JAK/STAT and toll activation. Cell Rep. 2020;33(6):108368.
88. Schymeinsky J, et al. Syk-mediated translocation of PI3Kdelta to the leading edge controls lamellipodium formation and migration of leukocytes. PLoS One. 2007;2(11):e1132.
89. Weavers H, et al. Corpse engulfment generates a molecular memory that primes the macrophage inflammatory response. Cell. 2016;165(7):1658–71.
90. Loo AE, Ho R, Halliwell B. Mechanism of hydrogen peroxide-induced keratinocyte migration in a scratch-wound model. Free Radic Biol Med. 2011;51(4):884–92.
91. Shen WL, et al. NAD(P)H oxidase-derived reactive oxygen species regulate angiotensin-II induced adventitial fibroblast phenotypic differentiation. Biochem Biophys Res Commun. 2006;339(1):337–43.
92. Lambeth JD. Nox enzymes, ROS, and chronic disease: an example of antagonistic pleiotropy. Free Radic Biol Med. 2007;43(3):332–47.
93. Schafer M, Werner S. Oxidative stress in normal and impaired wound repair. Pharmacol Res. 2008;58(2):165–71.
94. Weavers H, Wood W, Martin P. Injury activates a dynamic cytoprotective network to confer stress resilience and drive repair. Curr Biol. 2019;29(22):3851–62 e4.
95. Andre-Levigne D, et al. Reactive oxygen species and NOX enzymes are emerging as key players in cutaneous wound repair. Int J Mol Sci. 2017;18(10)
96. Bryan N, et al. Reactive oxygen species (ROS)—a family of fate deciding molecules pivotal in constructive inflammation and wound healing. Eur Cell Mater. 2012;24:249–65.
97. Murray RZ, et al. Development and use of biomaterials as wound healing therapies. Burns Trauma. 2019;7:2.
98. Huang J, et al. Tunable sequential drug delivery system based on chitosan/hyaluronic acid hydrogels and PLGA microspheres for management of non-healing infected wounds. Mater Sci Eng C Mater Biol Appl. 2018;89:213–22.
99. Schonfelder U, et al. Influence of selected wound dressings on PMN elastase in chronic wound fluid and their antioxidative potential in vitro. Biomaterials. 2005;26(33):6664–73.
100. Nishikawa M, Hashida M, Takakura Y. Catalase delivery for inhibiting ROS-mediated tissue injury and tumor metastasis. Adv Drug Deliv Rev. 2009;61(4):319–26.
101. He S-Q, Zhang Y-H, Venugopal SK, Dicus CW, Perez RV, Ramsamooj R, Nantz MH, Zern MA, Jian W. Delivery of antioxidative enzyme genes protects against ischemia/reperfusion-induced liver injury in mice. Liver Transpl. 2006;12:1869–79.
102. Goodwin GH, Johns EW. Isolation and characterisation of two calf-thymus chromatin non-histone proteins with high contents of acidic and basic amino acids. Eur J Biochem. 1973;40(1):215–9.
103. Hori O, et al. The receptor for advanced glycation end products (RAGE) is a cellular binding site for amphoterin. Mediation of neurite outgrowth and co-expression of rage and amphoterin in the developing nervous system. J Biol Chem. 1995;270(43):25752–61.

104. Park JS, et al. Involvement of toll-like receptors 2 and 4 in cellular activation by high mobility group box 1 protein. J Biol Chem. 2004;279(9):7370–7.
105. Huebener P, et al. The HMGB1/RAGE axis triggers neutrophil-mediated injury amplification following necrosis. J Clin Invest. 2015;125(2):539–50.
106. Fiuza C, et al. Inflammation-promoting activity of HMGB1 on human microvascular endothelial cells. Blood. 2003;101(7):2652–60.
107. Lee G, et al. Fully reduced HMGB1 accelerates the regeneration of multiple tissues by transitioning stem cells to GAlert. Proc Natl Acad Sci U S A. 2018;115(19):E4463–72.
108. Limana F, et al. Exogenous high-mobility group box 1 protein induces myocardial regeneration after infarction via enhanced cardiac C-kit+ cell proliferation and differentiation. Circ Res. 2005;97(8):e73–83.
109. Izuishi K, et al. Cutting edge: high-mobility group box 1 preconditioning protects against liver ischemia-reperfusion injury. J Immunol. 2006;176(12):7154–8.
110. Fei L, et al. Preconditioning with rHMGB1 ameliorates lung ischemia-reperfusion injury by inhibiting alveolar macrophage pyroptosis via the Keap1/Nrf2/HO-1 signaling pathway. J Transl Med. 2020;18(1):301.
111. Wu H, et al. Preconditioning with recombinant high-mobility group box 1 protein protects the kidney against ischemia-reperfusion injury in mice. Kidney Int. 2014;85(4):824–32.
112. Mitola S, et al. Cutting edge: extracellular high mobility group box-1 protein is a proangiogenic cytokine. J Immunol. 2006;176(1):12–5.
113. Lin Q, et al. High-mobility group box-1 mediates toll-like receptor 4-dependent angiogenesis. Arterioscler Thromb Vasc Biol. 2011;31(5):1024–32.
114. Schwiebert EM, Zsembery A. Extracellular ATP as a signaling molecule for epithelial cells. Biochim Biophys Acta. 2003;1615(1–2):7–32.
115. Chen Y, et al. ATP release guides neutrophil chemotaxis via P2Y2 and A3 receptors. Science. 2006;314(5806):1792–5.
116. McDonald B, et al. Intravascular danger signals guide neutrophils to sites of sterile inflammation. Science. 2010;330(6002):362–6.
117. Weavers H, et al. Systems analysis of the dynamic inflammatory response to tissue damage reveals spatiotemporal properties of the wound attractant gradient. Curr Biol. 2016;26(15):1975–89.
118. Thuma L, et al. Drosophila immune cells extravasate from vessels to wounds using Tre1 GPCR and Rho signaling. J Cell Biol. 2018;217(9):3045–56.
119. LeBlanc MG, Lehmann R. Domain-specific control of germ cell polarity and migration by multifunction Tre1 GPCR. J Cell Biol. 2017;216(9):2945–58.
120. de Oliveira S, Rosowski EE, Huttenlocher A. Neutrophil migration in infection and wound repair: going forward in reverse. Nat Rev Immunol. 2016;16(6):378–91.
121. Lammermann T, et al. Neutrophil swarms require LTB4 and integrins at sites of cell death in vivo. Nature. 2013;498(7454):371–5.
122. Afonso PV, et al. LTB4 is a signal-relay molecule during neutrophil chemotaxis. Dev Cell. 2012;22(5):1079–91.
123. Poplimont H, et al. Neutrophil swarming in damaged tissue is orchestrated by connexins and cooperative calcium alarm signals. Curr Biol. 2020;30(14):2761–76 e7.
124. Wilgus TA, Roy S, McDaniel JC. Neutrophils and wound repair: positive actions and negative reactions. Adv Wound Care (New Rochelle). 2013;2(7):379–88.
125. Wright HL, et al. Neutrophil function in inflammation and inflammatory diseases. Rheumatology (Oxford). 2010;49(9):1618–31.
126. Dovi JV, He LK, DiPietro LA. Accelerated wound closure in neutrophil-depleted mice. J Leukoc Biol. 2003;73(4):448–55.

127. Li L, et al. Live imaging reveals differing roles of macrophages and neutrophils during zebrafish tail fin regeneration. J Biol Chem. 2012;287(30):25353–60.
128. Hesketh M, et al. Macrophage phenotypes regulate scar formation and chronic wound healing. Int J Mol Sci. 2017;18(7)
129. Mathias JR, et al. Resolution of inflammation by retrograde chemotaxis of neutrophils in transgenic zebrafish. J Leukoc Biol. 2006;80(6):1281–8.
130. Wang J, et al. Visualizing the function and fate of neutrophils in sterile injury and repair. Science. 2017;358(6359):111–6.
131. Tauzin S, et al. Redox and Src family kinase signaling control leukocyte wound attraction and neutrophil reverse migration. J Cell Biol. 2014;207(5):589–98.
132. Gurevich DB, et al. Live imaging of wound angiogenesis reveals macrophage orchestrated vessel sprouting and regression. EMBO J. 2018;37(13)
133. Larouche J, et al. Immune regulation of skin wound healing: mechanisms and novel therapeutic targets. Adv Wound Care (New Rochelle). 2018;7(7):209–31.
134. Sherratt JA, Murray JD. Models of epidermal wound healing. Proc Biol Sci. 1990;241(1300):29–36.
135. Saiko G, Cross K, Douplik A. Mathematical model of an innate immune response to cutaneous wound in the presence of local hypoxia. Adv Exp Med Biol. 2016;923:427–33.
136. Kao LS, et al. Should perioperative supplemental oxygen be routinely recommended for surgery patients? A Bayesian meta-analysis. Ann Surg. 2012;256(6):894–901.
137. Maggelakis S. Modeling the role of angiogenesis in epidermal wound healing. Appl Math Model. 2003;27:189–96.
138. Schugart RC, et al. Wound angiogenesis as a function of tissue oxygen tension: a mathematical model. Proc Natl Acad Sci U S A. 2008;105(7):2628–33.
139. Dale PD, Maini PK, Sherratt JA. Mathematical modeling of corneal epithelial wound healing. Math Biosci. 1994;124(2):127–47.
140. Sheardown H, Cheng YL. Tear EGF concentration following corneal epithelial wound creation. J Ocul Pharmacol Ther. 1996;12(3):239–43.
141. Dallon JC, Sherratt JA, Maini PK. Modeling the effects of transforming growth factor-beta on extracellular matrix alignment in dermal wound repair. Wound Repair Regen. 2001;9(4):278–86.
142. Cumming BD, McElwain DL, Upton Z. A mathematical model of wound healing and subsequent scarring. J R Soc Interface. 2010;7(42):19–34.
143. Ferguson MW, O'Kane S. Scar-free healing: from embryonic mechanisms to adult therapeutic intervention. Philos Trans R Soc Lond B Biol Sci. 2004;359(1445):839–50.
144. Virchow R. Cellular pathology. As based upon physiological and pathological histology. Lecture XVI—Atheromatous affection of arteries. 1858. Nutr Rev. 1989;47(1):23–5.
145. Husemann Y, et al. Systemic spread is an early step in breast cancer. Cancer Cell. 2008;13(1):58–68.
146. Hu Z, et al. Quantitative evidence for early metastatic seeding in colorectal cancer. Nat Genet. 2019;51(7):1113–22.
147. Park SY, Nam JS. The force awakens: metastatic dormant cancer cells. Exp Mol Med. 2020;52(4):569–81.
148. Vesely MD, et al. Natural innate and adaptive immunity to cancer. Annu Rev Immunol. 2011;29:235–71.
149. van den Broek ME, et al. Decreased tumor surveillance in perforin-deficient mice. J Exp Med. 1996;184(5):1781–90.
150. Rangwala S, Tsai KY. Roles of the immune system in skin cancer. Br J Dermatol. 2011;165(5):953–65.

151. Chow MT, Moller A, Smyth MJ. Inflammation and immune surveillance in cancer. Semin Cancer Biol. 2012;22(1):23–32.
152. Gonzalez H, Hagerling C, Werb Z. Roles of the immune system in cancer: from tumor initiation to metastatic progression. Genes Dev. 2018;32(19-20):1267–84.
153. Dvorak HF. Tumors: wounds that do not heal. Similarities between tumor stroma generation and wound healing. N Engl J Med. 1986;315(26):1650–9.
154. Schafer M, Werner S. Cancer as an overhealing wound: an old hypothesis revisited. Nat Rev Mol Cell Biol. 2008;9(8):628–38.
155. Fogarty CE, et al. Extracellular reactive oxygen species drive apoptosis-induced proliferation via Drosophila macrophages. Curr Biol. 2016;26(5):575–84.
156. Bald T, et al. Ultraviolet-radiation-induced inflammation promotes angiotropism and metastasis in melanoma. Nature. 2014;507(7490):109–13.
157. Feng Y, Martin P. Imaging innate immune responses at tumour initiation: new insights from fish and flies. Nat Rev Cancer. 2015;15(9):556–62.
158. Kohli K, Pillarisetty VG, Kim TS. Key chemokines direct migration of immune cells in solid tumors. Cancer Gene Ther. 2022;29(1):10–21.
159. Freisinger CM, Huttenlocher A. Live imaging and gene expression analysis in zebrafish identifies a link between neutrophils and epithelial to mesenchymal transition. PLoS One. 2014;9(11):e112183.
160. Powell D, et al. Cxcr1 mediates recruitment of neutrophils and supports proliferation of tumor-initiating astrocytes in vivo. Sci Rep. 2018;8(1):13285.
161. Pastor-Pareja JC, Wu M, Xu T. An innate immune response of blood cells to tumors and tissue damage in Drosophila. Dis Model Mech. 2008;1(2–3):144–54. Discussion 153
162. Srivastava A, et al. Basement membrane remodeling is essential for Drosophila disc eversion and tumor invasion. Proc Natl Acad Sci U S A. 2007;104(8):2721–6.
163. Cordero JB, et al. Oncogenic Ras diverts a host TNF tumor suppressor activity into tumor promoter. Dev Cell. 2010;18(6):999–1011.
164. Diwanji N, Bergmann A. Basement membrane damage by ROS- and JNK-mediated Mmp2 activation drives macrophage recruitment to overgrown tissue. Nat Commun. 2020;11(1):3631.
165. van den Berg MCW, et al. Proteolytic and opportunistic breaching of the basement membrane zone by immune cells during tumor initiation. Cell Rep. 2019;27(10):2837–46 e4.
166. Lopez-Cuevas P, et al. Transformed notochordal cells trigger chronic wounds in zebrafish, destabilizing the vertebral column and bone homeostasis. Dis Model Mech. 2021;14(3)
167. Sherwood DR, Sternberg PW. Anchor cell invasion into the vulval epithelium in C. elegans. Dev Cell. 2003;5(1):21–31.
168. Voisin MB, Probstl D, Nourshargh S. Venular basement membranes ubiquitously express matrix protein low-expression regions: characterization in multiple tissues and remodeling during inflammation. Am J Pathol. 2010;176(1):482–95.
169. Chauveau A, et al. Membrane nanotubes facilitate long-distance interactions between natural killer cells and target cells. Proc Natl Acad Sci U S A. 2010;107(12):5545–50.
170. Feng Y, Renshaw S, Martin P. Live imaging of tumor initiation in zebrafish larvae reveals a trophic role for leukocyte-derived PGE(2). Curr Biol. 2012;22(13):1253–9.
171. Liao LM, et al. Nonsteroidal anti-inflammatory drug use reduces risk of adenocarcinomas of the esophagus and esophagogastric junction in a pooled analysis. Gastroenterology. 2012;142(3):442–52 e5. Quiz e22–3
172. Rothwell PM, et al. Effect of daily aspirin on long-term risk of death due to cancer: analysis of individual patient data from randomised trials. Lancet. 2011;377(9759):31–41.
173. Tsoi KKF, et al. Long-term use of low-dose aspirin for cancer prevention: A 10-year population cohort study in Hong Kong. Int J Cancer. 2019;145(1):267–73.

174. Sansom OJ, et al. Suppression of intestinal and mammary neoplasia by lifetime administration of aspirin in Apc(Min/+) and Apc(Min/+), Msh2(-/-) mice. Cancer Res. 2001;61(19):7060–4.
175. Heusinkveld M, et al. M2 macrophages induced by prostaglandin E2 and IL-6 from cervical carcinoma are switched to activated M1 macrophages by CD4+ Th1 cells. J Immunol. 2011;187(3):1157–65.
176. Coffelt SB, et al. Angiopoietin-2 regulates gene expression in TIE2-expressing monocytes and augments their inherent proangiogenic functions. Cancer Res. 2010;70(13):5270–80.
177. Folkman J. Tumor angiogenesis: therapeutic implications. N Engl J Med. 1971;285(21):1182–6.
178. Fernandez-Cortes M, Delgado-Bellido D, Oliver FJ. Vasculogenic mimicry: become an endothelial cell "but not so much". Front Oncol. 2019;9:803.
179. Li Y, et al. Targeting myeloid-derived suppressor cells to attenuate vasculogenic mimicry and synergistically enhance the anti-tumor effect of PD-1 inhibitor. iScience. 2021;24(12):103392.
180. Veglia F, Sanseviero E, Gabrilovich DI. Myeloid-derived suppressor cells in the era of increasing myeloid cell diversity. Nat Rev Immunol. 2021;21(8):485–98.
181. Steeg PS. Tumor metastasis: mechanistic insights and clinical challenges. Nat Med. 2006;12(8):895–904.
182. Mlecnik B, et al. The tumor microenvironment and Immunoscore are critical determinants of dissemination to distant metastasis. Sci Transl Med. 2016;8(327):327ra26.
183. Condeelis J, Pollard JW. Macrophages: obligate partners for tumor cell migration, invasion, and metastasis. Cell. 2006;124(2):263–6.
184. Wyckoff JB, et al. Direct visualization of macrophage-assisted tumor cell intravasation in mammary tumors. Cancer Res. 2007;67(6):2649–56.
185. Wyckoff J, et al. A paracrine loop between tumor cells and macrophages is required for tumor cell migration in mammary tumors. Cancer Res. 2004;64(19):7022–9.
186. Lin EY, et al. The macrophage growth factor CSF-1 in mammary gland development and tumor progression. J Mammary Gland Biol Neoplasia. 2002;7(2):147–62.
187. Schuster E, et al. Better together: circulating tumor cell clustering in metastatic cancer. Trends Cancer. 2021;7(11):1020–32.
188. Leach J, Morton JP, Sansom OJ. Neutrophils: Homing in on the myeloid mechanisms of metastasis. Mol Immunol. 2019;110:69–76.
189. Szczerba BM, et al. Neutrophils escort circulating tumour cells to enable cell cycle progression. Nature. 2019;566(7745):553–7.
190. Gay LJ, Felding-Habermann B. Platelets alter tumor cell attributes to propel metastasis: programming in transit. Cancer Cell. 2011;20(5):553–4.
191. Placke T, et al. Platelet-derived MHC class I confers a pseudonormal phenotype to cancer cells that subverts the antitumor reactivity of natural killer immune cells. Cancer Res. 2012;72(2):440–8.
192. Anderson NM, Simon MC. The tumor microenvironment. Curr Biol. 2020;30(16):R921–5.
193. Hanahan D, Coussens LM. Accessories to the crime: functions of cells recruited to the tumor microenvironment. Cancer Cell. 2012;21(3):309–22.
194. Truffi M, Sorrentino L, Corsi F. Fibroblasts in the tumor microenvironment. Adv Exp Med Biol. 2020;1234:15–29.
195. Roma-Rodrigues C, et al. Targeting tumor microenvironment for cancer therapy. Int J Mol Sci. 2019;20(4)
196. Raskov H, et al. Cytotoxic CD8(+) T cells in cancer and cancer immunotherapy. Br J Cancer. 2021;124(2):359–67.
197. Song G, Tian T, Zhang X. A mathematical model of cell-mediated immune response to tumor. Math Biosci Eng. 2020;18(1):373–85.

198. Idrees M, Sohail A. Bio-algorithms for the modeling and simulation of cancer cells and the immune response. Bio-Algorithms Med-Syst. 2021;17(1):55–63.
199. Folkman J, Kalluri R. Cancer without disease. Nature. 2004;427(6977):787.
200. Olson DJ, Luke JJ. The T-cell-inflamed tumor microenvironment as a paradigm for immunotherapy drug development. Immunotherapy. 2019;11(3):155–9.
201. Josefowicz SZ, Lu LF, Rudensky AY. Regulatory T cells: mechanisms of differentiation and function. Annu Rev Immunol. 2012;30:531–64.
202. Kim SH, et al. Adoptive immunotherapy with transient anti-CD4 treatment enhances anti-tumor response by increasing IL-18Ralpha(hi) CD8(+) T cells. Nat Commun. 2021;12(1):5314.
203. Ahrends T, Borst J. The opposing roles of CD4(+) T cells in anti-tumour immunity. Immunology. 2018;154(4):582–92.
204. Makhlouf AM, El-Shennawy L, Elkaranshawy HA. Mathematical modelling for the role of CD4(+)T cells in tumor-immune interactions. Comput Math Methods Med. 2020;2020:7187602.
205. Vafaei S, et al. Combination therapy with immune checkpoint inhibitors (ICIs); a new frontier. Cancer Cell Int. 2022;22(1):2.
206. Herbst RS, et al. Predictive correlates of response to the anti-PD-L1 antibody MPDL3280A in cancer patients. Nature. 2014;515(7528):563–7.
207. Duan Q, et al. Turning cold into hot: firing up the tumor microenvironment. Trends Cancer. 2020;6(7):605–18.
208. Chen Q, Sun T, Jiang C. Recent advancements in nanomedicine for 'cold' tumor immunotherapy. Nanomicro Lett. 2021;13(1):92.
209. Dogra P, et al. Mathematical modeling in cancer nanomedicine: a review. Biomed Microdevices. 2019;21(2):40.
210. Mantovani A, et al. Role of tumor-associated macrophages in tumor progression and invasion. Cancer Metastasis Rev. 2006;25(3):315–22.
211. Lan C, et al. Expression of M2-polarized macrophages is associated with poor prognosis for advanced epithelial ovarian cancer. Technol Cancer Res Treat. 2013;12(3):259–67.
212. Anfray C, et al. Current strategies to target tumor-associated-macrophages to improve anti-tumor immune responses. Cells. 2019;9(1)
213. Norton KA, Jin K, Popel AS. Modeling triple-negative breast cancer heterogeneity: Effects of stromal macrophages, fibroblasts and tumor vasculature. J Theor Biol. 2018;452:56–68.
214. Kather JN, et al. In silico modeling of immunotherapy and stroma-targeting therapies in human colorectal cancer. Cancer Res. 2017;77(22):6442–52.
215. Li X, et al. Computational modeling of the crosstalk between macrophage polarization and tumor cell plasticity in the tumor microenvironment. Front Oncol. 2019;9:10.
216. den Breems NY, Eftimie R. The re-polarisation of M2 and M1 macrophages and its role on cancer outcomes. J Theor Biol. 2016;390:23–39.
217. Mahlbacher G, et al. Mathematical modeling of tumor-associated macrophage interactions with the cancer microenvironment. J Immunother Cancer. 2018;6(1):10.
218. Ding X, et al. Engineering macrophages via nanotechnology and genetic manipulation for cancer therapy. Front Oncol. 2021;11:786913.
219. Feng Y, et al. Cell relay-delivery improves targeting and therapeutic efficacy in tumors. Bioact Mater. 2021;6(6):1528–40.
220. Zhao J, et al. Anti-tumor macrophages activated by ferumoxytol combined or surface-functionalized with the TLR3 agonist poly (I : C) promote melanoma regression. Theranostics. 2018;8(22):6307–21.
221. Muthana M, et al. Directing cell therapy to anatomic target sites in vivo with magnetic resonance targeting. Nat Commun. 2015;6:8009.

222. Meng QF, et al. Macrophage membrane-coated iron oxide nanoparticles for enhanced photothermal tumor therapy. Nanotechnology. 2018;29(13):134004.
223. Owen MR, et al. Mathematical modeling predicts synergistic antitumor effects of combining a macrophage-based, hypoxia-targeted gene therapy with chemotherapy. Cancer Res. 2011;71(8):2826–37.

Cell Migration in Cancer; Cell Migration in 2D and 3D

A. Brüning-Richardson and C. Kirby

What You Will Learn in This Chapter
The migration and invasion of cancer cells are central to metastatic disease, the main cause of death in cancer patients. One of the hallmarks of cancer is that tumour cells can leave the original tumour, migrate and invade distant tissues. Tumour cells can migrate via several different mechanisms. Solid epithelial-derived tumours initially disseminate via the epithelial–mesenchymal transition (EMT) program. This allows cells to acquire a mesenchymal phenotype and leave the original tumour via mesenchymal migration. Another mode of migration is amoeboid migration and diffuse tumours, e.g. high-grade gliomas, are thought to be able to switch between the two pathways. This is known as mesenchymal–amoeboid transition (MAT). Finally, groups or sheets of cells can migrate collectively, and the cells cooperate as a unit to migrate. This is known as collective cell migration. In vitro two-dimensional (2D) and three-dimensional (3D) models are routinely used as systems for investigating cell migration in cancer. This chapter will also explore the different mechanisms by which cancer cells can migrate and how using 2D and 3D supports these investigations.

A. Brüning-Richardson (✉) · C. Kirby
Department of Biological and Geographical Sciences, University of Huddersfield, Huddersfield, UK
e-mail: a.bruning-richardson@hud.ac.uk

A. Brüning-Richardson, S. Knipp (eds.), *Cell Migration in Development, Health and Disease*, Learning Materials in Biosciences,
https://doi.org/10.1007/978-3-031-64532-7_5

1 The Mechanisms of Tumour Cell Migration

One of the characteristics of cancer is that individual cells can migrate, meaning that they can detach from the original, or primary tumour, become motile and translocate to distant sites. The ability of a cell to migrate is central to tumour progression and is an important feature by which tumour cells disseminate and the progression to tumour metastasis begins. The epithelial–mesenchymal transition (EMT) signalling pathway is the main mechanism by which epithelial cells lose their cell-to-cell adhesions and cellular polarity and become cells displaying mesenchymal characteristics. This step is the first step of tumour cell dissemination in epithelial-derived carcinomas.

The main modes of cell migration which are adopted by cancer cells are (a) mesenchymal (crawling) migration of individual cells, which is typical for many cell types of mesenchymal origin and cells that have undergone a transition to the mesenchymal phenotype; (b) amoeboid migration of individual cells, which is usually exhibited by leukocytes and embryonic progenitor cells and (c) collective cell migration, which is typical of multiple epithelial cells that retain cell–cell connections and migrate coordinately as a multicellular group. The migration modes of tumour cells are plastic, i.e. they are not fixed, and tumour cells embarking on migration can exhibit different modes of cell migration in response to changes in external or internal conditions. This chapter will consider firstly the EMT signalling pathway and then the main modes of cell migration.

1.1 Epithelial–Mesenchymal Transition

Epithelial–Mesenchymal Transition (EMT) is a highly conserved process in which molecular alterations cause polarized, immobile epithelial cells to change their polarity through loss of tight junctions and cell adherence and become migratory mesenchymal cells. This change in cell morphology is illustrated in Fig. 1. Epithelial cells exhibit an apical and basal polarity whereby the plasma membranes are different in terms of their location, structure and composition. The apical membrane is the membrane towards the lumen in an epithelial duct, cavity or surface and is separated from the basolateral membrane by tight junctions. These junctions, also known as zona occludens, are areas where the membranes of two adjacent cells join to form a barrier that regulates the diffusion of substances between cells and allows epithelia to form cellular barriers that separate body compartments of differing compositions. The tight junctions also play a role in determining the composition of the apical and basolateral membranes and so also play a role in determining the cellular polarity. Loss of the tight junctions causes a loss of the ability of epithelial cells to maintain the separate compartments but also loss of the polarity of the epithelial cells.

EMT is an important process in embryonic development and is essential for many developmental processes such as gastrulation when the embryo transforms from a one-dimensional layer of epithelial cells (blastula) into a multilayered and multidimensional

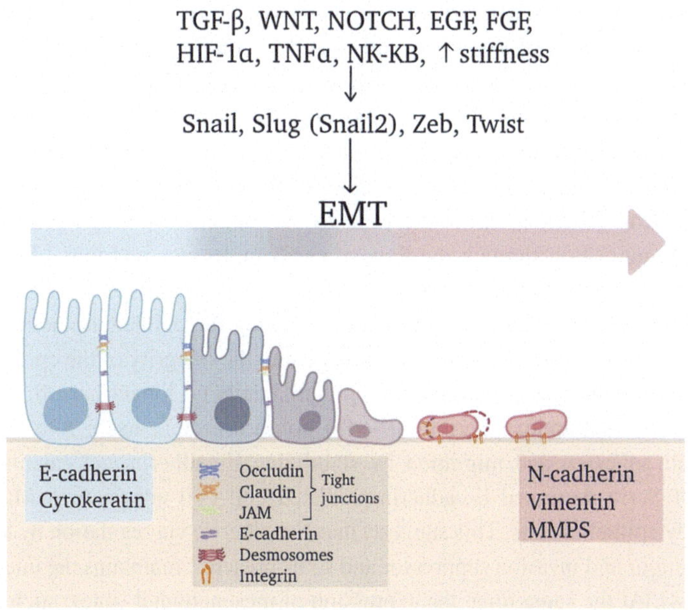

Fig. 1 EMT and mesenchymal cell migration. The cell progression through EMT is shown by colour change (blue to pink) highlighting the signalling transduction pathways and the transcription regulators involved. The loss of cell-to-cell adhesions is shown as the cell progresses from an epithelial to a mesenchymal phenotype. The turnover of the integrin-cell adhesion is shown on the migrating mesenchymal cell (red). ↑ stiffness refers to increased matrix stiffening in the tumour microenvironment. Figure drawn by D Sirnikova

structure called the gastrula. It is also central in the formation of neural crest cells, which arise from the neural tube during development. These cells migrate extensively to various parts of the growing embryo, where they differentiate into the peripheral nervous system and contribute to the formation of other tissues. The formation of the heart, musculoskeletal system and craniofacial structures are also examples where EMT plays a key role. However, a similar process has been observed by invasive and metastatic cancer cells, and suggests that tumour cells exploit the developmental EMT program for their metastatic dissemination and becomes the mechanism by which cancer cells spread. Here, epithelial-derived tumour cells can acquire an invasive and migratory phenotype and travel from the original site of the tumour and establish themselves at distant sites [1]. This program is due to molecular alterations causing the loss of the epithelial features, such as cell adherence junctions, which are anchoring structures that allow the cells to adhere to each other and support connective tissue. As stated previously, they also loose the associated apical–basal polarity, which means that the epithelial cells no longer have the characteristic apical and basolateral membranes that are typical of epithelial tissue. The loss of these features results in characteristics that are more typical of mesenchymal cells and the cells gain a motile and invasive phenotype [2], where they are able to migrate to distant sites and invade tissues. The cell-to-cell interactions in epithelial cells are due to the presence of adhesion molecules such as E-cadherin which are found within tight junctions, adherens junctions, desmosomes and gap junctions. The adherens junctions, which also contain actin filaments, and desmosomes are strong anchoring junctions. The catenins, which are cytoplasmic proteins, connect the E-cadherin complexes to the actin cytoskeleton. Gap junctions allow the cells to be connected through protein channels allowing communication through the flow of small molecules and ions [3]. These cell-to-cell adhesions are illustrated in Fig. 1.

EMT leads to the downregulation of genes encoding the epithelial junctional proteins causing the junctions to be dismantled and weakening the integrity of the epithelium. This causes a loss of apical–basal polarity and epithelial cells no longer have distinguishable membranes. The mammalian Scribble (Scrib) polarity protein has a role in regulating epithelial cell adhesion and migration by stabilizing E-cadherin and catenin coupling. Depletion of Scrib disrupted E-cadherin-mediated cell–cell adhesion in Madin–Darby canine kidney epithelial cells. This suggests that E-cadherin, via regulation by Scrib, plays a role as a tumour and invasion suppressor and so its presence maintains the integrity of the epithelium [4]. At the same time, the expression of mesenchymal genes, such as proteins N-cadherin, fibronectin and vimentin, are upregulated in these cells indicating that the cells are gaining characteristics of migratory cells [5]. As a result, these cells gain the ability to leave the original tumour and migrate into the surrounding connective tissue stroma. To be able to do this they will also have to digest through the basement membrane, which is the structure supporting the epithelium. This enzymatic digestion is mediated proteolysis by matrix metalloproteinases (MMPs). The process by which the cells move through the blood vessel wall and enter the blood circulation is called intravasation. Some circulating tumour cells and small cell clusters survive in the blood vessels and invade organs, such as the lungs or liver by a process known as extravasation. This is when the migrating cells leave the blood circulation to enter the tissues. These spreading tumour cells can eventually undergo the reverse mesenchymal–epithelial transition (MET) which promotes cell proliferation and metastasis at a distant site from the original tumour. An example of EMT is shown in Fig. 2.

1.1.1 Signalling Pathways Regulating EMT

The process of EMT occurs during normal development of the embryo, hence the signalling pathways involved in embryonic development are likely to play a role in the increased motility and invasiveness of tumour cells. There are many evolutionarily conserved pathways such as transforming growth factor beta (TGF-β), which mediates a diverse range of signalling functions involved in control of differentiation, proliferation and cell- or tissue-specific motility [6]. Wnt signalling plays a role in pattern formation during embryogenesis and studies have demonstrated a strong correlation between deregulation of Wnt signalling and human diseases [7]. The Notch pathway regulates cell proliferation, cell fate, differentiation and cell death in the embryo. It is a cell surface

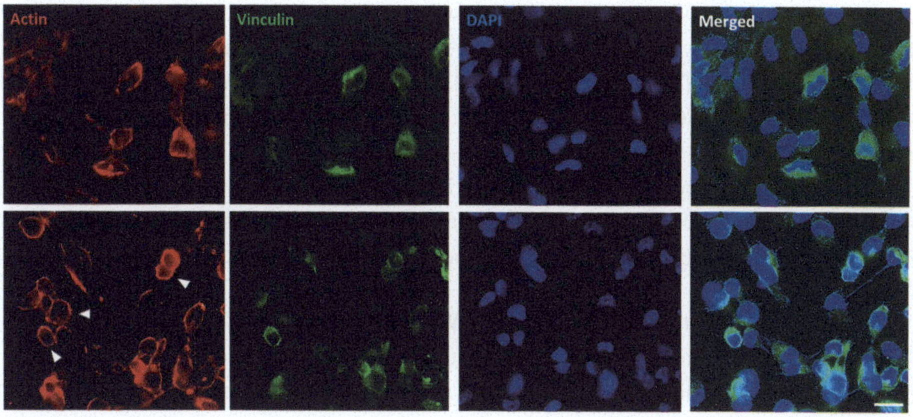

Fig. 2 Top: Epithelial ovarian cancer cells 1847 display partial EMT after plating at sub-confluence. Cells were fixed and stained for actin (red), vinculin (green) and with the nuclear marker DAPI (blue). Bottom: When treated with the anti-migratory inhibitor Bio-Indirubin, EMT is reversed and the cells assume their more epithelial morphology as indicated by the arrowheads. Vinculin staining becomes more prominent in the treated cells indicating a change in focal adhesion dynamics. Scale bar = 20 μm

receptor that transduces short-range signals by interacting with transmembrane ligands such as Delta (termed Delta-like in humans) and Serrate (termed Jagged in humans) [8]. In addition, growth factors that act via receptor tyrosine kinases, e.g. epithelial growth factor (EGF) and fibroblast growth factor (FGF), are essential components of signal transduction pathways that mediate cell-to-cell communication and also play key roles in processes such as cellular growth, differentiation, metabolism and motility [9]. A powerful stimulator of EMT is TGF-β, which is a cytokine secreted by tumour cells and the supporting stromal fibroblasts in the area surrounding the tumour, known as the tumour microenvironment [10]. Signalling pathways that are stimulated by hypoxia can also play a role. Hypoxia, or oxidative stress, is a lack of supply of oxygen in the tissues and hypoxia frequently exists in the tumour mass as the blood supply is often inadequate. Hypoxia causes overexpression of the transcription factor HIF-1α which can stimulate EMT. It was found that hypoxia, or HIF-1α regulates the expression of a transcription factor (Twist) by binding directly to the hypoxia-response element (HRE) in the proximal promoter of the gene [11]. Inflammatory cytokines, such as tumour necrosis factor alpha (TNFα), secreted by inflammatory tumour-associated macrophages (TAMS), activate the major inflammatory nuclear factor-κ light chain enhancer of activated beta cells (NF-κB) pathway which has led to similarities being made between the process of inflammation and EMT [12]. NF-κB represents a family of inducible transcription factors, which regulate genes involved in different processes of the immune and inflammatory response. Cancer is often characterized by the presence of hypoxia and inflammation thus involving both the HIF-1α and NF-κB pathways and it is now known that HIF-1α and NF-κB undergo crosstalk in pathological situations such as cancer. This means that these signalling pathways can influence each other, and evidence shows that HIF is activated by inflammation and NF-κB is modulated by hypoxia [13]. Finally, the presence of dense clusters of collagen fibrils can increase the stiffness of the extracellular matrix (ECM) and correlate with poor survival in breast cancer. It is not completely known how the mechanical signals are transduced into changes in transcription factors, but Twist has been identified as playing a role in the mechano-transduction that promotes EMT [14]. This list of signalling pathways involved in stimulating the process of EMT is not complete, but it does outline the main candidates involved.

1.1.2 Transcriptional Regulators of EMT

The previous section has outlined many of the extracellular signalling pathways involved in initiating EMT. One process by which a cell can respond to a variety of extra- and intracellular signals is called transcriptional regulation. Cancer is characterized by transcriptional dysregulation as most of the signalling pathways ultimately target the transcription machinery [15]. There are several transcription factors identified that respond to stimuli in the tumour microenvironment and that function as molecular switches for the EMT program. These include Snail, Slug (Snail 2), Twist and Zeb, and they all downregulate the expression of E-cadherin, inducing EMT and promoting tumour invasion [16, 17].

The transcriptional repressors, Snail and Slug, are members of the Snail family. These are proteins that bind to specific sites on DNA and prevent transcription of nearby genes. Snail and Slug play a pivotal role in several signalling pathways regulating EMT. The Snail family members all have a similar structure and have a specific zinc finger sequence which can combine with an enhancer box (E-box) in the promoter region of the target gene. The E-box is a DNA response element that acts as a protein-binding site and has been found to regulate gene expression of the target gene [18]. Snail can cause changes in *E-cadherin* gene (CDH1) expression by interacting with repressors, which are proteins that cause transcriptional repression and so lead to a reduced expression of the CDH1 gene. Or it can interact with epigenetic remodelling complexes, which are complexes that can modify gene expression without altering the DNA sequence [19]. Downregulation of the expression of E-cadherin is associated with the migratory properties and invasion of epithelial cells and can be correlated with the grade and stage of the tumour [20].

Slug has also been characterized as a strong E-cadherin repressor and a major EMT inducer and is correlated with distant tumour metastasis [21, 22]. Slug also cooperates with another transcriptional regulator, Twist, to promote EMT and metastasis. Slug mediates Twist-induced EMT, and Twist is required for Slug-induced suppression of EMT [23]. Twist also belongs to the evolutionarily conserved basic helix-loop-helix (bHLH) family which plays an essential role in embryonic development and tumour metastasis [24]. Like Snail, Twist binds to the E-box element in the promoter region of the E-cadherin gene, inhibiting its activity and reducing the transcription of E-cadherin. The subsequent induction of EMT, increases the migration, invasion and metastasis of tumour cells cancer [25]. Twist is known to play a role in a variety of invasive cancers, including breast [26], lung [27] and prostate cancers [28]. Snail and Slug may play important but distinct roles at each stage of the tumour dissemination process. This could involve a role in tumour cell dissemination, intravasation, transport to a distant site, extravasation, or adaptation and growth of the relocated tumour cells. Links between Snail, Slug and lymph node metastasis in a large sampling of mammary carcinomas, suggested a role for Slug in the maintenance of semi-differentiated structures and tumour invasion in the progression of breast cancer [29]. Transcription factors that regulate E-cadherin were expressed differentially in metastatic ovarian and breast carcinoma, with a stronger suppression of E-cadherin by Snail in breast carcinoma effusions, a clinical condition associated with extremely poor survival. Snail may predict a poor outcome in patients who have breast carcinoma metastatic effusions [30].

The TGF-β family of signalling molecules is proposed to have a role in cell migration in EMT [31]. It has been implicated in the upregulation of many of the transcription factors involved in regulating EMT including Snail [32, 33] and Slug [34]. Additionally, TGF-β-dependent activation of Snail and Slug is thought to be mutually exclusive, e.g. when Snail is activated by TGF-β, Slug is not, and vice versa. Also, the TGF-β dependent activation of Snail and Slug is dependent on the type of epithelial cell involved [35].

Zeb1 (also known as δEF1) and Zeb2 (also known as Sip1) can repress transcription by directly binding to 5-CACCT sequences located in various gene promotors. Hypoxic conditions occur when the uncontrollable growth of malignant tumour cells exceeds their blood supply, and this can stimulate the EMT process. It has been proposed that Zeb1/δefF1 and Zeb2/Sip1 are involved in the hypoxic (HIF-1α-dependent) downregulation of E-cadherin in von Hippel-Lindau (VHL)-negative renal clear cell carcinoma (RCC) cells. This mechanism is thought to contribute to the aggressive nature of these tumours [36, 37]. Zeb2/Sip1 is also involved in the upregulation of mesenchymal markers including vimentin and N-cadherin, indicating the transition from the epithelial to mesenchymal phenotype. Vimentin is a type-Ill intermediate filament normally expressed in cells of mesenchymal origin, so its increased expression is indicative of EMT [38, 39]. Snail and Zeb2/SIP1 have been associated not only with E-cadherin repression but also with acceleration of cancer invasion. The increased expression of MMPs, enzymes responsible for the protein degradation of the extracellular matrix (ECM) has also been associated with the loss of differentiation, tissue invasion and metastasis and tumour recurrence. A role for Snail Zeb2/Sip1 in the progression of hepatocellular carcinoma (HCC) cell lines was proposed to be due to transcription repression of the E-cadherin gene and upregulation of the MMP gene family promoting tumour invasion [40].

In this section, the signalling pathways and the role of transcriptional regulators that trigger EMT have been summarized. The following sections will summarize the different modes by which a cell can migrate.

1.2 Modes of Cell Migration

For cells to become motile and be able to migrate, they need to be able to change their shape and interact with the surrounding ECM. The ability of the cytoskeleton to reorganize and adapt to changes in the surrounding ECM will determine its migration ability. There are three key processes that impact cell motility, and these are actin dynamics, cell–substrate adhesion and ECM remodelling. The cytoskeleton is made up of three structures: (a) microtubules, which are composed of alpha- and beta-tubulin proteins and (b) the intermediate filaments, which include the keratins, lamins, neurofilaments, desmin and vimentin. Finally (c), the microfilaments, which are the smallest components of the cytoskeleton, and form a dynamic network made up of actin polymers and associated actin-binding proteins to form the actin cytoskeleton. The actin cytoskeleton can create force by the formation of cross bridges with myosin, this produces movement within the cell leading to changes in its shape [41]. The changes in cell shape can be converted into movement if the cells interact with the surrounding environment. Transmembrane receptors, such as integrins and growth factor receptors, cluster and recruit signalling molecules, e.g. small Rho GTPases and kinases, to assemble cell adhesion junctions with the ECM. These cell adhesions can provide friction, or traction, to facilitate cell movement. Also, as the cells attempt to move through their environment, they need to make space in the ECM. The MMPs enable digestion and remodelling of the ECM. These key features determine the mode by which a cell can migrate. Individual cells can migrate via mesenchymal and amoeboid migration. Collective cell migration involves migration of groups of intact cells [42]. These are considered below.

1.2.1 Mesenchymal Cell Migration

Mesenchymal cell migration is characterized by well-developed cell protrusions and cell adhesion. The invading cells are elongated and spindle shaped. They have a strong cell-to-matrix adhesion and the ability to degrade ECM [43]. For a cell to be able to move, it must attach itself to the ECM and use its front edge to move forward, exerting force to form a protrusion. The rear part of the cell must detach from the ECM, allowing retraction of the trailing edge and the cell to move forward. Cells move by repeated cycles of protrusion and attachment of the cell front, followed by detachment and retraction of the rear. The polymerization of actin filaments produces forces that drive the extension of the plasma membrane at the cell's leading edge as the polymers grow below the plasma membrane. To generate the pushing force for protrusion, the cell uses the energy of actin polymerization, mediated by proteins such as formins and regulated by the family of small Rho GTPases, e.g. Rac1 and Cdc42 [44]. F-actin is polymerized from the barbed, or spiked, end and this generates actin cables that provide support for the lamellipod area. These protrusions attach to focal adhesions and generate force. Retraction is caused by contractile forces produced by myosin motors pulling on actin filaments in the trailing end of the cell [45]. Mesenchymal cell migration requires strong cell adhesion and the timely attachment and detachment of the cell to the ECM via integrin-cell adhesions [46].

There are two types of cell protrusions that occur in mesenchymal cell migration, filopodia and lamellipodia. Filopodia are thin protrusions that extend beyond the edge of the cell. They are found in the leading edge of migrating cells such as fibroblasts or cells that extend long processes, such as neurons, and their role is to reach out and sense targets. Actin filaments are organized into parallel bundles in filopodia and can protrude by continuously releasing monomers from the back for reuse at the front of the polymer. The actin filaments are branched in lamellipodia, which are two-dimensional (2D) planar fan-shaped protrusions that generate the large pushing forces responsible for the progression of the leading edge. They can also play a role in cell navigation, sensing and probing the cell environment [41]. They have been extensively investigated in 2D cell culture on flat solid surfaces, however, lamellipodia have also been identified in cells migrating in three-dimensional (3D) matrixes that mimic more closely physiological conditions [47].

As discussed above, the polymerization of actin and the myosin motors create pushing and contracting forces and to convert these forces into movement the cell requires traction. Traction forces are provided by cell adhesion molecules, which provide the 'grip' on the ECM that the cell requires to transfer the energy into movement. Commonly it involves an interaction with cell adhesion receptors of the integrin family, which interact with specific types of ECM depending on integrin specificity [48, 49]. Cell adhesive receptors of the cadherin family mediate cell–cell interaction and the intracellular domains of the integrin and cadherin adhesion receptors are linked to the actin cytoskeleton [50, 51].

Enzymes that can degrade the ECM are upregulated and activated in tumour cells. Mesenchymal cells recruit proteases, such as the MMPs, to facilitate their movement through the ECM. On the leading edge of advancing cells, surface MMPs cause proteolytic degradation of the ECM cleaving many components, such as collagens, fibronectin and laminin. The local ECM degradation creates physicochemical trails for the migrating cells [52]. Mesenchymal migrating tumour cells usually originate from connective tissue tumours, e.g. sarcomas. They also originate from other tumour types after EMT allowing the cells to dedifferentiate and lose cell–cell junctions permitting dissemination into the ECM [53]. An example of mesenchymal cell migration is shown in Fig. 3.

1.2.2 Amoeboid Cell Migration

Cells migrate with low cell adhesion and/or high actomyosin-mediated contractility have little grip on the substrate and can assume spherical shapes and adopt amoeboid cell migration. Amoeboid cell migration is named after the type of movement observed in single-celled amoebae, known as *Dictyostelium discoideum* [54]. Amoeboid cells can change their shape to allow permeation through gaps in the surrounding ECM that can be as narrow as 3 μm [55]. The movement is rapid, and cells can move at velocities between 2 and 30 μm/min. There are intermittent interactions, or adhesions, between the cell membrane and surrounding ECM allowing the cells to "crawl" or creep along [54]. Amoeboid movement results from changes in the cytoskeleton of the cell that produces spherical protrusions that are loosely attached to the ECM, called blebs. The blebs are formed as the cell membrane expands and contracts. This process does not involve the

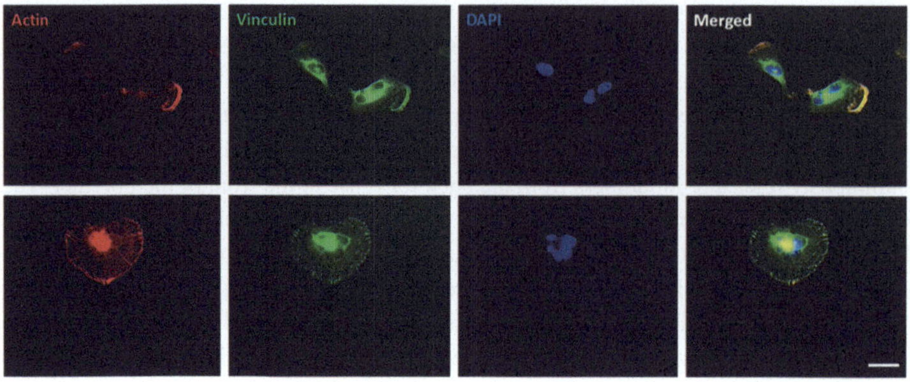

Fig. 3 Glioblastoma cell line U251 cells display a mesenchymal phenotype indicated by an elaborate migration front after plating at sub confluence. Cells were fixed and stained for actin (red), vinculin (green) and with the nuclear marker DAPI (blue). Bottom: When treated with the anti-migratory inhibitor Bio-Indirubin the mesenchymal phenotype is reversed and the cells assume a more rounded, epithelial morphology. Actin staining is relocated from a cell front localization to a distribution over the whole cell surface. A similar shift is observed for vinculin in the treated cells indicating a change in focal adhesion dynamics. Scale bar = 20 μm

polymerization of actin, as in mesenchymal cell migration, but contraction due to cross-bridge cycling between actin filaments and non-muscle myosin II (NMII), known as the contractile actomyosin cortex. During the formation of a bleb, the plasma membrane becomes separated from the actomyosin cortex and the NMII plays an essential role in this [56]. After expansion, the contractile actomyosin cortex reassembles under the membrane to enable retraction of the bleb [57]. The retraction can propel the cell forward if there is sufficient traction, or interaction, with the ECM [58]. Amoeboid cell migration is dependent on the ability of the cell to change its shape and form weak interaction with the underlying ECM so that the cell can move forward. Movement through the ECM does not require degradation and reorganization of the ECM and so is not dependent on proteolysis by MMPs [59]. Amoeboid movement is not prevented by integrin inhibition and so is independent of cell integrin adhesions [60]. The origin of amoeboid migrating tumour cells is often from the haematopoietic cells, e.g. in leukaemia and lymphoma, and small cell lung carcinoma. However, amoeboid cell migration can also be observed in cell subgroups in many types of tumours [61].

1.2.3 Collective Cell Migration

In collective cell migration, numerous cells can move together as a group, for example as cell sheets, strands or clusters. Friedl et al. [62] showed that groups of clustered cells, 5 to more than 100 cells, can detach from the primary tumour and migrate into the adjacent ECM using 3D collagen lattices [62]. Strong intercellular connections are necessary for the cells to be able to move as a cohesive group [63]. This type of cell migration is important in embryogenesis, wound healing but also plays a role in the migration of cancer cells [64]. The cells maintain a hierarchical structure and polarity; and they have a distinct leading front edge and a rear trailing edge [65]. It is proposed that the leader cells can sense and migrate in response to an external signalling gradient, and that adjacent cells follow due to the strong cell–cell contacts [64]. This type of cell motility also depends on actin dynamics, integrin-dependent adhesion to the ECM and MMP proteolysis and reorganization of the surrounding ECM [60]. Collective migration is present in most epithelial and mesenchymal tumour types [62].

1.2.4 Transition of the Mode of Cancer Cell Migration

The mode of migration in cancer cells is not fixed and cells can transition from one mode to another depending on the different environmental challenges and cues. MAT explains how a single mesenchymal cancer cell can accumulate amoeboid characteristics and migrate through narrow confinements in the surrounding ECM promoting tumour cell metastasis [55].

The transition from one mode of migration to another is brought about by the activation of transmembrane receptors such as integrin and growth factor receptors. These receptors recruit intracellular signalling molecules, e.g. Rho GTPases, to regulate the activity of the actin cytoskeleton and assemble adhesion complexes at sites of cell–ECM interaction, thus allowing the cells to change their characteristics. These Rho GTPases act as intracellular molecular switches, being active when bound to guanine triphosphate (GTP) and inactive when they are bound to guanine diphosphate (GDP) (see Fig. 4). Guanine nucleotide exchange factors (GEFs) cause the release of GDP and so promote GTP binding to activate Rho GTPase. GTPase activating proteins (GAPs) stimulate the hydrolysis of GTP and so inactivate Rho GTPase. The inactive Rho GTPases remain in the cytoplasm due to the activity of guanine nucleotide dissociation inhibitor (GDI) [46].

Members of the Rho family of small GTPases have been linked to mesenchymal and amoeboid cell movement. The GTPase Rac1 stimulates the formation of actin filaments driving cell motility by promoting lamellipodia, and is associated with mesenchymal cell migration [66]. Fibrosarcoma and carcinoma cells exhibited a mesenchymal-type migration which included the clustering of integrins and MMP together at fibre bindings sites and the generation of tube-like proteolytic degradation tracks [67]. This proteolysis of the ECM by the MMP, enables the actin protrusions to be pushed through channels in the ECM, as was evident by the degradation tracks. Rac1 activity can be linked to these observations. Overexpression of Rac1 is known to induce an increase in cellular reactive oxygen species (ROS). ROS itself leads to the expression of transcription factors including SNAIL which itself induces the upregulation of MMP expression. Interestingly, Rac1 also promoted the generation of 'invadopodia', F-actin enriched structures which are characterized by the presence of cortactin, cofilin, neural Wiskott–Aldrich syndrome protein (N-WASP) and Arp2/3, which are themselves surrounded by a ring of actin regulatory and adhesive molecules, such as integrins, talin, vinculin and paxillin. Upon elongation, maturation and stabilization of these complexes enabled by Rac1-driven actin

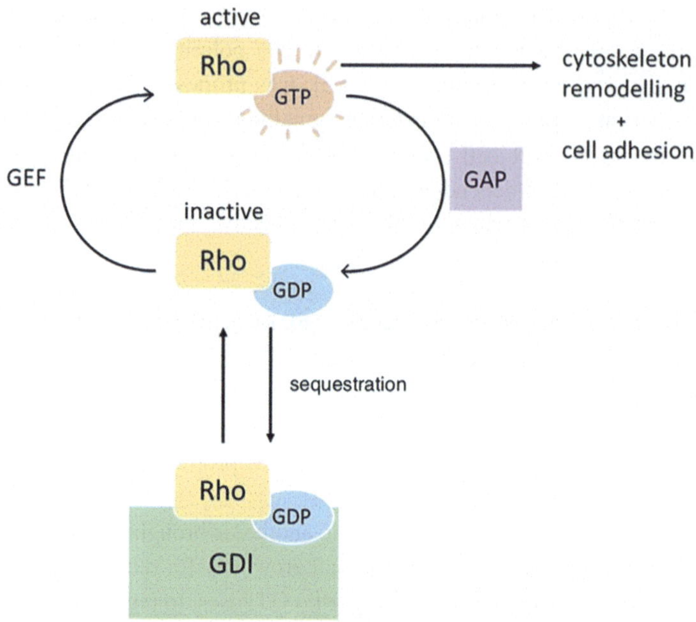

Fig. 4 Activation of Rho GTPase. Rho GTPase is active when bound to guanine triphosphate (GTP) and inactive when bound to guanine diphosphate (GDP). Guanine nucleotide exchange factors (GEFs) cause the release of GDP and promote GTP binding, activating Rho GTPase. GTPase activating proteins (GAPs) stimulate GTP hydrolysis, inactivating Rho GTPase. Inactive Rho GTPases remain in the cytoplasm due to guanine nucleotide dissociation inhibitor (GDI). Figure drawn by D. Sirnikova

polymerization, recruitment of microtubules and intermediate filaments secretion and recruitment of proteases was achieved. A group of MMPs are associated with invadopodia including MMP-2, MMP-9 and MT1-MMP, the ADAM family members, membrane-bound serine protease and the urokinase plasminogen activator receptor (uPAR). Inhibiting MMPs induces a transition from the mesenchymal to amoeboid phenotype that allows invasion in the absence of proteolysis and degradation of the ECM [67]. Furthermore, this transition is associated with weakened integrin-dependent adhesion in fibrosarcoma cells, which is also typical of amoeboid cell motility [68]. In contrast, RhoA signals to ROCK kinase, stimulating the formation of actin fibres and generation of the actomyosin contractility required for amoeboid cell motility [69]. This motility is ROCK dependent but independent of the MMPs, which are not required as the ROCK-dependent activation of myosin light chain (MLC) generates sufficient force to deform the ECM and thereby facilitate amoeboid cell movement [70]. In human fibrosarcoma cells, the mesenchymal amoeboid transition is associated with weakened integrin-dependent adhesion, which is also typical of amoeboid cell migrations [68]. The tight interplay between Rho and Rac determines the different modes of tumour cell movement and reveals how tumour cells can transition between mesenchymal and amoeboid movements [53].

Collective cell migration dependent on β-integrin-cell adhesions to maintain cohesion of cell clusters was impaired using a β-integrin antibody. The cell adhesions were disrupted, and detachment and dissemination of single amoeboid cells occurred indicating a transition from collective cell migration to amoeboid cell migration [60]. This emphasized the plasticity of tumour cell migration that depends on different environmental challenges and cues. Transition to amoeboid cell migration is associated with tumour progression and metastasis, but it can also play a role in the tumour evasion of medical treatment by enabling tumour cells to adopt alternative mechanisms for cell migration. Therefore, it is important to determine the mechanisms that regulate this transition and mechanisms permitting the plasticity of migration [71].

2 Two-Dimensional (2D) Cell Migration

2.1 2D Cell Culture Techniques

Cell culture is a process of growing cells in a controlled setting outside the natural environment. 2D cell culture has been practiced since the early 1900s and it is still widely used. The advantages are that it is inexpensive, the techniques of cell growth and downstream analyses are well established and there is a plethora of comparative published literature. The cells typically grow as monolayers on flat glass or plastic surfaces allowing them to multiply and spread. It may be necessary to coat the plastic surface to facilitate the adherence of the cells.

2.2 Cell Migration in 2D

A characteristic of mesenchymal cell migration in 2D is the establishment of cellular protrusion such as filopodia (thin membrane protrusions) and lamellipodia (broad, flattened cell extensions) cell protrusions (see Fig. 5). Cell advancement is predominantly driven by the lamellipodia, with the leading edge of the cell containing branched F-actin complexes that are formed by actin polymerization. This polymerization is controlled by signalling molecules, particularly the small GTPase Rac 1. In filopodia, the actin filaments run parallel towards the tip, and this allows molecular motors, such as unconventional myosin-X (MYO10), to move along and accumulate at the tip. These motors are thought to transport various proteins, including integrins, along the actin filaments to the tips of filopodia. The integrin–adhesion complexes tether the filopodia to the ECM [71].

The emerging adhesions in the advancing lamellipodium mature into larger focal adhesion structures which form the mechanical links between the intracellular actin complexes and the ECM [72]. Filopodia play a role in the progress of the lamellipodium

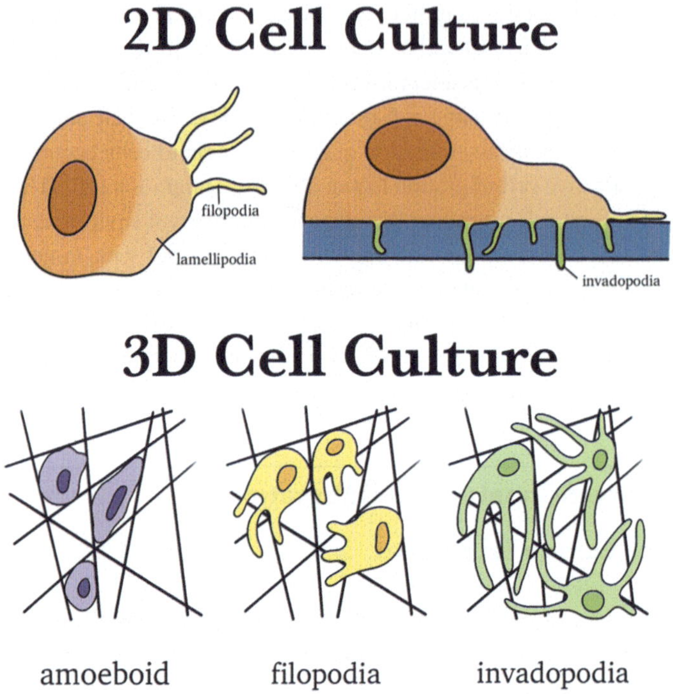

Fig. 5 2D versus 3D cell culture. Cells migrating on hard flat 2D surfaces display fan-shaped lamellipodia that exhibit thin filopodia on the leading edge. Invadopdia are present, though are limited by the hard 2D surface. Cells migrating in 3D scaffolds can display amoeboid cell migration, whereby they squeeze through the scaffold. The large firm spheroid nucleus may present an obstacle but some cells have flexible nuclei that can squeeze through narrow spaces. 3D cells can also display filopodia and invadopodia. Figure drawn by D Sirnikova

and project out of the lamellipodia. They have adhesion sites located at the base, shaft and tip. The base adhesions associate with the focal adhesions and play a role in the development of the filopodia. The *shaft* adhesions control the advance of the lamellipodia though not associated with focal adhesions. The *tip* adhesions play a role in motility and affect the advance of the lamellipodia. Both tip and shaft adhesion sites are integrin mediated [73, 74] and actin spikes can also be present on some filopodia [75]. Filopodia can also play a role in sensing the environment. They are required for guiding the migration of fibroblasts along external chemical gradients, this is known as chemotaxis, suggesting a role in probing the ECM. In this situation, the filopodia has been proposed to have a rigidity sensing mechanism allowing them to probe the stiffness of the substrate as the leading edge lamellipodia progresses [73]. In platelets, filopodia have been proposed to play a key role as antennae in sensing and adhering to the local environment during cell spreading [76].

Invadopodia are similar to filopodia, and they share common mechanisms with the actin-based dynamic protrusions of the plasma membrane. They are localized beneath the cell body and their progress will be limited in 2D cell culture as they cannot invade the plastic tissue culture surfaces. Hence to study them, the coverslips are frequently coated with a layer of artificial ECM, and increasing the ECM rigidity increases the formation of invadopodia (see Fig. 5). Their presence in 2D cell culture correlates with invasive characteristics as they contain proteases such as the MMPs which assist in degrading the ECM and so will contribute to the invasive nature of cells in vivo [42].

In addition, some cell types including cancer cells can adopt an amoeboid cell migration. Here rather than relying on generating cell movement by generation of cell protrusions, cells can move by enhanced contractility. ROCK-dependent myosin light chain phosphorylation and myosin II activation via the GTPase RhoA drive the adoption of a rounded or amoeboid phenotype. This is concomitant with the occurrence of cellular blebs, which are caused by myosin II activity and are effects of a local rise in hydrostatic pressure. As a result, the cortical actin network may rupture or the plasma membrane can detach from the cortical cytoskeleton. Cytosol can flow freely to induce the formation of blebs in the plasma membrane. Bleb retractions occur after recruitment of F-actin and actin-binding proteins into the bleb alongside myosin II activity.

2.3 Limitations of 2D Cell Culture

2D cell culture systems are limited by the fact that the cells grow on solid flat surfaces and so the expansion is limited to a sideways direction. This does not represent in vivo cell environments as the cells cannot expand in the third dimension, where cells are surrounded and influenced by other cells. As growth is limited in 2D, there is a forced membrane polarity of the cells in terms of an apical and basal surface. There is little cell-to-cell communication, and abnormal distribution of integrin adhesion ligand complexes, and the solid surface provides high matrix stiffness which influences cell growth [77] (see Fig. 5). The lamellipodia-based migration mode of migration in 2D represents only one of several mechanisms of cell motility in 3D environments. Cells grown in cell culture, both 2D and 3D, will consume the media, deplete nutrients and produce waste products damaging the environment the cells are in. However, 2D cell culture still provides important insights into the mechanisms of cell protrusions, chemotaxis and cell polarity. Although flat, 2D cell culture has predominated, recent research has shifted towards culture using three-dimensional (3D) structures, particularly in recognizing the important role of the ECM in regulating cell behaviour.

3 Three-Dimensional (3D) Cell Migration

Cell culture systems have evolved to grow cells in 3D. This method uses an extracellular matrix as a supporting scaffold and allows the cells to expand and interact with each other and the scaffold (see Fig. 5). The advantages of 3D cell culture are that the matrix can be more flexible, cells can grow with surrounding connections to the matrix and other cells. The ability of cancer cells to migrate from the primary tumour and metastasize to distant sites plays a role in the progression of cancer in a patient. For this reason, cell migration in vivo is important to elucidate the signalling mechanisms and structural changes that occur in a cell during cell migration. The disadvantages are that it is more expensive, and the cell culture techniques are more complex. It may also be difficult to mimic in vivo environments, however, in vivo techniques are progressing all the time.

3.1 3D Cell Culture Methods

Cells in 2D cell culture flatten out and have a forced apical–basal polarity though they usually regain their characteristics when placed back in appropriate 3D culture conditions. Epithelial cells can regain their natural apical–basal polarity and glandular cells are able to form a lumen allowing secretion. The different modes of 3D migration are dependent on the physical properties of the ECM [78].

3.1.1 Intact Tissue slices

Intact tissue slices cultured in vitro will retain their original 3D architecture. A limiting factor is the thickness of the slice which must be less than 0.3 mm thickness to allow for adequate delivery of oxygen and nutrients. Alternatively, a model that mimics the internal hypoxic conditions commonly present in a tumour due to the lack of nutrients and oxygen, may be valuable in understanding how these factors contribute to the growth of new blood vessels by angiogenesis and possible resistance to chemotherapeutic drugs [78].

3.1.2 Scaffold-Free Techniques

Cell growth in scaffold-free environments involves the spontaneous aggregation of cells into a mass of cells known as a spheroid. Spheroids are multilayered 3D aggregates whose cellular architecture recapitulates the original tumour's structural characteristics. As such they consist of distinct layers including an inner hypoxic and/or necrotic region, a layer of non-proliferative cells and towards the edge a region of highly proliferative cells. Such multicellular spheroids allow the investigation of the tumour microenvironment, cell-to-cell interactions, proliferation and absorption of nutrients in a more tumour-relevant setting. They are also characterized by diffusion gradients including nutrients, waste and oxygen. As such, the application of 3D spheroids for the development and testing of cytotoxic or anti-migratory drugs allows validation of effects on cellular responses or infiltration in preclinical drug screens. This is advantageous over the commonly used testing on 2D monolayers which lack a 3D structure, have only limited cell-to-cell contacts and diffusion gradients are non-existent [79, 80].

3.1.3 Scaffold-Based Techniques

Cell growth is supported by structures that anchor the cell and replicate the ECM. Hydrogels, which have a large water content, can be engineered to provide a more realistic microenvironment to the cells. They can have mechanics like those of many soft tissues and can support cell adhesion and protein sequestration. Hydrogels can be broadly classified as either natural or synthetic materials. Three naturally derived hydrogels are collagen, fibrin and alginate. Two synthetic hydrogels are polyacrylamide (PA) and polyethylene glycol (PEG) and there are two hybrid materials that combine elements of synthetic and natural polymers, and these are hyaluronic acid and polypeptides [81]. Examples for different migratory activity of cancer cells grown as spheroids and embedded in a collagen matrix are shown in Figs. 6 and 7.

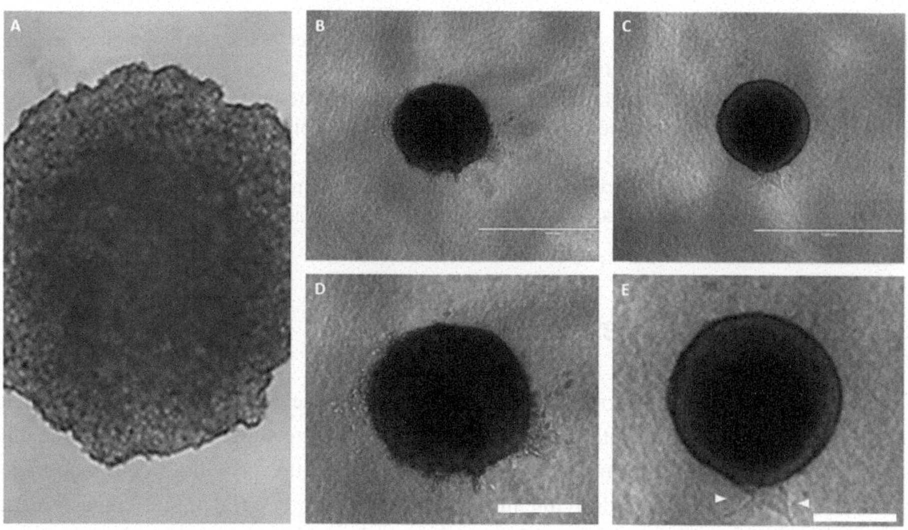

Fig. 6 The epithelial ovarian cancer cell line 1847 produces spheroids after plating in low adherence 96-well plates (**a**). When embedded in a matrix such as collagen, cell migration is observed in a sheet-like manner, away from the original spheroid (**b**); this is targeted after treatment with the anti-migratory inhibitor BIO-indirubin. After treatment, the epithelial sheets disappear and only single appendages are observed (**c**). Enhanced images of the control (**d**) and the treated spheroid (**e**) are shown and the changes in cell migration are indicated by the arrowheads. Scale bar = 1000 μm

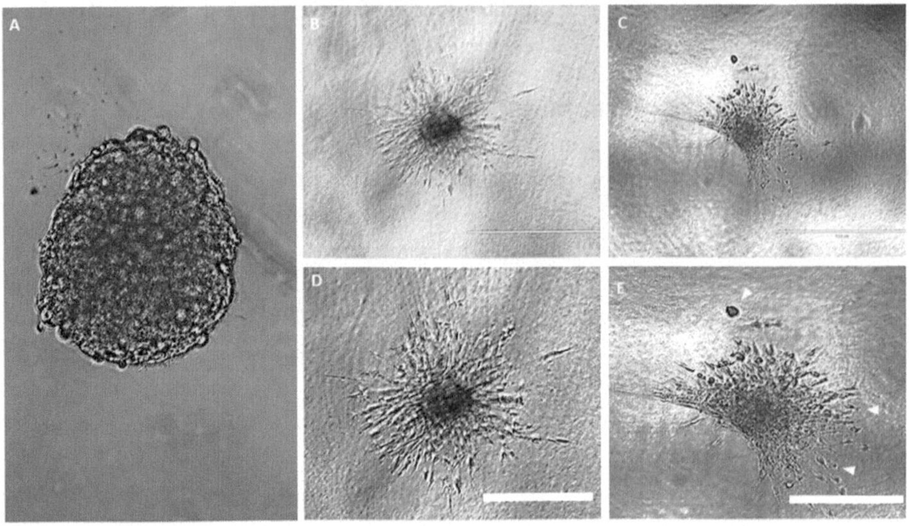

Fig. 7 The Glioblastoma cell line U251 produces spheroids after plating in low adherence 96-well plates (**a**). When embedded in a matrix such as collagen, cell migration is observed in a star-like manner with connective cellular appendages and single mesenchymal cells away from the original spheroid (**b**); this is targeted after treatment with the anti-migratory inhibitor BIO-indirubin. After treatment, the mesenchymal morphology starts to disappear, and individual rounded cells are observed in the vicinity of the original spheroid (**c**). Enhanced images of the control (**d**) and the treated spheroid (**e**) are shown and the changes in cell migration are indicated by the arrowheads. Scale bar = 1000 μm

3.2 Cell Migration in 3D

The characteristics of cell migration in 3D are shaped by the 3D structure through which the cell navigates. In vivo, epithelial and endothelial cells are supported by the basement membrane which forms a thin sheet-like supporting structure providing cell anchorage and a boundary from the underlying interstitial matrix. Interstitial tissue consists of a dense network of collagen fibres that results in many pores of varying magnitude that can allow the passage of migrating cells. This varying microenvironment will ensure that migrating cells adopt several modes with which to propel themselves through this changing landscape. In addition, the cells are also able to transition from one form of migration mode to another depending on the cues from the microenvironment [43].

3.2.1 Modes of Cell Migration in 3D

Tumour cells can adopt characteristics of amoeboid cell movement in 3D which is independent of proteases and involves the formation of membrane blebs due to the contractility of the actomyosin cortex. By forming blebs, tumour cells can squeeze themselves through gaps in the 3D-ECM. The contractility at the rear of the migrating cell generates a hydrostatic pressure and causes the flow of cytoplasm forward. This is a feature of amoeboid migration that facilitates forward movement of cells and has been observed in human breast adenocarcinoma cells [82]. Lobopodia are another type of blunt protrusions that can be observed in 3D cell migration. Lobopodial migration can be viewed as a mixture of amoeboid and mesenchymal migration. It uses asymmetric intracellular pressure to generate the bleb-like protrusions on the leading edge called lobopodia. Like mesenchymal migration, the cells adhere tightly to the ECM to exert traction and result in movement. The nucleus is able to move forward in a 'piston' like manner to generate pressure in the front half of the cell, moving the cell forward [83]. To form the lobopodia, the nucleus physically divides the cytoplasm into two parts, a posterior and anterior compartment. An increase in the hydrostatic pressure in the anterior compartment that lies between the nucleus and the leading edge drives the formation of the lobopodia. A protein that links the nucleoskeleton and cytoskeleton is located in the nuclear membrane and enables the nucleus to act like a piston, determining the migration of the leading edge of the migrating cells. By this mechanism, the cells migrating in 3D-ECM can switch from low-to high-pressure protrusions in response to the surrounding 3D matrix. This is typically observed in primary human fibroblasts and cancerous fibrosarcoma cells. The formation of lobopodia is also via actomyosin contractility in a similar way to the blebs in amoeboid transport [84, 85].

Filopodia are also commonly observed in migrating cells growing in in vivo models and may play a role in directional cell migration. Filopodia play a role in angiogenesis, the growth of new blood vessels from existing ones, and this has been investigated in zebrafish embryos. Here, vascular endothelial growth factor (VEGF) induced filopodial protrusions in the endothelial tip cell, and these protrusions invaded the surrounding ECM [86]. However, in the zebrafish embryo, inhibition of filopodia appeared to decrease the migration speed but not to alter the direction of the endothelial tip cell. This data suggested that filopodia may not be required for directional migration or chemotaxis, but play a role in probing the ECM [87]. Filopodia are present in the edges of migrating epithelial cell sheets and play a role in the fusion of these structures both in wound healing and in embryonic development. The small GTPase Rac1 leads to the assembly of filopodia and lamellipodia identifying a key role of Rac in organization of actin in these leading edge protrusions [88]. In the Drosophila embryo, the hole in the dorsal epithelium is zipped closed along the midline during the later stages of embryogenesis. Here filopodia play a role in the correct matching of cells in the opposing epithelial edges, maintaining the correct alignment of cells at the midline seam [89]. Actin-rich filopodia-like structures have been observed in cancer cells, as they invade through 3D microenvironments. In ovarian cancer cells, an increased number of these filopodia-like structures, called actin spikes, were induced by activation of the small GTPase RhoA. They appear at the cell front and lead forward movement rather than lamellipodia protrusions. The actin spikes correlate with the invasiveness of ovarian carcinoma [90].

Invasive cancer cells cultured on 3D ECMs form actin-rich membrane protrusions extended vertically from the leading ventral cell membrane. These are the invadopodia and they can extend vertically through the 3D-ECM. This contrasts with their limited invasiveness observed in 2D cell culture. These protrusions play an important role in the invasiveness of cancer, and they have the ability to remodel and degrade ECM. They contain actin regulatory proteins and adhesion molecules [91]. Cancer cell migration in breast cancer tumours in 2D and 3D has been described to involve mesenchymal-like migration. The cell protrusions described are lamellipodia and invadopodia/podosomes [92]. Podosomes are dynamic actin-rich adhesion structures that have a remarkable similarity to invadopodia in their structure [92]. The precise relationship between these two structures is confusing as most studies investigating podosomes have been carried out in 2D on glass or plastic hard surfaces. Studies on 3D identify invadopodia as the structures involved and these are carried out on thick 3D ECMs. The distinction between these two structures may be because different culture strategies have been used. It has been proposed that podosomes may be the forerunner structures that mature into invadopodia under certain physiological conditions, especially as podosomes have a lifespan of minutes, whereas invadopodia last from minutes to hours [92].

Glioblastoma cell invasion through 3D matrices is dependent on MMPs and thus indicates a mesenchymal-like migration. Treatment with MMP inhibitor (MMPI), Ilomastat, induced glioma cells to have an amoeboid-like morphology with invasive ability [59]. It was found that the ROCK/MLC signal was involved in the MMPI-induced switch to amoeboid morphology. The MMPI-treated glioma cells were able to compensate for the loss of their invasive mesenchymal proteolytic-dependent migration by developing an amoeboid-like migration. It was suggested that a combination of an MMP inhibitor, that inhibits proteolysis, and a ROCK inhibitor, that inhibits amoeboid movement, could be used as an attractive antitumour drug candidate for the treatment of glioblastoma. Treatment of glioblastomas with MMPI alone has not been very successful due to the ability of the tumour cells to undergo MAT and adopt an amoeboid-like migration [59].

3.2.2 The Nucleus in 3D Cell Migration

The dense collagen fibres present in the ECM can provide a significant obstacle to the migration of cells. Whereas the cytoplasm is flexible, the passage of the nucleus through constrained spaces is a significant physical challenge. The nucleus is a large, stiff and oval-shaped organelle that must be deformed and manoeuvred through small spaces due to the structural arrangement of collagen fibres in the ECM. However, the nucleus can also actively participate in driving migrations in the 3D-ECM, as previously discussed in the formation of lopodia. Excessive curvature of the nucleus and shape distortion need to be prevented to avoid rupture of the nuclear membrane and subsequent cell death [83]. Wolf et al. [93] showed that cell movement in a 3D matrix is limited by pore size due to the dimensions of the nucleus. Typically, the limit is 10% of the nuclear cross-sectional area. Neutrophils have flexible nuclei that can deform into thin cables due to flexible laminin networks and they can move through small pores in the ECM down to 2 μm^2. However, nuclei of invading tumour cells are less flexible due to different properties of the nuclear laminin. Tumour cells can perforate through pores with a limit of 7 μm^2, beyond these dimensions, tumour cells must digest the ECM to be able to migrate [93].

4 Conclusions

The EMT program is an important process whereby epithelial cells can become motile, lose their cell adhesions and apical–basal polarity, and migrate to distant sites [1]. The resulting mesenchymal cell migrates using actin polymerization to form filopodia and invadopodia in 3D. These structures are also observed in 2D cell culture. Fan-shaped flat lamellipodia are also a typical feature in 2D, though this may be a consequence of the properties of the 2D cell culture plastic, which presents a hard surface, preventing invasive invadopodia and encouraging cell spreading and the presence of lamellipodia. The role of these protrusions is to aid in the progress of the cells through their environment, though filopodia may also have a 'feeling' sense role. In mesenchymal cell migration, the cells have strong cell adhesion, usually integrin-dependent and they have ability to digest the ECM to enable movement through the environment [41, 42, 44]. There is flexibility in the mode of migration and the transition from mesenchymal to amoeboid migration involves the formation of membrane blebs due to the actinomyosin contraction. In this mode, cells interact weakly with their environment and can squeeze through the spaces in the ECM and so do not require proteolytic digestion of the ECM with MMPs. The cell-to-cell interactions during migration are crucial for the progress of the cell, as the contractile force generated by actin polymerization or the actinomyosin contractility will only produce cell movement if it can create traction with the ECM. In addition, the composition of the ECM, in 3D and in vivo, presents many obstacles for migrating cells, due to the dense network of collagen fibres. So mesenchymal migration is dependent on the proteolytic digestion of the ECM by MMPs, whereby amoeboid migration requires the cells to squeeze through the spaces between the collagen fibres in the ECM. Collective cell migration involves the migration of intact groups of cells in which the group displays strong cell-to-cell interaction and proteolytic capabilities at its leading edge and transition to other modes, e.g. amoeboid, is possible [45, 46, 51].

How the nucleus behaves during cell migration is an important consideration and it can become a limiting factor. It is not so important during mesenchymal migration as the cells make space in the ECM for the cells to move, but in amoeboid migration the size of the solid nucleus can be a limitation. However, some cells have developed mechanisms whereby the nucleus can participate in the migration, and this is evident in the formation of lobopodia when the nucleus can be forcibly moved in a 'piston'-like manner due to the creation of force in the anterior part of the cell [83]. Also, it is inherent in some cell types, e.g. leucocytes, for the nucleus to have more flexible properties. Both mechanisms will give information on the possible means employed by tumour cells to overcome this obstacle. The transition between migration modes is plastic and plays an essential role in the dissemination of tumour cells and progression of cancer. The Rho GTPases play an important role in cell migration and the transition between migration modes and there is an interaction between Rho and Rac in determining different modes of tumour cell movement [88].

Once a cell has escaped from the constraints of the tissue, then the modes of cell migration are plastic and cells can transition between mesenchymal and amoeboid migration. The composition of the ECM plays an essential role in determining the mode by which a cell can migrate [91]. Hence, it follows that the cell culture techniques employed in studies to investigate migration are important in determining the outcome. Over the years cell migration has been extensively studied in 2D cell culture and as studies move into 3D it has led to a better understanding of cell migration and the determining factors. Ultimately, this will help determine the mechanisms of cell migration in physiological tissues; the role played by the protrusions, cell-to-cell adhesions and the proteolytic degradation of the ECM.

Take-Home Message

- EMT is the process by which epithelial cells leave solid tumours and change their characteristics and become motile mesenchymal cells.
- Cancer cells can migrate as a single cell or as groups of cells. Single cells move via mesenchymal or amoeboid cell migration, and it is possible for cells to transition between these two modes. Collective cell migration is the mechanism by which groups of cells migrate.
- Amoeboid migration involves actinomyosin contractility and the formation of membrane bleb-like small protrusions of the cell membrane and low cell adhesion and MMP activity.
- Mesenchymal migration involves filopodia and invadopodium with substantial integrin-cell adhesion and MMPs activity.
- Investigating cell migration involves both 2D and 3D cell cultures and a combination of both is beneficial.

Acknowledgement The authors would like to sincerely thank Diana Sirnikova for her creative expertise in developing the figures.

References

1. Yeung KT, Yang J. Epithelial-mesenchymal transition in tumor metastasis. Mol Oncol. 2017;11(1):28–39.

2. Nieto MA. Epithelial plasticity: a common theme in embryonic and cancer cells. Science. 2013;342(6159):1234850.
3. Hartsock A, Nelson WJ. Adherens and tight junctions: structure, function and connections to the actin cytoskeleton. Biochim Biophys Acta. 2008;1778(3):660–9.
4. Qin Y, Capaldo C, Gumbiner B, Macara I. The mammalian Scribble polarity protein regulates epithelial cell adhesion and migration through E-cadherin. J Cell Biol. 2006;171:1061–71.
5. Porta-de-la-Riva M, Stanisavljevic J, Curto J, Francí C, Díaz VM, García de Herreros A, et al. TFCP2c/LSF/LBP-1c is required for Snail1-induced fibronectin gene expression. Biochem J. 2011;435(3):563–8.
6. Tzavlaki K, Moustakas A. TGF-β signaling. Biomolecules. 2020;10:3.
7. Komiya Y, Habas R. Wnt signal transduction pathways. Organogenesis. 2008;4(2):68–75.
8. Kopan R. Notch signaling. Cold Spring Harb Perspect Biol. 2012;4(10)
9. Hubbard SR, Miller WT. Receptor tyrosine kinases: mechanisms of activation and signaling. Curr Opin Cell Biol. 2007;19(2):117–23.
10. Katsuno Y, Lamouille S, Derynck R. TGF-β signaling and epithelial–mesenchymal transition in cancer progression. Curr Opin Oncol. 2013;25(1):76–84.
11. Yang M-H, Wu M-Z, Chiou S-H, Chen P-M, Chang S-Y, Liu C-J, et al. Direct regulation of TWIST by HIF-1α promotes metastasis. Nat Cell Biol. 2008;10(3):295–305.
12. Wu Y, Deng J, Rychahou PG, Qiu S, Evers BM, Zhou BP. Stabilization of snail by NF-kappaB is required for inflammation-induced cell migration and invasion. Cancer Cell. 2009;15(5):416–28.
13. D'Ignazio L, Batie M, Rocha S. Hypoxia and inflammation in cancer, focus on HIF and NF-κB. Biomedicines. 2017;5(2)
14. Wei SC, Fattet L, Tsai JH, Guo Y, Pai VH, Majeski HE, et al. Matrix stiffness drives epithelial-mesenchymal transition and tumour metastasis through a TWIST1-G3BP2 mechanotransduction pathway. Nat Cell Biol. 2015;17(5):678–88.
15. Casamassimi A, Ciccodicola A. Transcriptional regulation: molecules, involved mechanisms, and misregulation. Int J Mol Sci. 2019;20(6)
16. Xu S, Zhou Y, Biekemitoufu H, Wang H, Li C, Zhang W, et al. Expression of Twist, Slug and Snail in esophageal squamous cell carcinoma and their prognostic significance. Oncol Lett. 2021;21(3):184.
17. Siar CH, Ng KH. Differential expression of transcription factors Snail, Slug, SIP1, and Twist in ameloblastoma. J Oral Pathol Med. 2014;43(1):45–52.
18. Palma CS, Grassi ML, Thomé CH, Ferreira GA, Albuquerque D, Pinto MT, et al. Proteomic analysis of epithelial to mesenchymal transition (EMT) reveals cross-talk between SNAIL and HDAC1 proteins in breast cancer cells. Mol Cell Proteomics. 2016;15(3):906–17.
19. Wang Y, Shi J, Chai K, Ying X, Zhou BP. The role of snail in EMT and tumorigenesis. Curr Cancer Drug Targets. 2013;13(9):963–72.
20. Cowin P, Rowlands TM, Hatsell SJ. Cadherins and catenins in breast cancer. Curr Opin Cell Biol. 2005;17(5):499–508.
21. Gu A, Jie Y, Yao Q, Zhang Y, Mingyan E. Slug is associated with tumor metastasis and angiogenesis in ovarian cancer. Reprod Sci. 2017;24(2):291–9.
22. Shih J-Y, Yang P-C. The EMT regulator slug and lung carcinogenesis. Carcinogenesis. 2011;32(9):1299–304.
23. Casas E, Kim J, Bendesky A, Ohno-Machado L, Wolfe CJ, Yang J. Snail2 is an essential mediator of Twist1-induced epithelial mesenchymal transition and metastasis. Cancer Res. 2011;71(1):245–54.
24. Zhu QQ, Ma C, Wang Q, Song Y, Lv T. The role of TWIST1 in epithelial-mesenchymal transition and cancers. Tumour Biol. 2016;37(1):185–97.

25. Yang J, Weinberg RA. Epithelial-mesenchymal transition: at the crossroads of development and tumor metastasis. Dev Cell. 2008;14(6):818–29.
26. Cao J, Wang X, Dai T, Wu Y, Zhang M, Cao R, et al. Twist promotes tumor metastasis in basal-like breast cancer by transcriptionally upregulating ROR1. Theranostics. 2018;8(10):2739–51.
27. Wang Y, Shi L, Li J, Wang H, Yang H. Involvement of twist in NNK exposure-promoted lung cancer cell migration and invasion. Toxicol in Vitro. 2020;63:104740.
28. Lyu P, Zhang SD, Yuen HF, McCrudden CM, Wen Q, Chan KW, et al. Identification of TWIST-interacting genes in prostate cancer. Sci China Life Sci. 2017;60(4):386–96.
29. Côme C, Magnino F, Bibeau F, De Santa Barbara P, Becker KF, Theillet C, et al. Snail and slug play distinct roles during breast carcinoma progression. Clin Cancer Res. 2006;12(18):5395–402.
30. Elloul S, Elstrand MB, Nesland JM, Tropé CG, Kvalheim G, Goldberg I, et al. Snail, Slug, and Smad-interacting protein 1 as novel parameters of disease aggressiveness in metastatic ovarian and breast carcinoma. Cancer. 2005;103(8):1631–43.
31. Dhasarathy A, Phadke D, Mav D, Shah RR, Wade PA. The transcription factors Snail and Slug activate the transforming growth factor-beta signaling pathway in breast cancer. PLoS One. 2011;6(10):e26514.
32. Peinado H, Quintanilla M, Cano A. Transforming growth factor beta-1 induces snail transcription factor in epithelial cell lines: mechanisms for epithelial mesenchymal transitions. J Biol Chem. 2003;278(23):21113–23.
33. Medici D, Hay ED, Goodenough DA. Cooperation between snail and LEF-1 transcription factors is essential for TGF-beta1-induced epithelial-mesenchymal transition. Mol Biol Cell. 2006;17(4):1871–9.
34. Choi J, Park SY, Joo CK. Transforming growth factor-beta1 represses E-cadherin production via slug expression in lens epithelial cells. Invest Ophthalmol Vis Sci. 2007;48(6):2708–18.
35. Zavadil J, Cermak L, Soto-Nieves N, Böttinger E. Zavadil J, Cermak L, Soto-Nieves N, Bottinger Integration of TGF-beta/Smad and Jagged1/Notch signalling in epithelial-to-mesenchymal transition. EMBO J. 2004;23(5):1155–65.
36. Krishnamachary B, Zagzag D, Nagasawa H, Rainey K, Okuyama H, Baek H-J, et al. Krishnamachary B, Zagzag D, Nagasawa H, Rainey K, Okuyama H, Baek JH, Semenza GLHypoxia-inducible factor-1-dependent repression of E-cadherin in von Hippel-Lindau tumor suppressor-null renal cell carcinoma mediated by TCF3, ZFHX1A, and ZFHX1B. Cancer Res. 2006;66:2725–31.
37. Evans AJ, Russell RC, Roche O, Burry TN, Fish JE, Chow VWK, et al. VHL promotes E2 box-dependent E-cadherin transcription by HIF-mediated regulation of SIP1 and Snail. Mol Cell Biol. 2006;27:157–69.
38. Bindels S, Mestdagt M, Vandewalle C, Jacobs N, Volders L, Noël A, et al. Regulation of vimentin by SIP1 in human epithelial breast tumor cells. Oncogene. 2006;25(36):4975–85.
39. Vandewalle C, Comijn J, De Craene B, Vermassen P, Bruyneel E, Andersen H, et al. SIP1/ZEB2 induces EMT by repressing genes of different epithelial cell-cell junctions. Nucleic Acids Res. 2005;33(20):6566–78.
40. Miyoshi A, Kitajima Y, Sumi K, Sato K, Hagiwara A, Koga Y, et al. Snail and SIP1 increase cancer invasion by upregulating MMP family in hepatocellular carcinoma cells. Br J Cancer. 2004;90(6):1265–73.
41. Svitkina T. The actin cytoskeleton and actin-based motility. Cold Spring Harb Perspect Biol. 2018;10(1)
42. Jacquemet G, Hamidi H, Ivaska J. Filopodia in cell adhesion, 3D migration and cancer cell invasion. Curr Opin Cell Biol. 2015;36:23–31.

43. Friedl P, Alexander S. Cancer invasion and the microenvironment: plasticity and reciprocity. Cell. 2011;147(5):992–1009.
44. Pollard TD, Borisy GG. Cellular motility driven by assembly and disassembly of actin filaments. Cell. 2003;112(4):453–65.
45. Pollard TD. Actin and actin-binding proteins. Cold Spring Harb Perspect Biol. 2016;8(8)
46. Sadok A, Marshall CJ. Rho GTPases: masters of cell migration. Small GTPases. 2014;5:e29710.
47. Petrie RJ, Yamada KM. At the leading edge of three-dimensional cell migration. J Cell Sci. 2012;125(Pt 24):5917–26.
48. Kechagia JZ, Ivaska J, Roca-Cusachs P. Integrins as biomechanical sensors of the microenvironment. Nat Rev Mol Cell Biol. 2019;20(8):457–73.
49. Seetharaman S, Etienne-Manneville S. Integrin diversity brings specificity in mechanotransduction. Biol Cell. 2018;110(3):49–64.
50. Bachir AI, Horwitz AR, Nelson WJ, Bianchini JM. Actin-based adhesion modules mediate cell interactions with the extracellular matrix and neighboring cells. Cold Spring Harb Perspect Biol. 2017;9(7)
51. Collins C, Nelson WJ. Running with neighbors: coordinating cell migration and cell-cell adhesion. Curr Opin Cell Biol. 2015;36:62–70.
52. Friedl P, Wolf K. Proteolytic and non-proteolytic migration of tumour cells and leucocytes. Biochem Soc Symp. 2003;70:277–85.
53. Sanz-Moreno V, Gadea G, Ahn J, Paterson HF, Marra P, Pinner SE, et al. Rac activation and inactivation control plasticity of tumor cell movement. Cell. 2008;135:510–23.
54. Friedl P, Borgmann S, Bröcker EB. Amoeboid leukocyte crawling through extracellular matrix: lessons from the Dictyostelium paradigm of cell movement. J Leukoc Biol. 2001;70(4):491–509.
55. Holle AW, Govindan Kutty Devi N, Clar K, Fan A, Saif T, Kemkemer R, et al. Cancer cells invade confined microchannels via a self-directed mesenchymal-to-amoeboid transition. Nano Lett. 2019;19(4):2280–90.
56. Shih W, Yamada S. Myosin IIA dependent retrograde flow drives 3D cell migration. Biophys J. 2010;98(8):L29–31.
57. Charras GT, Hu CK, Coughlin M, Mitchison TJ. Reassembly of contractile actin cortex in cell blebs. J Cell Biol. 2006;175(3):477–90.
58. Paluch EK, Raz E. The role and regulation of blebs in cell migration. Curr Opin Cell Biol. 2013;25(5):582–90.
59. Yan S, Xue H, Zhang P, Han X, Guo X, Yuan G, et al. MMP inhibitor Ilomastat induced amoeboid-like motility via activation of the Rho signaling pathway in glioblastoma cells. Tumour Biol. 2016;37(12):16177–86.
60. Hegerfeldt Y, Tusch M, Bröcker EB, Friedl P. Collective cell movement in primary melanoma explants: plasticity of cell-cell interaction, beta1-integrin function, and migration strategies. Cancer Res. 2002;62(7):2125–30.
61. Madsen CD, Sahai E. Cancer dissemination—lessons from leukocytes. Dev Cell. 2010;19(1):13–26.
62. Friedl P, Noble PB, Walton PA, Laird DW, Chauvin PJ, Tabah RJ, et al. Migration of coordinated cell clusters in mesenchymal and epithelial cancer explants in vitro. Cancer Res. 1995;55(20):4557–60.
63. Choi P-W, Yang J, Ng SK, Feltmate CM, Muto MG, Hasselblatt KT, et al. Loss of E-cadherin disrupts ovarian epithelial inclusion cyst formation and collective cell movement in ovarian cancer cells. Oncotarget. 2016;7:4110–21.
64. Weijer CJ. Collective cell migration in development. J Cell Sci. 2009;122(18):3215–23.

65. Hallou A, Jennings J, Kabla AJ. Tumour heterogeneity promotes collective invasion and cancer metastatic dissemination. R Soc Open Sci. 2017;4(8):161007.
66. Ridley AJ, Paterson HF, Johnston CL, Diekmann D, Hall A. The small GTP-binding protein rac regulates growth factor-induced membrane ruffling. Cell. 1992;70(3):401–10.
67. Wolf K, Mazo I, Leung H, Engelke K, von Andrian UH, Deryugina EI, et al. Compensation mechanism in tumor cell migration: mesenchymal-amoeboid transition after blocking of pericellular proteolysis. J Cell Biol. 2003;160(2):267–77.
68. Carragher NO, Walker SM, Scott Carragher LA, Harris F, Sawyer TK, Brunton VG, et al. Calpain 2 and Src dependence distinguishes mesenchymal and amoeboid modes of tumour cell invasion: a link to integrin function. Oncogene. 2006;25(42):5726–40.
69. Amano M, Chihara K, Kimura K, Fukata Y, Nakamura N, Matsuura Y, et al. Formation of actin stress fibers and focal adhesions enhanced by Rho-kinase. Science. 1997;275:1308–11.
70. Wyckoff JB, Pinner SE, Gschmeissner S, Condeelis JS, Sahai E. ROCK- and myosin-dependent matrix deformation enables protease-independent tumor-cell invasion in vivo. Curr Biol. 2006;16(15):1515–23.
71. Alexandrova AY, Chikina AS, Svitkina TM. Actin cytoskeleton in mesenchymal-to-amoeboid transition of cancer cells. Int Rev Cell Mol Biol. 2020;356:197–256.
72. Miihkinen M, Grönloh MLB, Popović A, Vihinen H, Jokitalo E, Goult BT, et al. Myosin-X and talin modulate integrin activity at filopodia tips. Cell Rep. 2021;36(11):109716.
73. Johnson HE, King SJ, Asokan SB, Rotty JD, Bear JE, Haugh JM. F-actin bundles direct the initiation and orientation of lamellipodia through adhesion-based signaling. J Cell Biol. 2015;208(4):443–55.
74. Wong S, Guo WH, Wang YL. Fibroblasts probe substrate rigidity with filopodia extensions before occupying an area. Proc Natl Acad Sci USA. 2014;111(48):17176–81.
75. Murphy DA, Courtneidge SA. The 'ins' and 'outs' of podosomes and invadopodia: characteristics, formation and function. Nat Rev Mol Cell Biol. 2011;12(7):413–26.
76. Steketee MB, Tosney KW. Three functionally distinct adhesions in filopodia: shaft adhesions control lamellar extension. J Neurosci. 2002;22(18):8071–83.
77. Lee D, Fong KP, King MR, Brass LF, Hammer DA. Differential dynamics of platelet contact and spreading. Biophys J. 2012;102(3):472–82.
78. Hicks KO, Pruijn FB, Secomb TW, Hay MP, Hsu R, Brown JM, et al. Use of three-dimensional tissue cultures to model extravascular transport and predict in vivo activity of hypoxia-targeted anticancer drugs. J Natl Cancer Inst. 2006;98(16):1118–28.
79. Habanjar O, Diab-Assaf M, Caldefie-Chezet F, Delort L. 3D cell culture systems: tumor application, advantages, and disadvantages. Int J Mol Sci. 2021;22(22)
80. Sant S, Johnston PA. The production of 3D tumor spheroids for cancer drug discovery. Drug Discov Today Technol. 2017;23:27–36.
81. Caliari SR, Burdick JA. A practical guide to hydrogels for cell culture. Nat Methods. 2016;13(5):405–14.
82. Poincloux R, Collin O, Lizárraga F, Romao M, Debray M, Piel M, et al. Contractility of the cell rear drives invasion of breast tumor cells in 3D Matrigel. Proc Natl Acad Sci USA. 2011;108(5):1943–8.
83. Yamada KM, Collins JW, Cruz Walma DA, Doyle AD, Morales SRG, Lu J, et al. Extracellular matrix dynamics in cell migration, invasion and tissue morphogenesis. Int J Exp Pathol. 2019;100:144–52.
84. Petrie RJ, Koo H, Yamada KM. Generation of compartmentalized pressure by a nuclear piston governs cell motility in a 3D matrix. Science. 2014;345(6200):1062–5.
85. Petrie RJ, Harlin HM, Korsak LI, Yamada KM. Activating the nuclear piston mechanism of 3D migration in tumor cells. J Cell Biol. 2017;216(1):93–100.

86. Wakayama Y, Fukuhara S, Ando K, Matsuda M, Mochizuki N. Cdc42 mediates Bmp-induced sprouting angiogenesis through Fmnl3-driven assembly of endothelial filopodia in zebrafish. Dev Cell. 2015;32(1):109–22.
87. Wacker A, Gerhardt H, Phng L-K. Tissue guidance without filopodia. Commun Integr Biol. 2014;7(3):e28820.
88. Woolner S, Jacinto A, Martin P. The small GTPase Rac plays multiple roles in epithelial sheet fusion—dynamic studies of Drosophila dorsal closure. Dev Biol. 2005;282(1):163–73.
89. Jacinto A, Wood W, Balayo T, Turmaine M, Martinez-Arias A, Martin P. Dynamic actin-based epithelial adhesion and cell matching during Drosophila dorsal closure. Curr Biol. 2000;10(22):1420–6.
90. Jacquemet G, Green D, Bridgewater R, Kriegsheim A, Humphries M, Norman J, et al. RCP-driven α5β1 recycling suppresses Rac and promotes RhoA activity via the RacGAP1–IQGAP1 complex. J Cell Biol. 2013;202(6):917–35.
91. Caswell PT, Zech T. Actin-based cell protrusion in a 3D matrix. Trends Cell Biol. 2018;28(10):823–34.
92. Yamaguchi H, Wyckoff J, Condeelis J. Cell migration in tumors. Curr Opin Cell Biol. 2005;17(5):559–64.
93. Wolf K, Te Lindert M, Krause M, Alexander S, Te Riet J, Willis AL, et al. Physical limits of cell migration: control by ECM space and nuclear deformation and tuning by proteolysis and traction force. J Cell Biol. 2013;201(7):1069–84.

Biophysical and Biochemical Foundations of Cell Migration

Jonathan E. Dawson and Abdul N. Malmi-Kakkada

What You Will Learn in This Chapter

Many cell types, whether isolated or in cohorts, are motile. The focus of this chapter is to understand from a computational modeling perspective the motility of single cells and their collective aspects, such as herding, as demonstrated in eukaryotic cells during development. After introducing metrics typically used to quantify and characterize cell motility, a simple quantitative model of cell migration is presented. Step-by-step introduction to the construction of the model will be given. Implementation of this model in Matlab is shown, and basic results of the simulation are discussed. This chapter will also provide an introduction to essential biochemical components of cell migration.

This chapter will introduce basic physical and mathematical concepts involved in the modeling of cell migration. One particular modeling approach, namely self-propelled particle model, its ingredients, mathematical formulation, and computational framework, will be presented. Using Matlab we will build a simple

(continued)

J. E. Dawson (✉)
Engineering and Physics Department, Whitworth University, Spokane, WA, USA
e-mail: jdawson@whitworth.edu

A. N. Malmi-Kakkada
Department of Physics and Biophysics, Augusta University, Augusta, GA, USA
e-mail: amalmikakkada@augusta.edu

A. Brüning-Richardson, S. Knipp (eds.), *Cell Migration in Development, Health and Disease*, Learning Materials in Biosciences,
https://doi.org/10.1007/978-3-031-64532-7_6

self-propelled particle model to simulate cell migration in two dimensions. You will learn how to think about a biological process in terms of physics and maths. You will also be introduced to some intracellular biochemical players that have a key role in cell migration.

1 Introduction

Cell migration, the process by which individual cells and cell groups move and translocate, plays a very important role in many biological processes such as embryonic development, cancer metastasis, and wound healing. Whether singly or in cohort, living cells actively respond to a variety of signals from their environment during migration. Generally, such stimuli are considered to consist of chemical (chemotaxis) or mechanical (adhesion and substrate contact; haptotaxis) mechanisms, as well as of temperature gradients (thermotaxis) or electric fields (electrotaxis) [1–4]. Experiments have shown that large colonies of densely packed migrating cells, such as in a tissue, display fascinating patterns of motion, almost like flock of birds or shoals of fish. How do groups of cells display such a coordinated motion which spans length scales much larger compared to the size of a single cell? Even though we understand a lot about how cells move individually, why cells move collectively in unison is not clear. This is where physical and computational models play a big role. Computational models can help to gain a better understanding of the behavior of a biological system, its underlying mechanisms, and make accurate predictions [5–7]. Self-propelled particle (SPP) models, based on agent-based modeling framework, have been quite successful in capturing some key features of collective cellular motion [8].

`Agent-based models` treat each individual cell as a particle or an agent whose movement and interactions with other particles or agents are described by an equation of motion.

This chapter will first introduce basic physical concepts necessary to understand cell motility. Thereafter, an SPP model for cell migration, its key ingredients, and the mathematical formulation will be presented. A step-by-step guide to building and simulating a computational SPP model using Matlab will be provided. Finally, key biochemical components of a cell which control its motility will be briefly discussed.

2 Basic Physics Concepts in Cell Motility

In general, a single cell migrates by a series of changes in its spatial position. The goal of this section is to describe the fundamental concepts that are essential to characterizing and quantifying cell movement. We work under the assumption that cell positions can be explicitly tracked in order to obtain individual cell trajectories using microscopy techniques (see Fig. 1a). Trajectory mapping, as illustrated in Fig. 1b, to characterize dynamics is used in contexts ranging from motion of tumor cells, to the motion of individual protein along DNA. This usually involves video microscopy where the movement of cells of interest is captured repeatedly at fixed time intervals to construct a trajectory (see Fig. 1c for a sample trajectory).

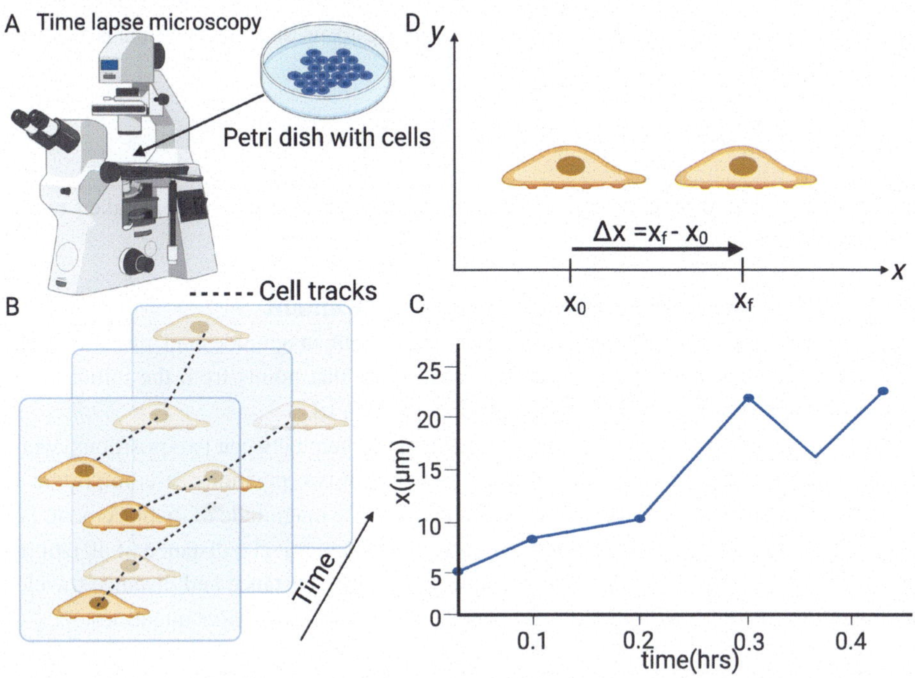

Fig. 1 (**a**) Video sequence of cells moving in a petri dish (or generally in other contexts) can be acquired using time lapse microscopy. (**b**) Cells are identified and tracked through the sequence of video time frames. (**c**) Plot of cell position vs. time will show a series of back and forth movements. (**d**) Cell displacement is visualized with an arrow pointing to the right

2.1 Displacement

The movement of a cell relative to a frame of reference leads to the change in the cell position. This change in position is referred to as displacement. Mathematically, for the simplest case of one-dimensional (1D) motion, displacement is defined as

$$\Delta x = x_f - x_0, \tag{1}$$

where x_f is the final position and x_0 the initial position. We use the symbol Δx to denote displacement. For example, consider the cell movement relative to the origin illustrated in Fig. 1d.

If a cell changes its position through a sequence of displacements Δx_i (where generally displacement is a vector with magnitude and direction of the ith change in position), then after a period of time, the cell position relative to its starting position is the sum of individual displacements,

$$R(N_\tau) = \Sigma_{i=1}^{N_\tau} \Delta x_i, \tag{2}$$

where N_τ is the total number of individual displacements that the cell has taken over a period of time τ.

2.1.1 Difference Between Distance and Displacement

Displacement is a *vector* quantity meaning it has both magnitude and direction. The direction of displacement is visualized using an arrow that points from the initial to the final position. For example, see cell displacement in Fig. 1d.

Distance is defined as the magnitude of the displacement between two positions, while distance traveled is the total length of the path between two positions. It is important to keep in mind that distance traveled can be greater than the magnitude of displacement. For example, a cell could move back and forth many times and travel a distance of 40 μm but could still end up only at 10 μm from its starting position. Distance and distance traveled are quantities that have no direction and hence are referred to as *scalar* quantities.

`Scalar` is any physical quantity that can be specified only by a magnitude, i.e., a single number.

`Vector` is any physical quantity specified by both magnitude and direction.

2.2 Velocity

It is easiest to visualize velocity in terms of our day-to-day experience of driving a car. A large displacement in a short amount of time implies a large velocity. Velocity has the units of distance divided by time. For cells, typical unit for distance is in micrometers (μm) and time unit in either seconds(s), minutes(m), or hours(hr).

Average velocity is defined as the change in position divided by the time elapsed.

$$v_{avg} = \frac{\Delta x}{\Delta t} = \frac{x_f - x_0}{t_f - t_0}. \tag{3}$$

In this equation, Δt is the change in time, and Δx, x_f, and x_0 are as defined in Sect. 2.1. If the starting time t_0 is considered to be zero, and $t = t_f$, the average velocity simplifies to

$$v_{avg} = \frac{\Delta x}{t}. \tag{4}$$

If we wish to know how fast a cell moves at any particular point along its trajectory, we need to calculate its instantaneous velocity. To find the instantaneous velocity, we refer back to the equation for the average velocity (Eq. (3)) and let $t_0 = t$ and $t_f = t + \delta t$, where δt is a small increment in time. Then taking the limit $\delta t \to 0$, we find the equation for instantaneous velocity,

$$v(t) = \lim_{\delta t \to 0} \frac{x(t + \delta t) - x(t)}{\delta t} = \frac{dx(t)}{dt}. \tag{5}$$

Just as with average velocity, instantaneous velocity is a vector quantity with dimension of length per time. Instantaneous velocity defined above can be generalized to three dimensions given the position vector $\mathbf{r}(t) = (x(t), y(t), z(t))$:

$$\mathbf{v}(t) = \left(\frac{dx(t)}{dt}, \frac{dy(t)}{dt}, \frac{dz(t)}{dt} \right). \tag{6}$$

Of course, movements are not only characteristics of objects such as cars. Many intriguing types of cell movements give rise to multicellular organisms and their shape. To give an idea of the different speeds at which cells move, a collection of cell speeds is presented in Table 1.

2.3 Basic Vector Operations

In one dimension, the direction of a vector can be specified by a either positive or negative sign. In two dimensions or higher, a vector has to be specified with respect to a reference

Table 1 Cell speeds of a variety of different cells given in μm per time unit. Measurements are based on time lapse microscopy. Speed depends on factors such as temperature, experimental conditions, etc.

Cell type	Speed	Reference [9][a]
Fish keratocytes	10–50 μm/min	106,807, 106,817
Dictyostelium discoideum	10 μm/min	106,825
Human neutrophil	9 μm/min	106,809
Glioma cells	50 μm/hour	106,810
Mouse fibroblastoid L929 cells	30 μm/hour	106,808
Human H69 small cell lung cancer cell	16 μm/hour	106,815

[a] BioNumbers database is accessible at https://bionumbers.hms.harvard.edu/search.aspx using the BioNumber ID given under reference column

frame using an arrow whose length is proportional to the vector magnitude and pointing in the direction of the vector specified by an angle. Here we describe some basic operations with vectors.

2.3.1 Vector Addition

A method for adding vectors is the **analytical method** using simple geometry and trigonometry. First step in the analytical method is to find the perpendicular components of a vector. If the perpendicular components A_x and A_y (see Fig. 2) are known, then the vector **A** can be calculated analytically. The following equations are used to calculate the magnitude and the direction of a vector given the components A_x and A_y,

$$A = \sqrt{A_x^2 + A_y^2}, \tag{7}$$

$$\theta_A = \tan^{-1}\left(\frac{A_y}{A_x}\right). \tag{8}$$

Note that Eq. (7) is the Pythagorean theorem relating the sides of a right triangle.

A visual representation of the analytical method is shown in Fig. 2, where two vectors **A** and **B** are added to obtain the resultant vector **C**. For analyzing cell dynamics, if **A** and **B** represent two separate displacements of a cell along its migratory path, then **C** is the total displacement (also referred to as a resultant displacement). Following are the steps required to use the analytical method:

- After identifying the x- and y-axis, find the components of each vector to be added along those axes. For example, the vector **A** has components $A_x = A\cos(\theta_A)$ and $A_y = A\sin(\theta_A)$ where θ_A is the angle that vector **A** makes with the x-axis.
- Add the components of the individual vectors separately along each axis to find the components of the resultant vector, i.e., $C_x = A_x + B_x$ and $C_y = A_y + B_y$.

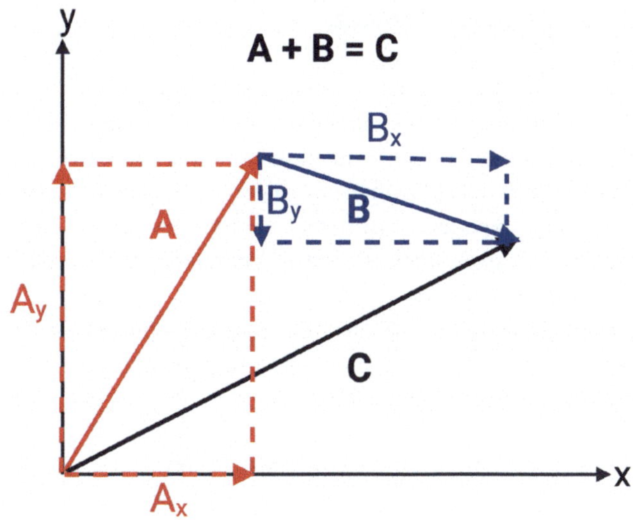

Fig. 2 Vectors **A** and **B** are two displacements of a cell, and **C** is the resultant or total displacement. The analytical method is used to determine the magnitude and direction of the resultant vector **C**

- To obtain the magnitude of the resultant vector **C**, use $C = \sqrt{C_x^2 + C_y^2}$.
- The direction of the resultant vector is given by $\theta_C = \tan^{-1}(\frac{C_y}{C_x})$.

2.3.2 Vector Subtraction

Subtraction of vectors, $\mathbf{A} - \mathbf{B} = \mathbf{A} + (-\mathbf{B})$, is the same as addition of a negative vector. Therefore, the method for subtracting vectors using the analytical method is identical to addition.

Analyzing vectors using the analytical method is useful in many areas of biophysics as perpendicular quantities are usually independent of one another.

Questions

1. Three displacement vectors **G**, **E**, and **F** are specified by their magnitudes $G = 10$, $E = 7$ and $F = 20$. These vectors make angles $\theta_G = 35°$, $\theta_E = -110°$, and $\theta_F = 110°$ with respect to the x-axis. Calculate the vector $\mathbf{H} = \mathbf{G} + 2\mathbf{E} - \mathbf{F}$ using the analytical method.

Answers

H = 28.15 and $\theta_H = -68.65°$

3 Mean Square Displacement

Diffusion at the scale of cells or at the subcellular scale is important for the functioning of organisms. Botanist Robert Brown observed the erratic motion of particles in pollen [10] and thought he was observing the swimming motion of tiny creatures. His namesake Brownian or diffusive motion is ubiquitous in nature and is now understood to originate from the thermal agitation experienced by particles at the microscopic scale. Mean squared displacement (MSD) is a measurement used to determine if a particle is undergoing diffusive motion, or directed motion.

Let us consider the simple case of one-dimensional motion in a system of N cells where cell positions $x_i(t)$ refer to the positions of a particular cell. We generalize the displacement (Eq. (1)) over a lag time δt, to obtain

$$\Delta x_i(t, \delta t) = x_i(t + \delta t) - x_i(t). \tag{9}$$

Assuming that there is no reason for a cell to prefer moving to the right or left, we expect the mean displacement $\langle \Delta x(\delta t) \rangle$ to be zero for any δt. Here, the angular bracket denotes average over all cells at every time point t. On the other hand, during directed motion, the mean displacement will grow linearly with time according to the relation,

$$\langle \Delta x(\delta t) \rangle = v \delta t. \tag{10}$$

Compare this to the familiar equation distance = velocity × time.

Because we expect $\langle \Delta x(\delta t) \rangle$ to be generally zero for random motion, we calculate the mean square displacement $\langle \Delta x^2(\delta t) \rangle$ which will be nonzero if the cell moves. For pure random movement such as Brownian motion, the MSD grows linearly with time:

$$\langle \Delta x^2(\delta t) \rangle = 2D\delta t, \tag{11}$$

where D is the diffusion constant which characterizes the range of motion. The quantity $\sqrt{\langle \Delta x^2(\delta t)}$ is the typical distance covered by cell motion, i.e., a typical distance traversed by a cell in the amount of time δt. Generalizing the motion to x, y, and z coordinates, the three-dimensional MSD equation for motion that is independent along each dimension is

$$\langle \Delta r^2(\delta t) \rangle = \langle \Delta x^2(\delta t) + \langle \Delta y^2(\delta t) + \langle \Delta z^2(\delta t) = 6D\delta t. \tag{12}$$

The diffusion coefficient for simple spherical particles is defined as

$$D = \frac{k_B T}{6\pi \eta a}, \tag{13}$$

where k_B is Boltzmann's constant (with $k_B T = 4.1 \times 10^{-21} J$ at room temperature), η is the viscosity of the fluid medium, and a the particle radius. As the diffusion coefficient is inversely proportional to the radius, we expect smaller particles to diffuse faster. Equation (13) is commonly referred to as the Stokes–Einstein equation.

Questions

1. Estimate the time τ it takes for a molecule to traverse a cell given the cell size $R = 20\,\mu\text{m}$ and typical diffusion coefficient for a protein in the cell cytoplasm $D = 10\,\mu\text{m}^2/\text{s}$. (*Hint: Use Eq. (12).*)

Answers

$\tau \sim \frac{R^2}{6D} \sim 6.7\,\text{s}$.

4 Modeling Single-Cell Motility and Collective Cell Migration

Here, we will discuss the self-propelled particle (SPP) modeling approach for cell migration. Particle-based models do not resolve the details of cell shape and simply treat cells migrating in two dimensions as circular disks of radius R with velocity $\mathbf{u}$, Fig. 3a–b. $\mathbf{u}$ is the vector denoting the cell's velocity. In high school, you might have learned that a vector has both a magnitude and a direction. Let us denote by u_0 the magnitude, and by θ the direction of the velocity vector $\mathbf{u}$, Fig. 3c. θ is the angle that the vector makes with the horizontal x axis. Now that we know the magnitude and direction of the cell's velocity vector, we can easily write down the x- and the y- component of the velocity vector as $u_0 \cos(\theta)$ and $u_0 \sin(\theta)$, respectively. A compact representation of the velocity vector which encapsulates all the information we discussed thus far is: $\mathbf{u} = u_0\hat{\mathbf{u}}$, where $\hat{\mathbf{u}} = (\cos(\theta), \sin(\theta))$ is the unit vector ($\hat{}$ denotes a unit vector) pointing in the direction of the particle's motion, Fig. 3d. SPP model describes a migrating cell as an active particle.

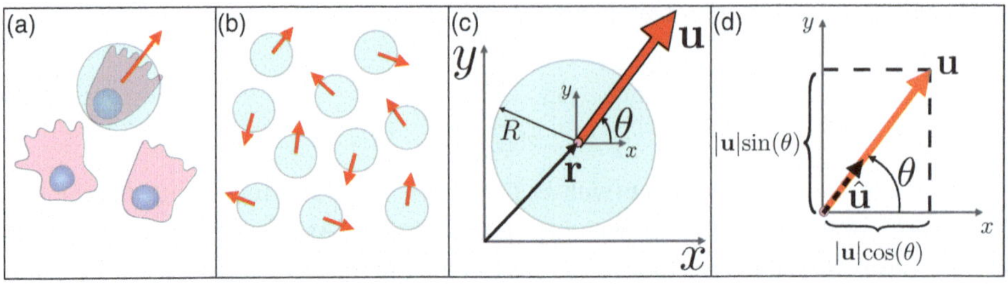

Fig. 3 (**a**) In a self-propelled particle (SPP) model, each individual cell migrating in two dimensions is described as an active particle and is represented by a circular disk. (**b**) An active system consists of a large number of active particles, interacting with each other and self-organizing with no leader and in the presence of noise. (**c**) A migrating cell, whose position coordinates are given by the vector **r**, is modeled as a circular disk of radius R moving with a velocity denoted by the **u** whose magnitude is $|\mathbf{u}| = u_0$ and the direction θ, where $|\cdot|$ denotes the length of a vector and θ is the angle with respect to the horizontal x axis. Since the motion of a cell is intrinsically random, the velocity of a migrating cell fluctuates. SPP model for cell migration takes this into account by adding small random fluctuations to the direction θ of the velocity vector **u**. (**d**) The x component of the velocity vector **u** is $u_0\cos(\theta)$, and the y component of the velocity vector **u** is $u_0\sin(\theta)$. The unit vector $\hat{\mathbf{u}} = \frac{\mathbf{u}}{|\mathbf{u}|}$ points in the same direction as **u** but has the length equal to 1

`Active particles` consume energy to generate their own persistent motion.

`Active system` is a collection of a large number of active particles, which interact with each other and self-organize with no leader and in the presence of noise.

What we mean here by the term "active" is that these particles are like "machines" that consume energy to generate their own motion, just like a biological cell has its own internal source of energy provided by adenosine triphosphate (ATP), the energy currency of the cell, which powers complex cellular processes, including cell motion. An active system is therefore a collection of a large number of active particles, which interact with each other and self-organize with no leader and in the presence of noise.

4.1 A Model for the Motility Dynamics of Self-Propelled Particles

Let us consider a group of motile cells as a system of particles that possess internally the ability to propel themselves forward. The motility dynamics of a group of self-propelled particles, each one representing an individual cell, is described by the following equation of motion:

$$m\mathbf{a}_i = m\frac{d^2\mathbf{r}_i(t)}{dt^2} = \mathbf{F}_i^{\text{drag}}(t) + \mathbf{F}_i^{\text{self-propelled}}(t). \tag{14}$$

Equation (14) is the Newton's equation of motion for the center of mass of a particle i, which states that at any given time t, the acceleration $\mathbf{a}$ of an object with mass m is equal to the net force $\mathbf{F}$ applied to that object at that time. The bold letters in Eq. (14) represent vector quantities, such as the position vector ($\mathbf{r}$), the acceleration ($\mathbf{a}$), and the force ($\mathbf{F}$). m is the mass of an individual active particle which represents an individual cell. The subscript i is the particle or cell index. The most basic contribution to the net force acting on the ith self-propelled particle, which we take into consideration here, comes from (i) the drag force $\mathbf{F}_i^{\text{drag}}$ and (ii) the self-propulsive force $\mathbf{F}_i^{\text{self-propelled}}$. These forces are described in detail in the following sections.

4.2 Motility of Cell Collective: Motion at Low Reynolds Number

Cells are typically small and their motions are slow. At the cellular length scale the motion takes place at low Reynolds number. The Reynolds number, Re, which describes the ratio of inertial to viscous forces in a system, is given by

$$\text{Re} = \frac{\rho u L}{\mu}, \tag{15}$$

where ρ is the fluid density, u is velocity of the object moving in the fluid, L is a relevant length scale describing the linear dimension of the object, and μ is viscosity of the fluid. The typical size of a human cell is about 10 μm across and moves at a speed no greater than 5 μm/s. Substituting these values for L and u in Eq. (15) and using $\rho \approx 10^3$ kg/m^3 and $\mu \approx 10^{-3}$ Pa · s, density and viscosity of water, respectively, we find that cells experience a Re of 10^{-4}, meaning that viscous forces on a cell are 10,000 times as great as those of inertia. Cells are, therefore, rarely subject to inertia. In the absence of inertia and without a constant input of directed force, cellular motion would essentially stop. In the regime of low Re, inertial effects (captured by the term $m\mathbf{a}$ in Eq. (14)) may be safely disregarded and therefore, the particle's acceleration $\mathbf{a}_i = 0$.

`Reynolds number` is the ratio of inertial to viscous forces within a fluid and is a good measure for evaluating whether the motion is dominated by inertial or viscous forces.

When Re is small, friction or drag forces dominate the motion of an object and inertial effects can be neglected.

When Re is large, inertial effects dominate.

Questions

1. Calculate the Reynolds number for a fish moving in the water. Let the density of water be 1000 kg/m^3, the length of the fish 10 cm, the swimming speed of the fish 100 cm/s, and the dynamic viscosity of water $\eta = 0.001$ Pa s. (*Hint: Use Eq. (15).*)
2. Determine whether the movement of the fish in Question #1 is inertia- or viscosity-dominant.

Answers

1. The Reynolds number Re for a fish swimming in the water is approximately equal to 10^5.
2. Since Re $\gg$ 1, the motion is inertia dominant.

By substituting $\mathbf{a} = 0$ (low Reynolds limit) in Eq. (14), the dynamic equation describing the change in the position of a self-propelled particle over time simplifies to

$$0 = \mathbf{F}_i^{\text{drag}}(t) + \mathbf{F}_i^{\text{self-propelled}}(t). \tag{16}$$

We will now discuss the nature of the above two forces included in our model, their possible sources, in the context of cell migration, and mathematical representations.

4.3 Drag

During migration, cells generally experience viscous drag from surrounding cells together with the confinement of cells within a stiff extracellular matrix [11]. The drag on the cell, denoted by the first term on right-hand side (RHS) of Eq. (14), is usually proportional to the cells velocity $\mathbf{u}$ and is defined as

$$\mathbf{F}_i^{\text{drag}}(t) = -\gamma \mathbf{u}_i(t), \tag{17}$$

where γ is the drag coefficient and $\mathbf{u}_i$ is the cell velocity.

4.4 Self-Propulsion

The first term on the right-hand side (RHS) of Eq. (14) is the force acting locally on a particle that gives rise to its motion and is defined as

$$\mathbf{F}_i^{\text{self-propelled}}(t) = F_0 \hat{\mathbf{u}}(\theta_i(t)). \tag{18}$$

This self-propulsive force of magnitude F_0 ensures that at any given time t, the particle moves persistently in a given direction $\theta_i(t)$, Fig. 4a. In order to maintain a constant velocity, the cell needs to consume energy, which is provided through cell internal processes like the phosphorylation of ATP. $\hat{\mathbf{u}}(\theta_i(t)) = (\cos(\theta_i(t)), \sin(\theta_i(t)))$ is the unit vector pointing in the direction of the particle's motion at time t. The x- component (in the Cartesian coordinates) of the ith particles' position at time t is $x_i(t) = |\mathbf{r}_i(t)| \cos(\theta_i(t))$, where $|\mathbf{r}_i(t)|$ is the magnitude of the vector $\mathbf{r}_i(t)$, and is defined as $|\mathbf{r}_i(t)| = \sqrt{(x_i(t))^2 + (y_i(t))^2}$. Similarly, the y- component (in the Cartesian coordinates) of the ith particles position is $y_i(t) = |\mathbf{r}_i(t)| \sin(\theta_i(t))$. $|\mathbf{b}|$ represents the magnitude of a vector $\mathbf{b}$.

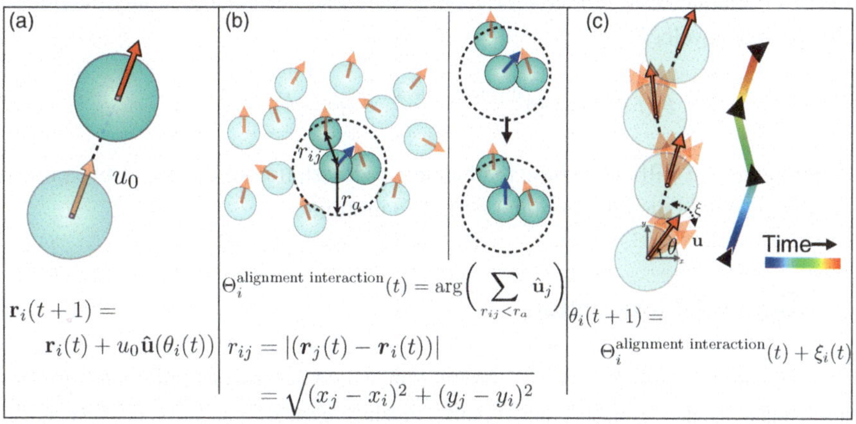

Fig. 4 (**a**) The self-propulsive force $\mathbf{F}_i^{\text{self-propelled}}$ acts locally on the individual particle i at each time t and propels the particle with a fixed speed u_0 in the direction θ_i. (**b**) In the presence of many cells, such as in the tissue, cell–cell interaction due to cell's adhesion to its neighbor might lead to realignment of the cell's velocity. Such an interaction is assumed to be short-ranged, extending to a distance r_a from the cell's center of mass, and only affects the direction of the cells velocity θ such that the direction of the motion of a particle reorients to an average direction that is more aligned with its neighbors. The average direction is obtained by vector addition of the velocity unit vectors $\hat{\mathbf{u}}_j$ of each cell that is a nearest neighbor of cell i. The function arg() returns the angle of the resultant vector $\sum \hat{\mathbf{u}}_j$. (**c**) The motion of a cell is intrinsically random. An SPP model for cell migration takes this into account by adding small random fluctuations to the direction θ of the velocity vector $\mathbf{u}$. As a result of these fluctuations, cell motion does not exhibit a straight trajectory, but rather a zigzag pattern. The distance a cell traverses between two time steps is however always equal

Putting together the mathematical forms of $\mathbf{F}^{\text{drag}}$ and $\mathbf{F}^{\text{self-propelled}}$ in Eq. (16), we have the following equation that describes the change in the position of a self-propelled particle over time:

$$\frac{d\mathbf{r}_i(t)}{dt} = u_0\hat{\mathbf{u}}_i(t), \tag{19}$$

where $\mathbf{r}_i$ is the vector representing the position of the ith self-propelled particle at time t and $u_0 = \frac{F_0}{\gamma}$ is the magnitude of the particles velocity. Note that $\mathbf{u}_i = d\mathbf{r}_i/dt$. We can write Eq. (19) in time discrete form for each component of the vector $\mathbf{r}$, so as to describe the change in the x and the y coordinate of a particle i's position,

$$x_i(t+1) = x_i(t) + u_0\cos(\theta_i(t)) \tag{20}$$

$$y_i(t+1) = y_i(t) + u_0\sin(\theta_i(t)), \tag{21}$$

where we have set the time step $\Delta t = 1$.

In addition to Eqs. (20) and (21), in a self-propelled particle model, the motility of a cell i is also influenced by cell–cell interaction and noise, which affect only the direction θ_i of the particle's velocity,

$$\theta_i(t+1) = \Theta_i^{\text{alignment interaction}}(t) + \Theta_i^{\text{noise}}(t). \tag{22}$$

In the following, we will present the mathematical description of the above two contributions to the change in the direction of a particle's motion.

4.5 Alignment Interaction

Here we consider a type of cell–cell interaction which represents cell's adhesion to its neighbor, such as in the tissue, and results in alignment of the cells velocity with its neighbors. This interaction is mediated primarily by cadherins on the cell surface [12]. A consequence of cell–cell adhesion on cell migration is that one cell's direction of motion is influenced by its neighbors, Fig. 4b. Such an interaction must be relatively short-ranged, extending to only nearest neighbors, and has the effect of cells aligning the direction of their velocity with another [13, 14]. We will consider the following to be the features of the cell–cell interaction-based velocity realignment:

- First of all, such an interaction must be short ranged. In other words, the direction of a cells velocity aligns with only those other cells which are in its close proximity or are its nearest neighbors. To identify the nearest neighbors of a cell i we will first need to measure r_{ij}, which is the distance between the cell i and its neighbor j and is defined as

$$r_{ij}(t) = \sqrt{(x_i(t) - x_j(t))^2 + (y_i(t) - y_j(t))^2}, \tag{23}$$

where $(x_i(t), y_i(t))$ and $(x_j(t), y_j(t))$ are the centers of cells i and j at time t. We will now introduce a parameter r_a such that if $r_{ij} < r_a$, then cell j is considered to be a nearest neighbor of cell i.

- Secondly, the velocity alignment due to cell–cell interaction only affects the direction θ of a cells motion, such that the direction of a cells velocity at the next time step is reoriented to an average direction which includes the cells' own direction and the direction of all its nearest neighbors.

Taking the above two features into account, the velocity realignment over time can be mathematically represented as

$$\Theta_i^{\text{alignment interaction}}(t) = \text{atan2}\left(\frac{\sum_{r_{ij}<r_a} \sin(\theta_j(t))}{\sum_{r_{ij}<r_a} \cos(\theta_j(t))}\right), \tag{24}$$

where atan2() represents the inverse of tangent. Basically, the right-hand side of Eq. (24) is the average of the velocity angles of all the particles, or the cells, which are within a distance of r_a from the center of mass of cell i, including cell i. In obtaining Eq. (24), we have used the concept of vector addition of two or more vectors, which gives us the mean or average direction of the vectors.

4.6 Noise

The motion of individual cells is intrinsically random. There are multiple possible sources of the observed randomness in cell motility. There is apparent stochasticity inherent to the intracellular dynamics, which in turn generates a stochastic force [15]. In addition, the motion of an individual cell takes place in a noisy environment surrounded by other randomly moving neighbors [16]. Shape changes during cell motility might also give rise to randomness in the direction of cell motion.

Generally, in the SPP model for cell migration, unlike in the case of a Brownian particle, the noise is considered to influence only the motility direction θ_i and not the translation of the particle. In other words, in the SPP model for migrating cells, the self-propelled particle likes to go in a certain direction for a certain persistence time before it veers off, Fig. 4c. The angular noise is defined as

$$\Theta_i^{\text{noise}} = \xi_i(t) \ . \tag{25}$$

In Eq. (25) $\xi_i(t)$ is the angular white noise which is uniformly distributed in $[-\pi, +\pi]$ with mean $\langle \xi_i(t) \rangle = 0$ and the autocorrelation function $\langle \xi_i(t)\xi_j(t') \rangle = D_r \delta_{ij} \delta(t - t')$, where D_r is the strength of the noise and first and second delta functions are Kronecker and Dirac deltas, respectively.

Substituting Eqs. (24) and (25). into Eq. (22), the time discrete form for the change in the direction θ_i of the velocity of a particle i is

$$\theta_i(t+1) = \text{atan2}\left(\frac{\sum_{r_{ij}<r_a} \sin(\theta_j(t))}{\sum_{r_{ij}<r_a} \cos(\theta_j(t))}\right) + \xi_i(t). \tag{26}$$

Note that the first term in Eq. (26) includes, by definition, the angle $\theta_i(t)$ of the particle, or cell, i at time t. Equations (20), (21), and (26) together govern the motility dynamics of single cells and the migration cell collectives.

4.7 Dynamic Equation for the Migration of Cell Collective

The full description of the motion of a cell collective, according to the model presented here, in time discrete form is given by the set of Eqs. (20), (21), and Eqs. (26).

5 Matlab Implementation of the Self-Propelled Particle Model for Cell Migration

In Sect. 4 a self-propelled particle model for cell migration was introduced. Now, we will build a simple Matlab code that will simulate this model.

5.1 Initial Conditions and System Initialization

We will use Matlab to develop the code for the self-propelled particle model we discussed in the preceding sections. In order to build the computational model for self-propelled particle in a computing platform, we first need to initialize the system. Let us assume that our system consists of N number of self-propelled point particles, each one representing the center of mass of an individual migrating cell. The particles move with a speed of u_0 in a two-dimensional domain whose length is Lx along the x- axis and Ly along the y-axis. We chose $N = 1000$, $u_0 = 0.05$, $Lx = Ly = 25$. All the parameters that we will use in our program code are in dimensionless form. In order to keep track of the positions and the velocities of each individual particle, we will define some arrays. Arrays are like containers that store information in a ordered manner. Each element of the array is uniquely indexed, and it can be accessed at any time using that index. We will use two arrays to store the coordinates of each particles location, one array each for x_i and y_i components, respectively. i is the index of a particle and $i \in [1 \ldots N]$. Since our particles are motile, we will need one array to store the angle θ_i of each particle's velocity. Finally, we will also use two arrays to store each particle's velocity, one array each for ux_i and uy_i, i.e., x and y components of the velocity, respectively.

Program Code: Initial Conditions and System Initialization

```
close all;
clear all;
N=1000; % number of particles or cells
Lx=25.; % length of domain in x-dir
Ly=25.; % length of domain in y-dir

uo=0.05;   % particle speed

xp=rand(N,1).*Lx; % random initial x-pos of each particle
yp=rand(N,1).*Ly; % random initial y-pos of each particle
theta=pi.*2.*rand(N,1);% rand initial vel. angle of...
                    %particles
uxp=uo.*cos(theta);
uyp=uo.*sin(theta);

onebd=ones(N,1); % an array of ones

Dr=0.05;  % strength of rotational noise
zeta=(2.*pi).*Dr;  % random fluctuation in angle per ...
               % time step (rads between 0 and 2pi)
```

The motion of each individual cell is governed by Eqs. (20), (21), and (26). The x and y coordinates of position and velocity angle θ of each particle are updated at each time step. We fix the time step to be $dt = 1$. We implement a periodic boundary condition, which means that if a particle's position is calculated in the next time step to be outside at one side of the domain, then its coordinates are shifted such that the particle enters the domain from the opposite side. The results of the simulation are shown in Fig. 5.

Program Code: Time-Dependent Dynamics of Cell Motility

```
Tmax=100000; % Total time for which the simulation should run
dt=1.;     % time step
ra=1.;      % interaction radius

for nsteps=1:Tmax
   nsteps

   xp=xp+uxp.*dt; % update position x-coord of each particle
```

```
    yp=yp+uyp.*dt; % update position y-coord of each particle

    for i=1:N
       %periodic boundary condition
       if(xp(i)<0);xp(i)=xp(i)+Lx;end
       if (yp(i)<0);yp(i)=yp(i)+Ly;end
       if (xp(i)>Lx);xp(i)=xp(i)-Lx;end
       if (yp(i)>Ly);yp(i)=yp(i)-Ly;end

       %find distance of a particle i with all other...
       %particles at a time t
       sep(1:N)=(onebd.*xp(i)-xp(1:N)).^2+...
       (onebd.*yp(i)-yp(1:N)).^2;

       %find angle of nearest neighbors (including i)
       nearang=theta(sep<ra);
       %find mean angle of neigbours (including i)
       mtheta(i)=...
       atan2(sum((sin(nearang))),sum((cos(nearang))));
    end

    theta=mtheta'+zeta.*(rand(N,1)-0.5); % the total angle ...
        %of the particle i is sum of the mean and rand noise

    uxp=uo.*cos(theta);
    uyp=uo.*sin(theta);

    cla;
    set(gcf,'doublebuffer','on');
    plot(xp,yp,'bo');
    drawnow;

end
```

6 Biochemical Foundations of Cell Migration

The computational modeling approach described above can be used to simulate cell migration and better understand various biophysical aspects of cell migration. We remind here that cell migration is a complex process involving a spectrum of biophysical and biochemical processes. Here, we discuss briefly some of the crucial biochemical aspects relevant for cell migration [17]. The process of migration of a single cell or a cell collective involves the following key components: (i) cell

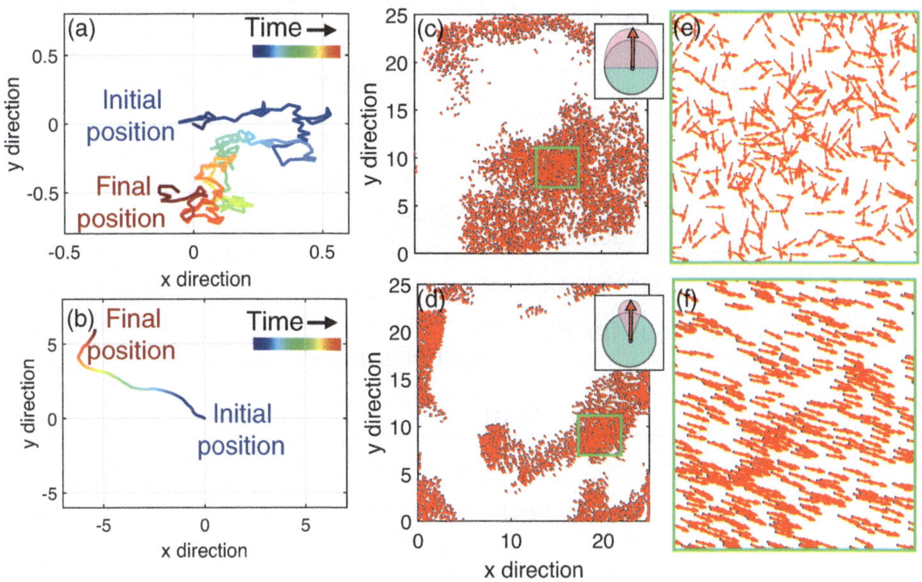

Fig. 5 (**a–b**) The trajectory of a self-propelled particle with (**a**) a strong angular noise of strength $D_r = 0.5$, and (**b**) a weak angular noise of strength $D_r = 0.05$. (**c**) Migratory behavior of a group of $N = 5000$ self-propelled particles. The angular strong noise strength for each particle $D_r = 0.5$, which is equivalent to angular fluctuation in the range $[\frac{-\pi}{2}, \frac{\pi}{2}]$, inset. The snapshot is the result of the simulation after 500 time steps. (**d**) Migratory behavior of a group of $N = 5000$ self-propelled particles. The angular strong noise strength for each particle $D_r = 0.05$, which is equivalent to angular fluctuations of $[\frac{-\pi}{20}, \frac{\pi}{20}]$, inset. The snapshot is the result of the simulation after 500 time steps. (**e**) In the case of a strong angular noise, the particles move in all directions and do not exhibit collective migration, as shown by the velocity vectors of a subgroup of particles all pointing in random directions. (**f**) In the case of a weak angular noise, the particles show velocity alignment with their nearest neighbors. The velocity vectors of a subgroup of particles all point in the same direction

polarization—the intrinsic asymmetry in cell properties such as shape, organization of subcellular components, etc., which distinguishes the front/back of cell during migration; (ii) extensions of lamellipodial or filopodial protrusions; (iii) establishment of adhesion between the cell and the surface or the substrate on which migration occurs; and (iv) the force generation by cells required for movement.

6.1 Cell Polarization

For directional cell migration to occur, cell needs to become polarized, i.e., the cell front facing the migratory direction becomes distinct from the back of the cell. Such breaking of symmetry occurs primarily through actin polymerization at the front edge,

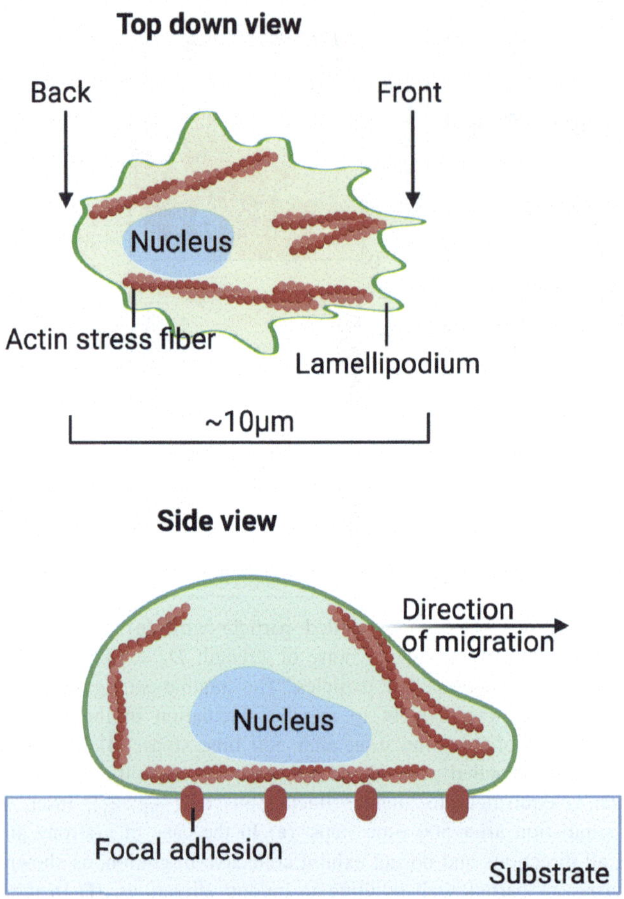

Fig. 6 Cell migration involves protrusions generated by lamellipodia, cell body contractions driven by actin stress fibers and adhesion to the substrate through focal adhesion sites

which results in membrane protrusions (Fig. 6). Actin is the most abundant protein in cell and is an important component of the cell cytoskeleton. Two states of actin exist with monomeric form (G-actin) or filamentous form (F-actin). The interaction between filamentous actin and myosin forms the basis of contraction [17].

6.2 Protrusions

Once polarized, the cell starts extending actin-based protrusions at its front edge such as lamellipodia or filopodia. Lamellipodia are branched networks of actin filaments that push a broader stretch of the cell membrane, while filopodia are thin, tubular protrusions composed of linear bundles of actin filaments. Lamellipodia are formed by the Arp2/3 complex proteins that bind to preexisting filaments and initiate the assembly of new filaments. Wasp/Wave family of proteins regulated by Rho GTPases controls the activity of Arp2/3 complex [17].

`Actin` is the most abundant protein in most eukaryotic cells, making it a critical player in many cellular functions ranging from cell shape, polarity, and cell motility.

Actin filaments are composed of globular actin (G-actin) and underlie cell membrane. Actin stress fibers are composed of many actin filaments.

6.3 Adhesion Formation

Molecular structures called focal adhesions (see Fig. 6) connect the actin cytoskeleton to the extracellular matrix (ECM)—the noncellular external component of tissue that the cells attach to while migrating [17]. Receptors on cell surfaces (e.g., integrins) interact with ligands in the ECM which in turn activate intracellular signaling pathways at focal adhesion sites. Focal adhesion serves as sites where cells generate traction forces required to move themselves forward.

6.4 Disassembly of Adhesions

At the front edge of a migrating cell, some of the older adhesions disassemble, while others grow into more mature molecular assemblies. Kinases such as FAK and Src and phosphatases regulate the disassembly of adhesions [17]. Finally, adhesions also undergo turnover at the rear, which is essential for retraction of the back end of the cell. Adhesion turnover is regulated mainly by myosin II-dependent actin filament contractility.

Take-Home Message

- Basic physical laws in mechanics can be used to computationally model cell migration.
- Reynolds number is an important measure for evaluating whether the cellular movement is dominated by inertial or viscous forces.

(continued)

- Physiologically, cellular movements typically take place in viscous environments characterized by low Reynolds number, and consequently inertia can be neglected.
- Computational modeling approach can be used to simulate cell migration and better understand its various biophysical aspects.
- Self-propelled particle (SPP) model is a computational model in which individual cells are represented as active particles which interact with each other.
- Collective self-organized behavior, in which individual cells form large groups moving in the same direction, is an emergent phenomenon described by the SPP model and is shown to depend on the noise strength and cell number density.
- For directional cell migration to occur, cell needs to become polarized, a process in which actin plays an important role.

References

1. Iglesias PA. Sci Signal. 2012;5(213):pe8.
2. Lara Rodriguez L, Schneider IC. Integr Biol. 2013;5(11):1306.
3. Roca-Cusachs P, Sunyer R, Trepat X. Current Opin Cell Biol. 2013;25(5):543.
4. Zajdel TJ, Shim G, Wang L, Rossello-Martinez A, Cohen TP, Scheepdog DJ.Cell Syst. 2020;10: P506–514.E3. https://doi.org/10.1016/j.cels.2020.05.009
5. Malmi-Kakkada AN, Li X, Samanta HS, Sinha S, Thirumalai D. Phys Rev X 2018;8(2):021025.
6. Malmi-Kakkada AN, Sinha S, Li X, Thirumalai D. Biophys J. 2022;121:3719–3729.
7. Marchant CL, Malmi-Kakkada AN, Espina JA, Barriga EH. Nat Mat. 2022;21(11):1314.
8. Camley BA, Rappel WJ. J Phys D: Appl Phys. 2017;50(11):113002.
9. Milo R, Jorgensen P, Moran U, Weber G, Springer M. Nucleic Acids Res. 2010;38 Suppl_1:D750.
10. Brown R, Edinburgh New Philosoph J. 1828;5:358.
11. Cai D, Dai W, Prasad M, Luo J, Gov NS, Montell DJ. Proc Natl Acad Sci U S A. 2016;113:E2134–E2141. https://doi.org/10.1073/pnas.1522656113
12. Méhes E, Vicsek T. Integr Biol (United Kingdom) 2014;6:. https://doi.org/10.1039/c4ib00115j
13. Vicsek T, Czirk A, Ben-Jacob E, Cohen I, Shochet O. Phys Rev Lett. 1995;75:1226. https://doi.org/10.1103/PhysRevLett.75.1226
14. Mayor R, Carmona-Fontaine C. Trends Cell Biol. 2010;20:319. https://doi.org/10.1016/J.TCB.2010.03.005
15. Naoki H, Sakumura Y, Ishii S. J Theoret Biol. 2008;255:. https://doi.org/10.1016/j.jtbi.2008.08.012
16. Yamao M, Naoki H, Ishii S. PLoS ONE 2011;6:e27950. https://doi.org/10.1371/journal.pone.0027950
17. Alberts B, Johnson A, Lewis J, Raff M, Roberts K, Walter P. Molecular Biology of the Cell. 4th Edition. New York: Garland Science; 2002. p. 906–911

Part II

Methods to Investigate Cell Migration

Methods to Investigate Cell Migration

A. Brüning-Richardson and S. E. Lawler

What You Will Learn in This Chapter

Cell migration is of great relevance to researchers studying development, wound healing, the immune system and its responses and diseases associated with deregulated cell migration. The study of cell migration has been made possible by the development of a huge variety of experimental set-ups in 2D and 3D focussing on specific aspects of cell migration. In this chapter, some of such specific aspects of cell migration will be discussed and the assays available to investigate the particular features of cell migration. Examples of traditional cell migration assays and most recent developments ranging from wound healing assays to microfluidics studies will be given, followed by commonly used assays in a 3D setting and finally highly specialised intravital in vivo research. The advantages and disadvantages of such assays are highlighted.

A. Brüning-Richardson (✉)
Department of Biological and Geographical Sciences, University of Huddersfield, Huddersfield, UK
e-mail: a.bruning-richardson@hud.ac.uk

S. E. Lawler
Department of Pathology and Laboratory Medicine, Brown University, Providence, RI, USA

A. Brüning-Richardson, S. Knipp (eds.), *Cell Migration in Development, Health and Disease*, Learning Materials in Biosciences,
https://doi.org/10.1007/978-3-031-64532-7_7

1 Cell Migration Assays

The discovery of the migratory ability of cells and the association of this cellular behaviour and disease is becoming increasingly of interest to understand the underlying mechanisms and to characterise cell migration in great detail at cellular and subcellular levels. To be able to study cell migration it is necessary to learn how to utilise the cells of interest. It follows that the development of in vitro cultures of cells was a big milestone in the study of cell migration. The first in vitro cell cultures were established by Ross Granville Harrison (1870–1959), an American embryologist, in the footsteps of Wilhelm Roux (1850–1924). Harrison removed living frog embryonic tissue and established with this growth conditions outside the original donor. He also realised that the growth of the tissue was dependent on a solid support, the most ideal he identified as being lymph from adult frogs. He also realised the need to keep his specimens sterile and decided to wash the embryo donors with sterile distilled water and as such being one of the first tissue culture users to adopt aseptic techniques. Having established this first in vitro system Harrison was able to observe extensions of long threads emanating from his specimens, with active movement resembling neurons [1]. The first human cells to be grown in vitro were established from cancer tissue from the cervical cancer sufferer, Henrietta Lacks (1920–1951), in 1951. Her surgeon had removed a womb tissue biopsy and passed this on to a cancer researcher, George Otto Gey (1926–1970). Gey found that with the right culture conditions the tumour cells continued to proliferate and grow indefinitely; he was the first as such to make an established cell line and named it after Henrietta Lacks, HeLa [2, 3]. HeLa cells were the first to be used for studies in cancer research but also played an important role in the development of the polio vaccine [4]. Dr. Jonas Salk developed an inactivated vaccine after global outbreaks of Polio. HeLa cells were found to be easily infected with the causative Poliomyelitis virus and therefore allowed the mass production of the virus. Years later, to increase production and characterisation of the human immunodeficiency virus (HIV) HeLa cells were used again to uncover that the cells were relatively resistant to infection, which allowed the study of the life cycle of the HIV virus [5]. Ethical issues with using HeLa cells eventually arose, as at the time of removal the original tissue had been taken without consent and then 'immortalised' as a research tool; even today HeLa cells are used, and it has been estimated that there are over 70,000 publications based around their use [1]. Posthumously, Henrietta Lack's descendants were notified of the unconsented removal and use of the tissue and as a consequence laws and regulations with regards to patient consent were changed in the USA [1].

Following on from the first success of growing cells indefinitely, many cancer cells are now routinely grown in vitro culture; although many are known as established cell lines having been grown in culture medium supplemented with foetal calf serum (FCS), it is nowadays preferred to especially grow cancer cells recently obtained from patients in their stem cell like state by maintaining them in serum free, and specific growth factor containing defined media [6]. The choice of growth medium is essential for successful cultivation and has its origins in J.F Morgan's (1918–1976) development of the first synthetic media for growing mammalian cells, M199 [6]. Following the in vitro culturing of cells in two-dimensional (2D) monolayers, recent developments include the generation of cellular spheroids from cell suspension by spinner methods, hanging drops or the use of low adherence plates. Spheroids are 3-dimensional (3D) structures resembling the original organ/tumour they originated from and as such resemble architecturally the original tissue. Advancements in these spheroid structures were the generation of organoids as described by Lancaster [7] where mini-brains were created from pluripotent human stem cells with the support of a 3D matrix and spinning bioreactor. As the created structures are composed of various types of nerve cells, they resemble the architecture of a human brain. The model was then taken forward to create brain tumour models after genetic manipulation of the brain organoids [8]. Technologies have even moved on from there with the possibility of 3D bioprinting whole organs [9, 10].

With expanding access to different cells there have also been huge advancements in the development of migration/invasion assays to analyse the migratory potential over the last 15 years. Depending on the scientific question, cells can be analysed for their migratory ability in terms of speed or directional migration (random cell migration), their ability to follow external cues such as chemokine trails (chemotaxis) and their ability to invade into 3D environments (invasion). For each of these particular interests, there are experimental set-ups that can be explored. As such, it is important to have a defined research question which then allows the choice of the most appropriate experimental set-up.

2 Two-Dimensional (2D) Random Cell Migration

A very basic way to analyse cell migration is to study random cell migration. Here, the cells of interest are plated at sub-confluence on a specialised imaging dish and then allowed to migrate within a certain timeframe. Live cell imaging is the choice of capturing the migratory ability and in our laboratories, we routinely carry out these experiments to determine the speed and directionality of glioma cells. Cells can be exposed to drugs, manipulated to overexpress genes of interest, or be silenced for genes of interest. Differences in comparison to controls can then be assessed by analysing speed, distance travelled, directionality of travel or visual morphological changes. The following set-up is based around our own experiences but may be adapted according to individual needs and available equipment and facilities.

For this experimental set up, the chosen cells to be investigated are trypsinised and a cell count is carried out. Depending on the proliferative rate of the cells and the time frame of the experiment, cells are diluted to a subconfluent level and then added to special, preferably, glass bottom dishes such as those available from Ibidi. We routinely use round dishes that are segmented into four quarters to allow the set-up of four different experimental conditions at the same time. The cells are added to the dish at their pre-determined cell density and then imaged over a certain time interval using live imaging systems such as the BioStation IM-Q (Nikon) or the Incucyte (Essen BioScience). We recently also had excellent results with the HoloMonitor (Phase Holographic Imaging PHI AB). These are all systems that combine the imaging of live cells whilst maintaining them under their optimum growth condition. Imaging intervals can be defined again based on the migratory ability of the cells and the particular scientific question in mind. For our own random cell migration studies, we allowed glioma cells to migrate over 72 h with imaging at 30-min intervals. The resulting movies can then be analysed using freely available software tools such as ImageJ [11] with the Tracker plugIn and the Cell migration analysis tool from Ibidi or software tools such as Volocity (Perkin Elmer) or the Incucyte system (Essen BioScience) which are fully automated. Parameters such as cell speed, direction, distance travelled can be determined and tracks of cells can be plotted to allow direct comparison of different cell populations. Newest developments also allow the imaging in 3D. Therefore, random cell migration studies can be assessed which is extremely informative in terms of basic migratory characteristics allowing comparisons between motile cells as well as investigating the effect of specific anti-migratory inhibitors. Live cell migration imaging is one of the most straightforward assays in terms of experimental design and set-up to consider, however, this requires access to, potentially, very expensive imaging equipment and software tools. In addition, special plastic dishes can also be expensive. Detailed methods can be found in [12, 13]. Figure 1 gives an overview of random cell migration and analysis.

2.1 Scratch Assay (Wound Healing)

To study wound healing or the ability of cells to migrate towards each other in a collective manner the wound healing assay or scratch assay is used. This assay is described in many variations in the scientific literature, but the underlying principle is to assess cell migration towards a wound that has been artificially created in a lawn of cells. The effect of, for example anti-migratory inhibitors is assessed by the ability of the cells to refill the gap that has been introduced. The percentage change of the gap can then be determined and used to analyse cell migration. Here is just one example of many how to set up and interpret the scratch assay.

Cells are seeded at a pre-determined cell density in 24 well plates such that after 24 h of growth they reach 80–90% as a monolayer. Prior to cell seeding, a line is drawn underside each well across the centre with a fine marker pen. After 24 h incubation (time points depend on the cell type used) at normal growth conditions, a gentle, slow scratch using a p200 pipette tip is applied across the centre of the cell monolayer, perpendicular to the marker line. The wells are then washed twice with culture medium to remove any detached cells and the medium is replaced with fresh medium with and without, for example drugs or inhibitors. Cell migration across the scratch is determined by imaging at 0, 4, 8 and 24 h using an imaging system such as the EVOS (ThermoFisher Scientific) or an automated live cell imager such as the HoloMonitor (Phase Holographic Imaging PHI AB) at ×4 magnification. Quantification of cell migration can be achieved either by handler

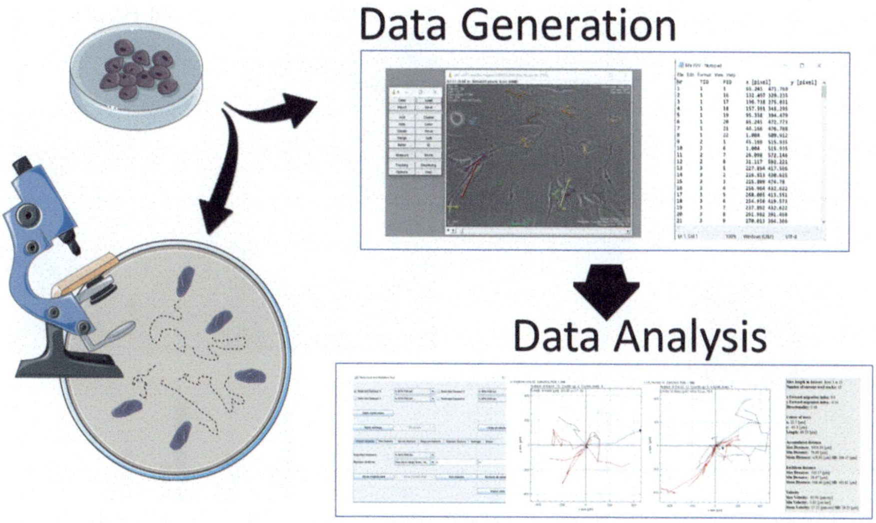

Fig. 1 Cell speed and directionality can be measured using the random cell migration assay. This follows several steps including (1) plating cells at sub-confluence, (2) microscopical imaging which can include a tracker dye and data acquisition, (3) and data analysis using a tracker software such as MTrackJ, which can be converted to parameters such as speed, distance travelled, directionality by additional softwares or PlugIns such as the Ibidi Migration and Chemotaxis tool [14]. Individual cell tracks can be visualised in generated rose plots by the same software. Images and graphical representation created by Dr. Anke Brüning-Richardson and Delnia Abed, University of Huddersfield. This diagram created with https://smart.servier.com/

analysis using ImageJ software [11] to measure the gap and to determine the percentage change in the area over time or by making use of the supportive software analysis tools that are offered as part of automated imaging systems. This method is straightforward and easy to set up at low cost and to repeat once the optimum cell densities and assay conditions have been determined. However, some uncertainties may remain over the role of cell proliferation to cover the wound, the importance of the extracellular matrix (ECM), potential cell damage when the wound is introduced and cell displacement. Advances in this technology have been made and recently assays are available that create a wound by stamping, with the help of a solid or liquid barrier, chemical droplets, microfluidics, electrical application or laser ablation (wound assay) [15]. For detailed methodology please refer to [16–19]. Figure 2 gives an overview of the scratch wound assay highlighting the different steps including plating of cells and introducing a scratch and images of cell migration into a scratch and graphical representation.

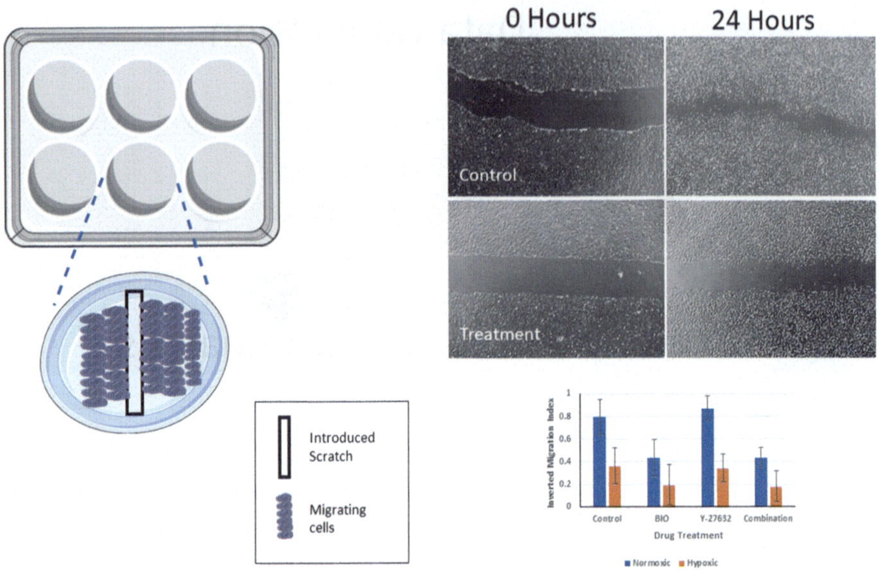

Fig. 2 Wound healing or the ability of cells to migrate towards each other in an epithelial-like sheet can be investigated using wound healing assays. This follows a setup of various steps including (1) the addition of cells to flat bottom plastic plates for the creation of a cell lawn; (2) the introduction of a wound across the cell lawn and (3) analysis of scratch wound closure. As an example, here, representative images are depicted showing the glioma cell line U251 without treatment ('Control') and after treatment with an anti-migratory drug ('Treatment') at 0 Hours and 24 Hours. The scratch area has hardly changed in the treated cells whereas in the control cells it is nearly covered by cells that had migrated into it. An example for data presentation is shown in graphical format where the glioma cell line U251 was treated with two inhibitors, 'BIO' and 'Y-27632', and in combination under normoxic and hypoxic conditions. The effect of the treatment is most pronounced when the cells are treated in combination. Images and graphical data were created by Becky Griggs, University of Huddersfield. This diagram created with https://smart.servier.com/

2.2 Transwell Assay

Following on from the scratch assay, one can also assess the ability of cells to migrate towards a chemoattractant, therefore, to assess their ability to follow or migrate towards an external cue such as a chemical 'trail'. This is established in the so-called 'Transwell assays' where cells are seeded within a plastic inlet that contains a porous membrane (available at different pore sizes) at its bottom and can be then placed within, for example, a 12- or 24-well flat bottom plastic plate. To the inlet containing the membrane, the cells of interest are added, potentially without, for example FCS or a ligand that stimulates migratory activity, whereas in the plate well the chemoattractant is added to the medium which then surrounds and wets the inlet. Cells will start to migrate towards their chemoattractant also having to pass through the membrane. Migratory activity can, therefore, be assessed by fixing and staining the membrane and counting the number of cells found to have migrated across the membrane. One of the advantages of this technique in cancer research is that cells then have to cross the transwell membrane which is a good representative of a leaky endothelium the cells would have to cross in the body. In terms of invasion, matrices can be added to the inlet to mimic migration across an ECM and then also the transwell membrane itself. The assay is cheap, simple to use and easy to set up and carry out. On the other hand, the assay only allows assessment of cell migration in single cells and the physiological relevance of the assay is not given.

A commonly used protocol for the determination of the ability of cells to migrate chemotactically is the transwell assay is shown here. To obtain meaningful results this assay should be performed in triplicate and then repeated three times over different days. Preparations are necessary for this assay, as the transwell inserts (which come in different pore sizes) may be pre-coated with an ECM such as fibronectin prior to use. An incubation step is then followed by some PBS washes. In the case of inhibition studies cells should be pre-incubated with inhibitor for at least one hour. To begin the assay, culture medium plus the chemoattractant, is added to the wells of the plate outside the inlet, this could be, for example FCS. To the inside of the insert, growth medium is added without the chemoattractant. Cells are trypsinised and counted and in this example approx. 50,000 cells in 50 μl of culture medium are added to the transwell inlet. The plate is then returned to the incubator and cells are allowed to migrate for 4–6 h towards the chemoattractant, depending on the particular cell type/line to be investigated. When the assay is completed, the inlets are dipped into PBS to wash. The remaining cells on the inside of the inlet are scraped off with a Q-tip stick. The transwell inlet is again dipped into PBS to wash any remaining unattached cells. The PBS is removed from the inlets and replaced with 1% glutaraldehyde in PBS per well. The membrane is allowed to fix for 10 min (or two minutes if methanol acetate is used). The inlets are placed into distilled water for one minute and then in DAPI diluted 1:500 in PBS; they are then incubated for 5–10 min. After incubation, the inlets are washed once with PBS and twice with water. They can be stored in the original plate and final wash with water wrapped in foil at 4 °C until analysis. For analysis of the membranes, cells are counted in three to four randomly selected fields within the inlet (three inlets/condition) by microscopy; automated software such as Velocity can be used for this. To determine migration the results are expressed as % control after all counts are averaged. For detailed methodology please refer to [20–23]. Figure 3 depicts an overview on the transwell assay. A photographic depiction of the transwell inlet is shown which then is placed inside the 24 well plate. Cells are added to the inlet and allowed to migrate towards the chemoattractant over several hours. The blue (DAPI) stained nuclei can be counted after fixing and staining the cells that have migrated through the membrane. The counts can then be shown in graphical format.

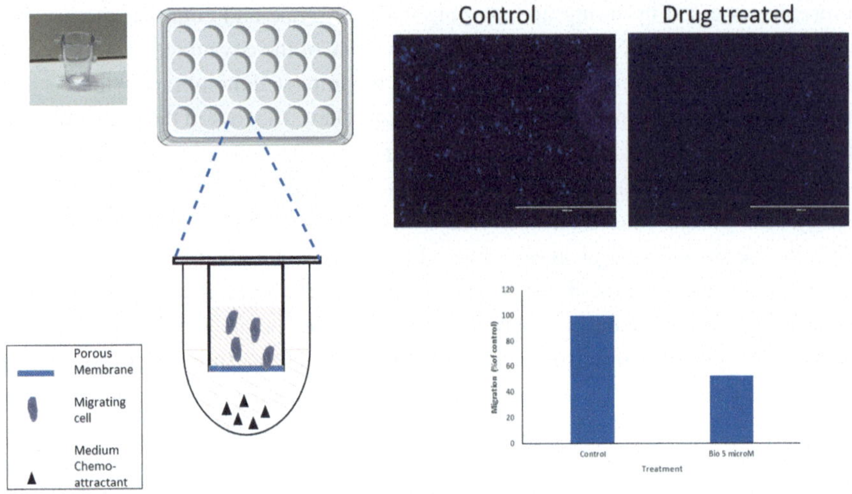

Fig. 3 Migration towards a chemoattractant can be assessed by transwell assays. This is achieved by (1) using transwell inlets in a flat bottom plastic plate allowing cells to migrate towards a chemoattractant over several hours, (2) fixing, staining and visualisation of migrated cells as indicated by the DAPI (blue) stain and (3) graphical representation of the results. An example of generated data is given here. In comparison to the control, the glioma cell line U251 pre-treated with an inhibitor, shows reduced migration. Images created by Dr. Anke Brüning-Richardson, University of Huddersfield. This diagram created with https://smart.servier.com/

2.3 Following Spatial Clues

Another way to analyse the ability of cancer cells to follow external spatial cues is the use of specialised plates which are manufactured to contain nanofibers which are either aligned or non-aligned. An example for the use of such plates is the investigation if glioma cells can follow neuronal tracks as part of their migratory activity. These plates therefore provide scaffolds that mimic neural topography. They allow better reproduction of native cell motility and elongated morphology than non-specific, random cell migration. The aligned nanofibers mimic neuronal tracks and therefore should be able to support cell migration along the tracks of aligned fibres. Cell spheroids can be implanted onto such tracks and by using a fluorescent life cell tracker it is possible to follow migration by imaging the fluorescent signals given off by the migrating cells. To determine migratory activity, it is then possible to, for example, measure the area covered by the migrating cells over time or determine the distance travelled by the migrating cells furthest away from the geometric centre of the original spheroid plated onto the nanofibre plate. We have used this assay successfully to assess our cells' ability to follow external cues in this way. Drawbacks of this technique are the costs associated with the nanofibre plates and the generation of spheroids if using specialised plates, as well as accessibility to microscope systems.

2.3.1 Nanofibre Assay

For this assay, spheroids of the cells to be investigated should be produced. This can be achieved by various means. To make aggregates of cells, several methods can be utilised. Here, aggregates are made by using a method utilising agar. Agarose is dissolved in high glucose medium. The agarose is completely dissolved by placing in a boiling waterbath until the agarose is fully dissolved. The agarose is then allowed to cool down and filter-sterilised under sterile conditions or autoclaved after which antibiotics can also be added. The filtered or autoclaved agarose is pipetted into 35 mm culture dishes before the agar solidifies. The dishes are allowed to solidify in a tissue culture hood to keep them sterile. The dishes are wrapped in parafilm and stored at 4 °C until use. They can be stored up to 3 weeks.

For use, the agar plates are then incubated at 37 °C for one hour with the appropriate culture medium. The medium is removed allowing to clean the agar plate from debris and new culture medium is added. Cells are trypsinised and counted and added at the desired cell number to the agar plate. For aggregate formation to occur it is advisable to rock the agar plate gently and to continue to do so every few hours. The forming aggregates are incubated for a minimum of 24 h under normal growth conditions. To aid visualisation the formed aggregates may be stained with fluorescent markers such as CellTRacker at pre-determined concentration for one hour in the incubator. The preferred CellTRacker dye is Green CellTRacker as others have been shown to affect cell motility. Excess dye is removed by transferring the aggregates gently with a pipette to two 35 mm dishes sequentially and adding culture medium to the final dish. For the assay, the commercially available nanofibre plates are gently washed three time with sterile water or PBS to remove any debris from the wells. The plates are allowed to airdry and then may be pre-coated with a substrate such as fibronectin, laminin, collagen or hyaluronan to aid attachment. When pre-coating the substrate is added at the pre-determined concentration to the plate (to each well) and then incubated for two hours at room temperature. After incubation, the wells are washed twice with sterile PBS and growth medium is added.

The aggregates are picked from the dishes using a dissecting microscope. Great care has to be taken to do this aseptically in a tissue culture hood. All equipment must be wiped down with ethanol prior to use. Aggregates are picked with 10 μl pipettes and are then transferred onto the nanofibre plates in the centre of the wells. Care must be taken not to injure the nanofibers aligned on the surface of the plates with the pipette tip. The plates are then covered and allowed to sit in the tissue culture hood up to 25 min to support attachment of the aggregates to the plate. Finally, growth medium is added (with or without inhibitor of interest) to the plates and cell migration is recorded by imaging the fluorescent signal at several time points over a pre-determined time frame, for example, 24 h, at low magnification such as ×4 or ×10. Imaging may be achieved by time lapse or taking snapshots with a fluorescent microscope.

Instead of using cell aggregates derived from agar plate, spheroids may be generated using the hanging drop method. Here, cells are trypsinised, counted and prepared at a pre-determined cell density. Working aseptically, petri dishes are prepared, by taking off the lid of the dish and adding 20 μl aliquots of the cell suspensions to the inner side of the lid using a 20 μl pipette. Droplets of the cell solutions are carefully added to the lid keeping some space between each drop. Ten ml sterile PBS is added to the bottom of the dish after which the lid with the hanging drops is carefully place back in its normal position onto the Petri dish. The dish is then carefully transferred to the incubator. Spheroid formation from the cells is achieved within 72 h or may take up to 1 week, depending on the nature of the cells in question. For use on the nanofibers, the lid is carefully taken off the petri dish, inverted, so the drops are upright and then the spheroids are picked from the original drop with a 2 μl pipette. The spheroids are usually detectable by their opaque appearance. For uniformity and consistent size, it is advisable to use an alternative way of making spheroids, such as low adherence plates, which will make this type of assay expensive due to the cost of such plates. For detailed methods please refer to [24–27].

Figure 4 depicts a schematic overview for the nanofibre assay. Specialised nanofibre plates are used to add spheroids to the aligned fibres. Cancer cells migrate away from the original spheroid following the fibre tracks; this can be visualised by the addition of

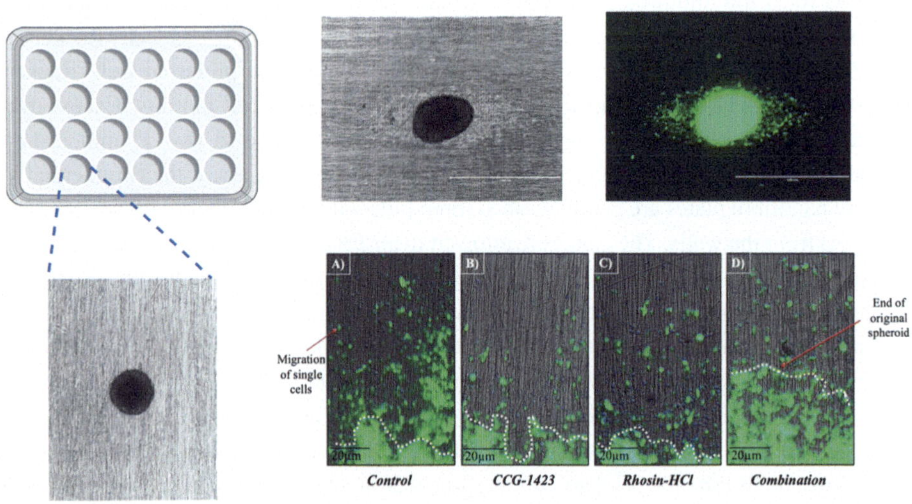

Fig. 4 Nanofibre plates. Migration along external spatial cues can be achieved using aligned nanofibre plates. This is achieved by (1) adding spheroids to the nanofibre plates allowing cell migration away from the original spheroid over time (2) cell migration is visualised by the addition of fluorescent dyes which can be used for assessing cell migration. Here, spheroids formed from the glioma cell line U87 under investigation are placed onto the fibres of the special nanofibre plates and then allowed to migrate along the tract. They had been treated with two anti-migratory inhibitors, 'CCG-1423' and 'Rhosin-HCL', and in combination, or left untreated ('untreated'). The green, fluorescent label clearly shows the migrating cells. The distance travelled away from the original spheroid and area covered can then be quantified. Images created by Dr. Anke Brüning-Richardson and Dr Jessica Senior, University of Huddersfield. This diagram created with https://smart.servier.com/

fluorescent markers. Effects on cell migration after addition of anti-migratory drugs can be assessed by the fluorescent signal and distance travelled by the cells.

3 Invasion in 3D

Invasion of cells into the surrounding matrix can be assessed by the introduction of an extracellular matrix the cells can invade into. Commonly used matrices include Matrigel, collagen and synthetic matrices such as PeptiGels (Manchester BioGel) and hydrogels. In the case of Matrigel we successfully used this in combination with the transwell assay where a layer of Matrigel is added on top of the membrane the cells migrate through to promote invasive activities of the cells under observation. Protrusions and cellular extensions from the cell or cell aggregates can then be counted by, for example, confocal microscopy after fluorescently labelling the protrusions emanating into the matrix.

Following from this particular example we also routinely use a collagen-based invasion assay to assess the effect of anti-migratory inhibitors on cell migration/invasion in 3D glioma spheroids. Here spheroids are embedded in a collagen-supported matrix and migrating cells are allowed to invade into the surrounding environment. Invasion is assessed microscopically over time and data can be analysed by establishing individual migration indices for the cells in question. Our standard migration assay protocol is given here.

Before the assay is initiated spheroids are generated using the aforementioned low adherence 96 well plates. We have tested plates from different manufacturers and have found that the Nunc low adherence plates gave the most consistent results in terms of spheroid homogeneity and also the different types and range of cancer cells tested. It is advisable to obtain a variety of plates from different manufacturers and do some initial assessment of which plate type may be the most appropriate for a particular cell type. Cells are again trypsinised and counted; cells are then added at a pre-determined cell density to the wells of the plate and incubated for at least 72 h to allow the cells to form visible spheroids. This may take longer, for example, in our experience when patient-derived cell lines are used. Once the spheroids are formed, they are ready for the invasion assay. For this, the matrix to be used is prepared, as mentioned above this can be an animal-derived product such as collagen, or equally applicable and in our experience, preferable, synthetic matrices such as PeptiGel or hydrogels. The chosen matrix is then prepared with the growth medium and a polymerising agent. Just before addition of this mixture, the growth medium in the 96 well plate is carefully removed from the plate avoiding at same time to also remove the spheroid. Once the polymerising agent has been mixed into the matrix this is then added to the spheroids from whom the growth medium has been removed. This has to be done swiftly but at the same time carefully to avoid the formation of bubbles as the matrix will start to polymerise. The plate is placed into the incubator to aid polymerisation which is usually achieved within 30 min. To the matrix plugs, growth medium with or without inhibitor is added and the individual spheroids are imaged at certain intervals, i.e., once daily over 72 h, by microscopy. Images can be analysed manually or recorded automatically using either the created images and software tools such as ImageJ to determine the migration index of the migratory cells away from the original spheroid or with an imaging analyser such as Holomonitor (Phase Holographic Imaging PHI AB). Advantages of imaging systems over manual analysis is the avoidance of user bias, data generation within minutes and accessibility in the form of graphs and read out. However, such systems are expensive and require an initially extensive budget.

In our experience the spheroids and migratory cells embedded in natural or synthetic matrices can also be prepared for further analysis by immunohistochemistry or immunofluorescence. We were able to obtain valuable information of the effect, of for example, novel anti-migratory inhibitors, on cytoskeletal components of the migratory cells. To do this, the spheroids in their matrix plugs are fixed and stained for these processes whilst in their original plate [28], making these first steps of processing very easy. For detailed methodology please refer to [12, 13, 29]. Figure 5 depicts a schematic overview for the 3D invasion assay which highlights the different steps including the generation of spheroids in plates, cell invasion into a collagen matrix, visualisation of cell invasion by immunofluorescence microscopy and graphical representation of results.

3.1 Organoids and Organs on a Chip

Organoids are 3D structures that are derived from primary cells or from induced pluripotent stem cells. Stem cells are able to differentiate and to grow together to form a variety of specific organoids. To start the production of organoids the stem cells are mixed with a matrix such as Matrigel and droplets are added to a 24 well plate at room temperature. The plates are then transferred to an incubator to allow the formation of domed droplets. After formation for at least 1 week media are added to promote proliferation and differentiation into specific tissues such as brain or lung. The media are composed of proteins associated with the ECM and various growth factors, which are fine-tuned and formulated to create specific tissues. Growing organoids is a multi-step and laborious process which must be monitored to ascertain that they contain the right tissue and differentiated state. This can be achieved by high-content imaging analysing growth and differentiations. In addition, 3D recapitulation of the formed structures in association with cell morphology, as well as expression of certain biomarkers is necessary to confirm the integrity of the organoid. As such this technology can be costly and time consuming, however, organoids are known to recapitulate the original tissue and organ from which they were generated.

An additional aspect of organoids is organ-on a chip assay. This mimics the physiological aspects of organs by co-culturing cells in a supportive 3D matrix with the help of microfluidics to allow the continued flow of nutrients and interactants passing the forming cellular structures. The biological relevance of such model makes it especially attractive for drug screens or toxicity tests.

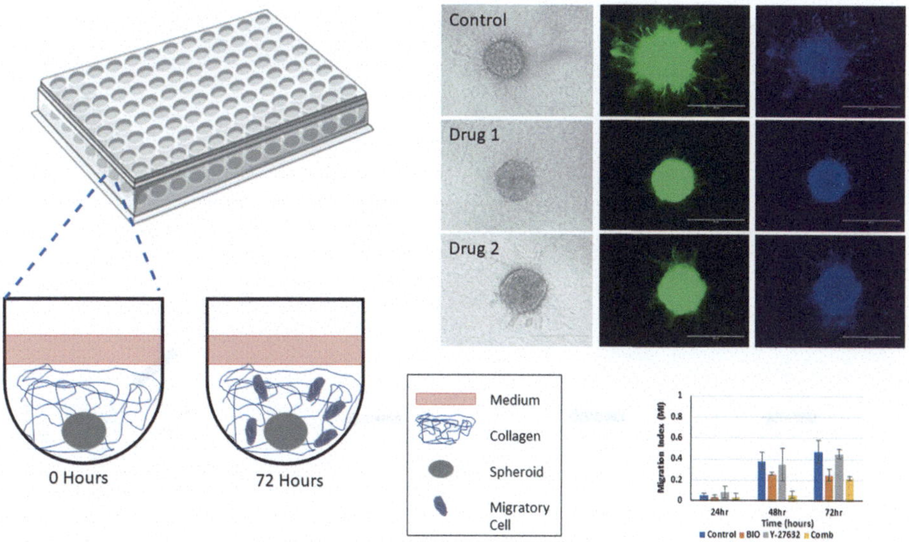

Fig. 5 Cell migration/invasion in a 3D environment can be achieved by 3D invasion assays. This is achieved by (1) the generation of spheroids from cell suspensions in special low adherence plates, (2) embedding in a matrix such as collagen and (3) assessing cell migration from microscopic images over several time points. Immunofluorescent dyes may also be added for visualisation as shown in this example for a patient-derived glioma cell line either untreated ('Control') or treated with Drug 1 or Drug 2. Once the experiment is completed, migration can be determined by calculating the migration index (MI), for example as published by Cockle et al. [12]. Here an example is shown for treatment of the glioma cell line U251 with two drugs, 'BIO' or 'Y-27632', or in combination over 72 h. The combination treatment of the cells has the most pronounced effect on cell migration/invasion. Images and graphical representation created by Dr. Anke Brüning-Richardson and Becky Griggs, University of Huddersfield. This diagram created with https://smart.servier.com/

3.2 Tissue Slice Invasion Model

Tissues may also be directly used for analysis of cell migration. This is based on the collection of a tissue in question, for example brain tissue, which is stored in a pre-determined growth medium with added antibiotics and fungicides. The tissue can be stored at 4 °C up to 24 h after collection. The tissue is cut into sections between 0.25 and 0.5 cm, which are then stored in the same growth medium as before. They are then embedded in 3% agar diluted in PBS; the cast for this is then placed in a refrigerator.

A Leica vibratome is set up and then 250 μm sections are cut. Speed settings may vary according to the tissue to be cut. The sections are then transferred into 6 well plates with the growth medium and cultured in the incubator at 37 °C and 5% CO2.

The growth medium is replaced, and spheroids can be placed on top of the slices for adhering and then migrating into the tissue. After ending of the experiment, the tissue slices are then fixed. This is achieved by removing the supernatants (these can also be kept,

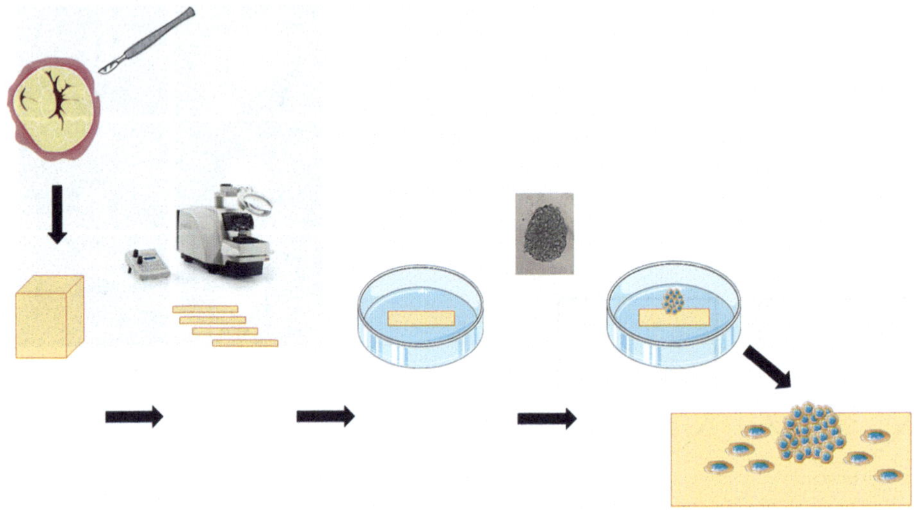

Fig. 6 Organotypic tissue slices. Migration/invasion within a cell-specific environment can be assessed using tissue slices. In this example, (1) brain tissue is obtained after biopsy, (2) the tissue is cut into small cubes and then sliced at a certain thickness with a vibratome. The slices are maintained within a pre-determined, growth-promoting medium immobilised on a platform in a culture dish such as a Petri dish (3). For brain tumour studies, glioma spheroids can then be added to the brain slice and allowed to interact with the brain tissue. Live cell imaging may be achieved, or the tissue can be prepared for immunohistochemistry after completion of the experiment. Image created by Dr. Anke Brüning-Richardson, University of Huddersfield. This diagram created with https://smart.servier.com/

spun down and stored at −80 °C for further analysis), and then washing the slices twice with DPBS; the final wash of PBS is then replaced with 10% formalin, which is incubated for two hours. The formalin is removed, and the washing process repeated twice. Keeping the slice still in the washing solution the cassettes for further immunohistological sample preparation are obtained and the slices transferred to the cassette for further processing. Processing can be done manually using routine immunohistochemistry (IHC) procedures or via a processing machine. The samples are then ready for embedding in paraffin and sectioning. For detailed methodology please refer to [30–33]. Figure 6 depicts a schematic overview for the tissue slicing assay. This overview highlights the cutting of the tissue to be investigated, preparation for sectioning, growing in culture medium and addition of spheroids of interest for tissue invasion assays.

4 In Vivo Imaging

Cell migration can also be imaged in vivo, however, ethical considerations and costs and appropriate facilities for housing animals and for the imaging must be carefully considered to opt for this particular approach. For in vivo imaging, animals may be used for non-invasive techniques, including magnetic resonance imaging (MRI) and positron emission topography (PET) imaging. More commonly used techniques include the application of luminescence/ fluorescence imaging. This allows the observation of processes within an entire organism. Intra-vital imaging (IVM) permits continuous monitoring of cellular processes in living organisms. This focuses on certain organs or tissues. It is a highly invasive method involving surgical exteriorisation of the observed part. The basis of this is the development of fluorescent probes and fluorescent microscopy utilising fluorophores which are excited at certain wavelengths and then emit photons that can be detected. Tumours can be labelled fluorescently by genetic manipulation and attempts were made to detect such labelled tumours in situ. It was soon realised that tissues are a barrier to imaging due to distortion, scattering and absorption of visible light whilst travelling. As a consequence, it is difficult to detect fluorescent images and other challenges of phototoxicity and signal loss arise. To overcome these problems multiphoton microscopy has been successfully applied to live imaging due to the use of specific fluorophores which are excited with low-energy infrared light thus avoiding the aforementioned problems associated with absorption of visible light. Depending on the site of the tumour, the experimental design and tissue thickness, IVT can be used to achieve good imaging contrast allowing high resolution. Another positive of this procedure is the ability of infrared light to produce additional label-free signals known as second harmonic generation (SHG) by which one can determine additional features of molecular structures, including collagen type I fibres [34]. Finally, third harmonic generation (THG) is also possible so that certain structures which are at an interface with different refraction indices may also be visualised including lipid bodies in an aqueous environment. Even though multiphoton excitation allows the detection of signals that are at a certain depth within the body, most sites are still not accessible due to the depth at which they are found. To circumvent this problem migratory behaviours of tumour cells can be imaged by exposing the tumour surgically, within a time frame of approx. 40 h. To improve on the time given to image such tumour behaviour intravital windows are created in the animal imaging over long periods of time up to several years and this has been achieved for sites such as the brain, breast and organs found in the abdomen. It has been therefore possible to study cancer cells under physiological conditions in their original site and migratory behaviour for various cancer types have been described; however, such experimental set-ups require access to expensive imaging equipment and facilities, animal facilities and should be considered carefully due to ethical issues around the use of animals for such invasive procedures. Examples of this include the work described by Beerling et al. [35] where the migratory behaviour of pancreatic cancer cells was studied. Although these cells were believed to be highly aggressive and motile it was shown that only a small proportion of the tumour cells were able to migrate away from tumour lobes and adopted migration strategies similar to those seen in other in vivo mouse models. The results from the study suggested that the various migration patterns ranging from collective streaming to single migration are dependent on the microenvironment. Therefore, it follows that the microenvironment will influence the ability of the cells to migrate and may be a novel way to determine the key regulators in migration in cancer cells which can be exploited to target cell migration and dissemination of cancers.

Take-Home Message

This chapter introduced concepts of in vitro and in vivo assays to investigate cell migration.

- Migration/invasion assays for cell migration should be carefully chosen to answer a particular scientific question.
- 2D assays such as random cell migration, scratch assays, transwell assays and nanofibre migration allow determination of different aspects such as speed, direction of travel, following external cues, of cell migration.
- 3D assays such as invasion assays and organotypic tissue slices allow investigation of cell migration in a more cell/tumour-representative environment.
- In vivo assays allow the investigation of cell migration in situ.

References

1. Abercrombie M. Ross Granville Harrison, 1870-1959. Biogr Mems Fell R Soc. 1961;7:110–26. https://doi.org/10.1098/rsbm.1961.0009.
2. Editorial. Henrietta Lacks: science must right a historical wrong. Nature. 2020;585:7. https://doi.org/10.1038/d41586-020-02494-z.
3. https://www.immunology.org/hela-cells-1951
4. Turner T. Development of the polio vaccine: a historical perspective of Tuskegee University's role in mass production and distribution of HeLa cells. J Health Care Poor Underserved. 2012;23(4 Suppl):5–10. https://doi.org/10.1353/hpu.2012.0151. PMID: 23124495; PMCID: PMC4458465
5. https://media.nature.com/original/magazine-assets/d41586-020-02494-z/d41586-020-02494-z.pdf
6. Morton HJ. Joseph F. Morgan: the man and his contributions. In Vitro. 1978;14:3–10.
7. Lancaster MA, Renner M, Martin CA, Wenzel D, Bicknell LS, Hurles ME, Homfray T, Penninger JM, Jackson AP, Knoblich JA. Cerebral organoids model human brain development and microcephaly. Nature. 2013;501(7467):373–9. https://doi.org/10.1038/nature12517. Epub 2013 Aug 28. PMID: 23995685; PMCID: PMC3817409
8. Bian S, Repic M, Guo Z, Kavirayani A, Burkard T, Bagley JA, Krauditsch C, Knoblich JA. Genetically engineered cerebral organoids model brain tumor formation. Nat Methods.

2018;15(8):631–9. https://doi.org/10.1038/s41592-018-0070-7. Epub 2018 July 23. Erratum in: Nat Methods. 2018;15(9):748. PMID: 30038414; PMCID: PMC6071863

9. Saini G, Segaran N, Mayer JL, Saini A, Albadawi H, Oklu R. Applications of 3D bioprinting in tissue engineering and regenerative medicine. J Clin Med. 2021;10(21):4966. https://doi.org/10.3390/jcm10214966. PMID: 34768485; PMCID: PMC8584432
10. Kabir A, Datta P, Oh J, Williams A, Ozbolat V, Unutmaz D, Ozbolat IT. 3D Bioprinting for fabrication of tissue models of COVID-19 infection. Essays Biochem. 2021;65(3):503–18. https://doi.org/10.1042/EBC20200129. PMID: 34028514
11. Collins TJ. ImageJ for microscopy. Biotechniques. 2007;43(1 Suppl):25–30. https://doi.org/10.2144/000112517. PMID: 17936939
12. Cockle JV, Picton S, Levesley J, Ilett E, Carcaboso AM, Short S, Steel LP, Melcher A, Lawler SE, Brüning-Richardson A. Cell migration in paediatric glioma, characterisation and potential therapeutic targeting. Br J Cancer. 2015;112(4):693–703. https://doi.org/10.1038/bjc.2015.16. Epub 2015 Jan 27. PMID: 25628092; PMCID: PMC4333505
13. Egbivwie N, Cockle JV, Humphries M, Ismail A, Esteves F, Taylor C, Karakoula K, Morton R, Warr T, Short SC, Brüning-Richardson A. FGFR1 expression and role in migration in low and high grade pediatric gliomas. Front Oncol. 2019;9:103. https://doi.org/10.3389/fonc.2019.00103. PMID: 30931252; PMCID: PMC6425865
14. https://ibidi.com/chemotaxis-analysis/171-chemotaxis-and-migration-tool.html?gad=1&gclid=EAIaIQobChMItN7E5-ef_wIV0tvtCh0VaQHWEAAYASAAEgI8FvD_BwE. Accessed 31 May 2023
15. Riahi R, Yang Y, Zhang DD, Wong PK. Advances in wound-healing assays for probing collective cell migration. J Lab Autom. 2012;17(1):59–65. Society for Laboratory Automation and Screening. https://doi.org/10.1177/2211068211426550.
16. Cory G. Scratch-wound assay. Methods Mol Biol. 2011;769:25–30. https://doi.org/10.1007/978-1-61779-207-6_2. PMID: 21748666
17. Cappiello F, Casciaro B, Mangoni ML. A novel in vitro wound healing assay to evaluate cell migration. J Vis Exp. 2018;133:56825. https://doi.org/10.3791/56825. PMID: 29608162; PMCID: PMC5931780
18. Pijuan J, Barceló C, Moreno DF, Maiques O, Sisó P, Marti RM, Macià A, Panosa A. In vitro cell migration, invasion, and adhesion assays: from cell imaging to data analysis. Front Cell Dev Biol. 2019;7:107. https://doi.org/10.3389/fcell.2019.00107. PMID: 31259172; PMCID: PMC6587234
19. Hossian AKMN, Mattheolabakis G. Cellular migration assay: an in vitro technique to simulate the wound repair mechanism. Methods Mol Biol. 2021;2193:77–83. https://doi.org/10.1007/978-1-0716-0845-6_8. PMID: 32808260
20. Marshall J. Transwell(®) invasion assays. Methods Mol Biol. 2011;769:97–110. https://doi.org/10.1007/978-1-61779-207-6_8. PMID: 21748672
21. Stoellinger HM, Alexanian AR. Modifications to the transwell migration/invasion assay method that eases assay performance and improves the accuracy. Assay Drug Dev Technol. 2022;20(2):75–82. https://doi.org/10.1089/adt.2021.140. Epub 2022 Feb 23. PMID: 35196113; PMCID: PMC8968842
22. Kenney RM, Loeser A, Whitman NA, Lockett MR. Paper-based Transwell assays: an inexpensive alternative to study cellular invasion. Analyst. 2018;144(1):206–11. https://doi.org/10.1039/c8an01157e. PMID: 30328422; PMCID: PMC6296866
23. Ponath PD, Wang J, Heath H. Transwell chemotaxis. Methods Mol Biol. 2000;138:113–20. https://doi.org/10.1385/1-59259-058-6:113. PMID: 10840748
24. Johnson J, Nowicki MO, Lee CH, Chiocca EA, Viapiano MS, Lawler SE, Lannutti JJ. Quantitative analysis of complex glioma cell migration on electrospun polycaprolactone using

time-lapse microscopy. Tissue Eng Part C Methods. 2009;15(4):531–40. https://doi.org/10.1089/ten.TEC.2008.0486. PMID: 19199562; PMCID: PMC3497888

25. Rao SS, Nelson MT, Xue R, DeJesus JK, Viapiano MS, Lannutti JJ, Sarkar A, Winter JO. Mimicking white matter tract topography using core-shell electrospun nanofibers to examine migration of malignant brain tumors. Biomaterials. 2013;34(21):5181–90. https://doi.org/10.1016/j.biomaterials.2013.03.069. Epub 2013 Apr 16. PMID: 23601662; PMCID: PMC4080638
26. Agudelo-Garcia PA, De Jesus JK, Williams SP, Nowicki MO, Chiocca EA, Liyanarachchi S, Li PK, Lannutti JJ, Johnson JK, Lawler SE, Viapiano MS. Glioma cell migration on three-dimensional nanofiber scaffolds is regulated by substrate topography and abolished by inhibition of STAT3 signaling. Neoplasia. 2011;13(9):831–40. https://doi.org/10.1593/neo.11612. PMID: 21969816; PMCID: PMC3182275
27. Roth AD, Elmer J, Harris DR, Huntley J, Palmer AF, Nelson T, Johnson JK, Xue R, Lannutti JJ, Viapiano MS. Hemoglobin regulates the migration of glioma cells along poly(ε-caprolactone)-aligned nanofibers. Biotechnol Prog. 2014;30(5):1214–20. https://doi.org/10.1002/btpr.1950. Epub 2014 July 24. PMID: 25044995
28. Harmer J, Struve N, Brüning-Richardson A. Characterization of the effects of migrastatic inhibitors on 3D tumor spheroid invasion by high-resolution confocal microscopy. J Vis Exp. 2019;151 https://doi.org/10.3791/60273. PMID: 31566595
29. Cheng V, Esteves F, Chakrabarty A, Cockle J, Short S, Brüning-Richardson A. High-content analysis of tumour cell invasion in three-dimensional spheroid assays. Oncoscience. 2015;2(6):596–606. https://doi.org/10.18632/oncoscience.171. PMID: 26244167; PMCID: PMC4506363
30. Weitz JR, Tiriac H, Hurtado de Mendoza T, Wascher A, Lowy AM. Using organotypic tissue slices to investigate the microenvironment of pancreatic cancer: pharmacotyping and beyond. Cancers (Basel). 2021;13(19):4991. https://doi.org/10.3390/cancers13194991. PMID: 34638476; PMCID: PMC8507648
31. Spennati G, Horowitz LF, McGarry DJ, Rudzka DA, Armstrong G, Olson MF, Folch A, Yin H. Organotypic platform for studying cancer cell metastasis. Exp Cell Res. 2021;401(2):112527. https://doi.org/10.1016/j.yexcr.2021.112527. Epub 2021 Mar 4. PMID: 33675807; PMCID: PMC8806469
32. Braun R, Lapshyna O, Eckelmann S, Honselmann K, Bolm L, Ten Winkel M, Deichmann S, Schilling O, Kruse C, Keck T, Wellner U, Bronsert P, Brandenburger M. Organotypic slice cultures as preclinical models of tumor microenvironment in primary pancreatic cancer and metastasis. J Vis Exp. 2021;172 https://doi.org/10.3791/62541. PMID: 34251366
33. Bakmand T, Troels-Smith AR, Dimaki M, Nissen JD, Andersen KB, Sasso L, Waagepetersen HS, Gramsbergen JB, Svendsen WE. Fluidic system for long-term in vitro culturing and monitoring of organotypic brain slices. Biomed Microdevices. 2015;17(4):71. https://doi.org/10.1007/s10544-015-9973-6. PMID: 26123417
34. Suijkerbuijk SJE, van Rheenen J. From good to bad: Intravital imaging of the hijack of physiological processes by cancer cells. Dev Biol. 2017;428(2):328–37. https://doi.org/10.1016/j.ydbio.2017.04.015. Epub 2017 May 1. PMID: 28473106
35. Beerling E, Oosterom I, Voest E, Lolkema M, van Rheenen J. Intravital characterization of tumor cell migration in pancreatic cancer. Intravital. 2016;5(3):e1261773. https://doi.org/10.1080/21659087.2016.1261773. PMID: 28243522; PMCID: PMC5226006

For Additional Learning Material We Also Recommend

Gatreau A, editor. Cell migration methods and protocols, Methods in molecular biology (MIMB), vol. 1749. New York: Humana; 2018.

Margadant C, editor. Cell migration in three dimensions, Methods in molecular biology (MIMB), vol. 2068. New York: Humana; 2023.

Monti M. Cell migration—methods and protocols. Eur J Histochem. 2019;63(1):3013. https://doi.org/10.4081/ejh.2019.3013. PMID: 30652435; PMCID: PMC6340306

Wells CM, Parsons M, editors. Cell migration. Developmental methods and protocols, Methods in molecular biology (MIMB), vol. 769. Totowa, NJ: Humana; 2017.

3D Bioprinting of Cell Migration

Jessica J. Senior

What You Will Learn in This Chapter

The ability of cells to migrate is a fundamental process that allows structural organisation and survival of human tissues and organs. One of the most effective ways of mimicking human tissues for the assessment of cell migration is through 3D bioprinting of tissue models. There exists a range of technologies that employ the principles of 3D bioprinting that allow the tailorability of bioinks and printing parameters so that both optimal cell viability and print fidelity can be achieved. Working through the typical bioprinting workflow, it is possible to fabricate 3D migratory models with the aim of establishing biological mechanisms of cell migration, studying novel candidate drugs to limit migration of cells in diseases such as cancer or even develop pro-migratory tissue models in a bid to restore tissue function following implantation. Multiple types of tissue have been investigated using various modalities of bioprinting, with some implanted models showing much promise ascribed by migration of exogenous and endogenous cells between host tissue and implants. As bioprinting is still very much a developing technology, it is anticipated that clinically biorelevant models that better capture in vivo microenvironments will become available in the very near future. This chapter

(continued)

J. J. Senior (✉)
University of Huddersfield, Huddersfield, UK
e-mail: j.j.senior2@hud.ac.uk

A. Brüning-Richardson, S. Knipp (eds.), *Cell Migration in Development, Health and Disease*, Learning Materials in Biosciences,
https://doi.org/10.1007/978-3-031-64532-7_8

will first introduce the relevance and principles of 3D bioprinting, with a detailed insight into the major bioprinting workflow stages consisting of model design, print design, model fabrication, model propagation and application. The chapter will then focus on the three predominant 3D bioprinting operatives; laser-induced forward transfer, inkjet printing and robotic dispensing and their subtypes, with their optimal operating specifications being indicated for each. The chapter will then go on to present pre-established 3D bioprinted migratory models in research with examples in cancer, skin and muscle tissue mimics and their advances in the field.

1 Introduction

Amongst all the techniques used in tissue engineering, 3D bioprinting is arguably one of the most sophisticated and functionally accurate strategies for achieving biorelevant models in vitro. 3D bioprinting requires a multidisciplinary approach for the fabrication of a successful tissue model, including expertise in engineering, cell biology, materials science and medicine and each component contributes to the optimisation of the final part. It allows the generation of models in 3D that closely mimic their intended in vivo environments when compared to culturing cells in 2D monolayers. In general, 3D bioprinting entails the layer-by-layer deposition of biological constituents (namely cells, biomaterials and media) for the fabrication of tissues through biomimicry [1]. Many applications have already been established for the utilisation of 3D bioprinted constructs in the assessment of cell migration, including the study of its biological pathways in health and disease, the use of drugs or other therapeutic strategies to correct cell migration and to assess the suitability of constructs destined for implantation [2, 3]. Although the process of 3D bioprinting is somewhat flexible and certain parameters may be adjusted, there is a typical workflow that can be followed in order to gain the best possible outcomes. The main focus of this chapter is to discuss each step in this workflow, to present different bioprinting techniques and when best to use them, and finally to highlight the current state-of-the-art in 3D printed migratory systems.

2 3D Bioprinting Workflow

Like other 3D biofabrication methods, 3D bioprinting begins with planning the theoretical design of the model by careful selection of cells and polymers to closely replicate the tissue of interest. During print design, a computerised 3D model is used for the generation of coding instructions that the printer can follow once the bioink has been prepared, and 3D bioprinting commences during the model fabrication step. After curing where necessary, the printed model is propagated under a series of defined, physiologically relevant

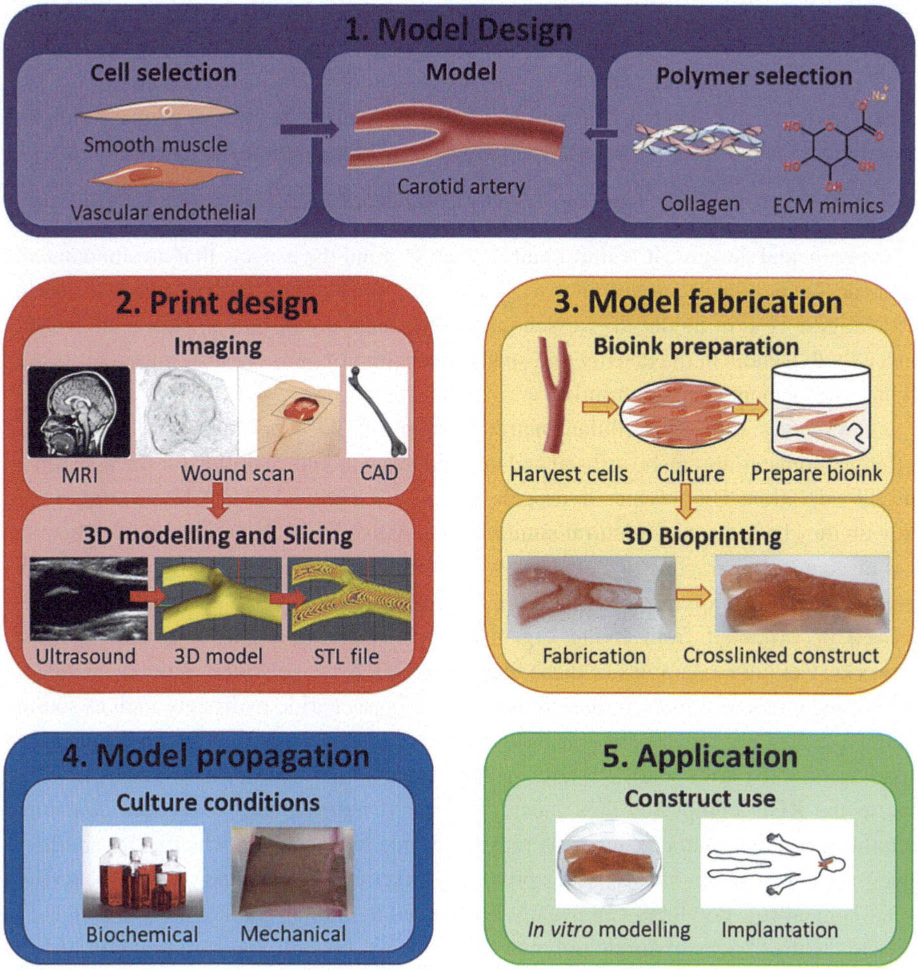

Fig. 1 Chronological stages for the design and production of a 3D bioprinted model by example of a carotid artery, including model design, print design, model fabrication, model propagation and application

conditions for enhanced functionality prior to being used for its intended application (Fig. 1). This section will discuss each portion of the workflow in detail with particular reference to the fabrication of 3D migratory models.

2.1 Model Design

The first critical step in the formation of a 3D bioprinted migratory model is selecting relevant cells and polymer materials to secure functional resemblance of the intended microenvironment. When designing the cell components within a model, many considerations need to be made with regards to cell source and the level of complexity one wishes to achieve with their model. Cells can be obtained from multiple different species (e.g. human, mouse and hamster), different origins (established cell lines or primary isolation from tissues) and with a given differentiation status (differentiated or uncommitted) [4]. It is also possible to dictate the number of different cell types incorporated into the model depending on the complexity of the desired research outcomes in question. Like all experimental designs, it is important to keep in mind the aspects that are fundamental to the design as consistent with in vivo systems as possible. For instance, when designing a 3D migratory skin model, it is essential that the anatomy of human skin is thoroughly understood. Details in the density and spatial allocation of cells are also important factors in the development of tissue mimics.

Representing the extracellular matrix portion of 3D bioprinted models, multiple different biomaterials can be utilised and tailored to act as cell scaffolds for the assessment of cell migration. Biopolymer hydrogels are often employed in 3D bioprinting as not only do they have strong structural similarities to native ECM, but their tunable chemical and mechanical characteristics also allow greater freedom and versatility during model fabrication [5]. Biopolymer hydrogels are defined as water-swollen polymeric substances derived or extracted from living organisms and featuring highly organised intrinsic networks. Common examples of these include protein hydrogels such as collagen, gelatin and Matrigel (derived from animal tissues) and polysaccharide hydrogels such as sodium alginate, gellan gum, pectin and agarose (extracted from seaweed, bacteria fermentation byproducts and citrus peel). Protein hydrogels are often conducive to cell attachment due to the RGD complexes they inherently bear (tri-amino acid sequences containing arginine–glycine–aspartate), whereas polysaccharide hydrogels are chemically similar to glycosaminoglycans in native tissues providing structural support and offering equivalent water content [6]. Polymers can usually be blended to exhibit qualities of the two and to obviate the drawbacks of single polymer functionality. Overall, the common aim of a 3D bioprinted model is to reflect the biocompatibility, biodegradability, mechanical properties and scaffold architecture of in vivo developing tissues [7] (Fig. 2). Therefore, for the successful manufacture of a 3D model where its in vivo counterpart is rich in collagenous ECM, the incorporation of collagen should be encouraged to generate biorelevant ECM-like environments. Tissue models may also be synthesised from synthetic hydrogels such as polyamides and polyethene glycol, however, they are typically less biocompatible due to a poorer resemblance of native tissue chemical structure.

2.2 Print Design

Preparing the print design for a 3D model often consists of (1) acquisition of imaging data and (2) 3D model reconstruction. During imaging, it is pertinent to interpret the target tissue architecture so that the structure's dimensions and model's performance are

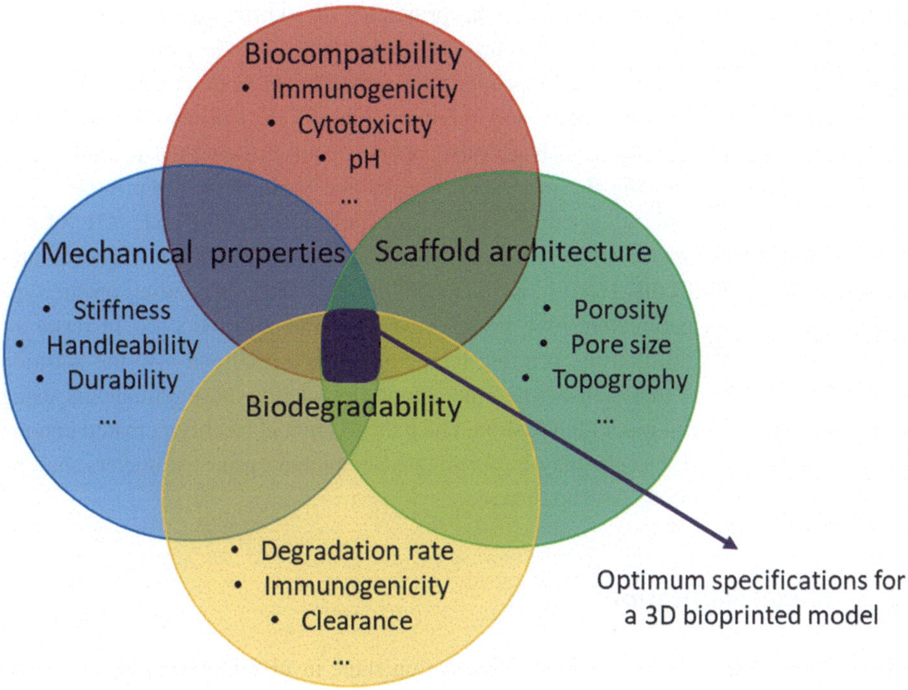

Fig. 2 Optimum specifications for a 3D bioprinted model, including biocompatibility, biodegradability, mechanical properties and scaffold architecture

functionally enhanced and as representative as possible. Anatomic data can be produced using a range of techniques and the selection of imaging software is tailored to the appropriate situation. For instance, when imaging hard tissues such as bone and tumour for the generation of a 3D bioprinted tissue mimic, computed tomography (CT) scanning may be more appropriate, whereas MRI imaging is better suited to assessing soft tissues. Other aspects of imaging to take into account include accessibility of the target tissue, imaging resolution and potential exposure to harmful radiation [8]. If the 3D migratory model design does not require imaging data as extensive as this and, instead, a simpler model is adequate, computer-aided design (CAD) software is easily accessible for generating 3D templates. Here models may be modified to introduce internal contours, pores, channels and other features to facilitate cell attachment, migration and nutrient flow [9].

Once the 3D model has been computerised, the data can then be transformed into a readable language that conforms with 3D bioprinter operation using modelling and slicing software. First, the 3D model is rendered to comply with stereolithography (STL) file format, describing the surface of the geometry using a series of surface tessellation arrangements for ease of recognition [10]. Following this, the STL model is translated into a stack of horizontal layers called 'slices' which the 3D bioprinter uses to extrude material in a layered fashion following the principles of additive layer manufacturing (ALM). Oftentimes, sliced stl files require conversion into G-Code format so that the languages between the file and the bioprinter are compatible. G-Code is made of a series of instructions manifested as letters and numbers that assign a printer function. This digital coding directs functions such as valve opening and closing time to orchestrate material extrusion, or moving the printhead to a defined location on the printbed [11]. For example, the code 'M760' opens the valve for printhead 1 and the code 'G1 X1.236 Y4.611 E0.01212; perimeter' linearly moves the extruder to these *X*, *Y* and *Z* coordinates when using the INKREDIBLE 3D bioprinter (Cellink, Sweden). Depending on the type of bioprinter, certain parameters may be selected and built into the code in terms of model infill pattern, infill density, printing temperature and speed and also the addition of overhang-supporting material, whilst other bioprinters define these parameters using manual operation prior to model fabrication. Once the 3D model has been created and the print path fully defined, bioinks may be formulated, remaining print parameters may be selected and 3D bioprinting can commence.

2.3 Model Fabrication

Two key steps that make up the model fabrication stage involve bioink preparation and 3D bioprinting. Preparation of the bioink is typically standard across all 3D bioprinter platforms, however, the operation of bioink deposition differs depending on the type of bioprinter and manufacturer. Initially, polymer materials and cells identified in the model design stage as appropriate ECM mimics are mixed at defined polymer concentrations and cell densities under sterile conditions. Multiple polymers and cell types can be mixed within a single bioink or, alternatively, several cartridges containing different bioink formulations may be installed within the bioprinting setup so that the complex heterogeneous nature of native tissue may be captured within a single migratory model [12]. Following placement of the bioink-containing cartridge into the printer, the deposition outlet is calibrated in reference to the printbed so that model dimensions can be accurately achieved. The deposition rate is then set as a function of flow rate or extrusion pressure and is directly correlated to bioink viscosity, print speed and extruding nozzle internal diameter. There are two main issues that are associated with bioprinting, however, in that maximising print fidelity often comes at a cost of compromising cell viability and vice versa. Using high-viscosity bioinks aids in improving the shape of a print yet exposes suspended cells to high local shear forces during deposition due to the high print pressures required for extrusion. Low viscosity bioinks therefore better perpetuate cell viability, however, printed models rapidly lose their shape following deposition onto a flat surface [13]. Several printing variables may be adjusted to best optimise material parameters (including hydrogel properties and gelation), printing parameters (including nozzle gauge, print pressure and print speed) and the collector platform (being a flat surface or a supporting gel suspension) (Fig. 3). Bioprinting offers versatility in the fabrication of

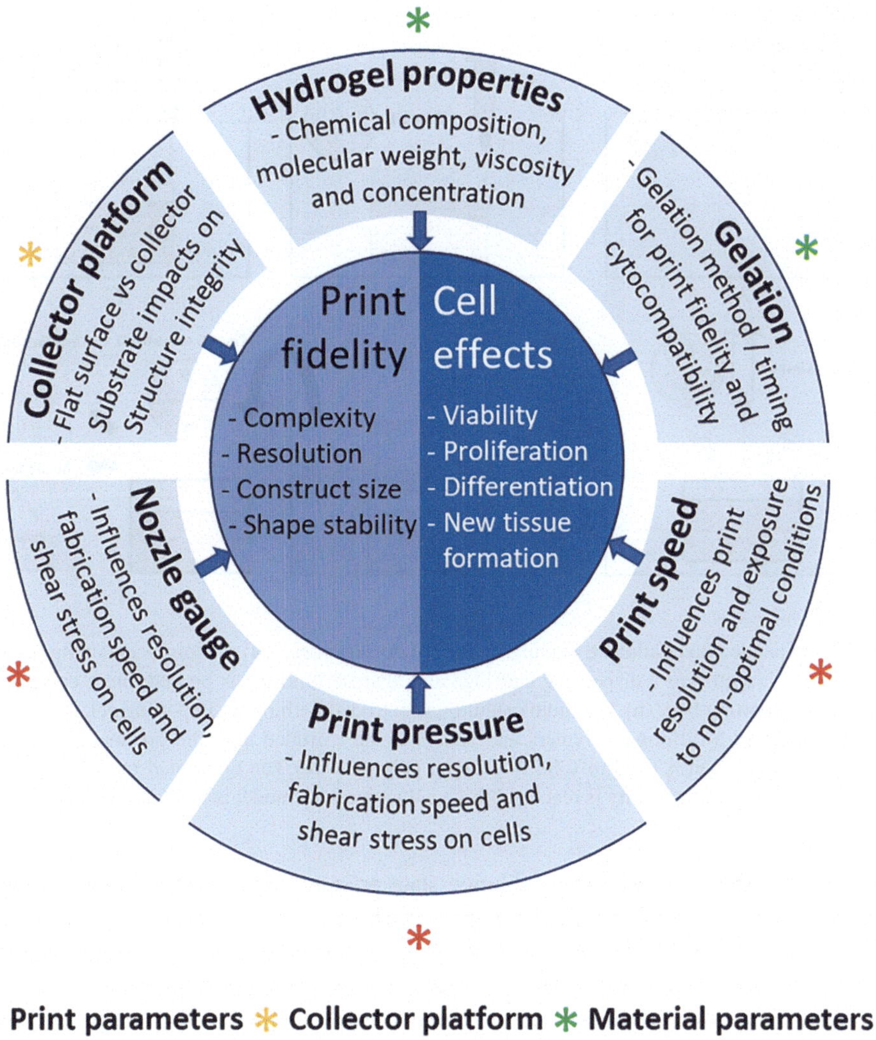

Fig. 3 Optimisation of 3D bioprinting parameters that are critical for successful biofabrication of migratory microenvironments

migratory models, and in the first instance, the print parameters and collector platform should be considered for adjustment prior to altering material parameters so that the intended model physiochemical and mechanical properties that have been designed to match a tissue of interest does not become compromised.

Perhaps the most effective way of enhancing both print fidelity and cell viability simultaneously is by using a supporting collector platform known as embedded bioprinting. This technique employs a sacrificial reservoir within which bioinks are extruded, retaining

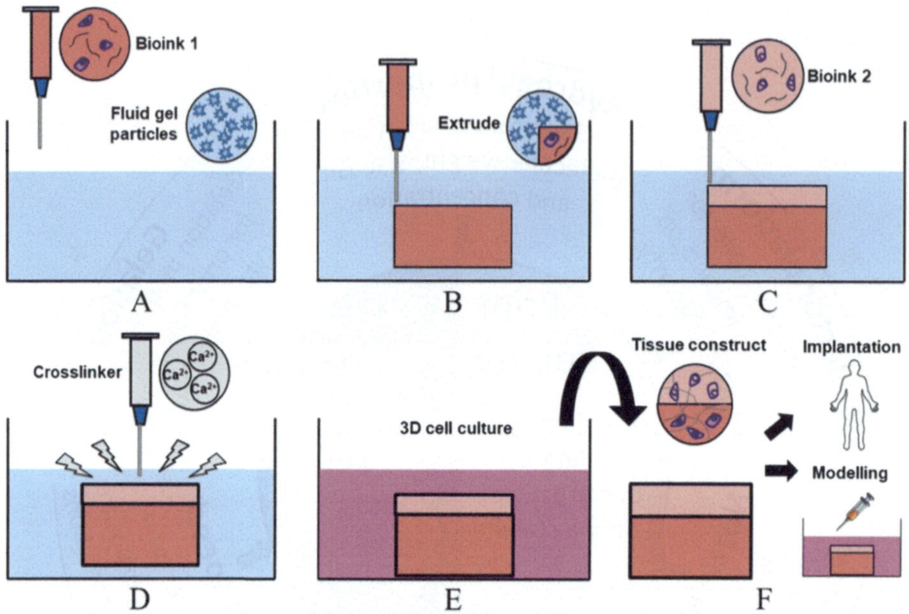

Fig. 4 Suspended layer additive manufacturing (SLAM) process. (**a**) A bioink-containing polymer and cells is loaded into a dispensing cartridge and agarose fluid gel is poured into a container to form the supporting bed. (**b**) The bioink solution is extruded within the supporting gel. (**c**) Multiple other bioinks with differing polymers and cells may be extruded atop the previous extrude. (**d**) Crosslinking solution is used to solidify charged polymers. (**e**) The bioprinted model is extracted and 3D cultured until maturity is reached. (**f**) The 3D migratory model is ready for assessment

the shape of the 3D model in its solution state prior to solidifying and removing from the support bed (Fig. 4). Charged polymers such as alginate, pectin and gellan can be crosslinked using divalent cations such as calcium solution by injection into the supporting bed and diffusion within the model [14]. This technology enables the printing of constructs from low-viscosity precursor solutions with greater complexity and shape specificity without limitation over build direction, supports or overhangs. Cell viability is preserved to a high degree and the method allows incorporation of multiple bioink layers, meaning that the complex hierarchical nature of native tissues can be realised effectively. Current embedded printing techniques use gelatin, Carbopol and agarose supporting gels, with agarose showing the most promise due to its inertness, rheological properties and ambient temperature ranges being optimal for cell culture [15].

2.4 Model Propagation

The culture conditions that 3D printed models are kept under are an additionally important aspect of the bioprinting process in that the conditions to which 3D prints are exposed can very much define the success of the part as a truly representative migratory model. Here, the tissue model is subjected to specific biochemical and physical cues to replicate the environments experienced within healthy or diseased states until the model is fit for use. Using mesenchymal stem cell culture as an example, when introduced to isobutylmethylxanthine (IBMX), indomethacin and dexamethasone, adipogenesis can be initiated through activation of PPARγ (peroxisome proliferator-activated receptors) [16]. Alternatively, when exposed to β-glycerophosphate, ascorbic acid and dexamethasone at critical levels, osteogenic gene expression is induced by extracellular-related kinase phosphorylation [17]. The physical environment may also be modulated by tailoring oxygen concentrations representative of those that native tissues face, such as the hypoxic environments within some tumours at the core or culturing keratinocytes at the air–liquid interface, simulating in vivo epidermal propagation. For tissues that naturally experience mechanical loading such as tendons and ligaments, application of mechanical forces on tenocytes and fibrocytes can secure cell phenotype and differentiation [18].

2.5 Application

The four main 3D bioprinting applications include drug discovery, cosmetics testing, tissue modelling/regeneration and medical devices. These applications can be applied towards studying 3D migration and bypasses the need for animal testing. Furthermore, using 3D bioprinting, it is possible to create representative 3D environments of tissue mimics, having the potential to overcome organ donor shortages and can been used to address the unmet need for medical devices. The major areas of study in 3D bioprinting migratory microenvironments include models of cancer, skin, angiogenesis and muscle. These areas will be discussed towards the end of the chapter as examples of 3D migratory models, but first, the main three techniques within the bioprinting arena will be discussed in detail.

3 3D Bioprinting Techniques

3D bioprinting employs the fundamentals of additive layer manufacturing (ALM) for the buildup of a 3D tissue model using polymer materials and cells. The main deposition-based biofabrication approaches include laser-induced forward transfer, inkjet printing and robotic dispensing, and each technique has associated advantages and drawbacks. The type of bioprinter to be used in a given application typically depends on the requirements of resolution, fabrication speed, polymer material viscosity, gelation speed and cost (Fig. 5).

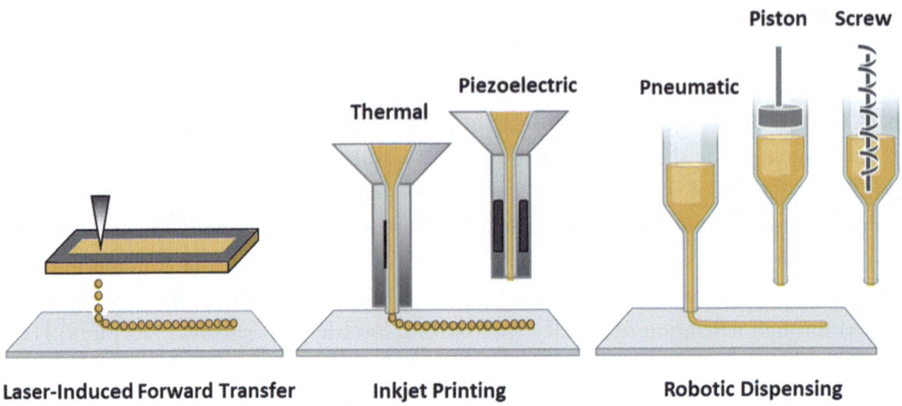

Fig. 5 3D bioprinting techniques including laser-induced forward transfer, inkjet printing and robotic dispensing

3.1 Laser-Induced Forward Transfer

Also known as laser-assisted 3D bioprinting or LIFT, this method utilises a laser pulse that is focused onto an adsorbing layer of glass ribbon that generates a high-pressure bubble and propels the bioink towards the collector substrate. The high precision afforded by laser-induced forward transfer enables the fabrication of micro- and even nano-scale tissue scaffolds making it the most favourable technique in terms of resolution. The timescales associated with this technology, however, can be potentially detrimental to the resulting tissue model, such that complex or larger structures require longer printing times, which may require crosslinking in situ to prevent loss of structure. This type of printing is conducive to a range of different material viscosities and is capable of operating both solid and liquid donor films in a single step. Despite advancements with LIFT, high laser powers can have an adverse effect on cell viability due to thermal damage. Overall, this technique offers a nozzle-free approach to depositing high densities of cells, yet this costly and time-consuming approach makes the generation of larger, clinically relevant 3D models a current constraint [19].

3.2 Inkjet Printing

Inkjet printing is a term used in biofabrication to describe the dispensing of a bioink through a small orifice onto a collector substrate with high precision of very small volumes (also known as drop on demand). This technique has been adapted from 2D ink-based commercial printers and can be further subcategorised into two modes of action, namely thermal inkjet printing and piezoelectric inkjet printing. During thermal inkjet printing, bioink droplets are generated through electrical heating of the printhead which produces vapour bubbles and subsequent deposition. Alternatively, using piezoelectric inkjet printing, a piezoelectric actuator propagates an alternating current, causing mechanical vibrations in the nozzle and inducing bioink droplet formation. Perhaps the greatest asset of inkjet printing is the rapid gelation time of 3D printed models. They also offer adequate resolution and fabrication speed yet are only conducive to printing low-viscosity materials. Cell viability can also be impacted due to the nozzle aperture being not much larger than the cells themselves, which is especially the case with thermal inkjet printing which further elicits heat stress on deposited cells. In summary, this method is attractive when producing smaller structures at a much lower cost than LIFT technology, yet the buildup of larger structures again influences the potential to produce models of clinically relevant sizes [20, 21].

3.3 Robotic Dispensing

Also recognised under the terminology 'extrusion bioprinting' due to the continuous extrude that these technologies produce, robotic dispensing 3D bioprinters operate using either pneumatic, piston or screw deposition. This mode of printing is arguably the most promising when it comes to printing models that are clinically translatable and within a suitable timeframe. A range of material viscosities are adoptable in robotic dispensing although resolution is generally lower than LIFT or inkjet means of 3D bioprinting. Piston-driven dispensing typically gives greater control over deposition, attributed to the fact that pneumatic systems may experience delays in the delivery of compressed gas, whilst screw dispensing can generate larger pressure drops at the nozzle, which may prove problematic for suspended cells [22, 23].

4 3D Bioprinted Models of Cell Migration

As 3D bioprinting has made staggering advances in recent years, the intelligent design and fabrication of complex migratory systems have successfully been realised across multiple applications. These developments have given us a greater appreciation of suitable materials, cells and bioactive molecules to be used in bioinks, as well as advancing techniques that address the issues associated with bioprinting to make model fabrication more accessible and feasible. Within this chapter section, the current state-of-the-art 3D bioprinting cell migration will be presented, with representative examples in cancer, skin and muscle 3D migratory systems all developed using LIFT, inkjet or robotic dispensing technologies.

4.1 Cancer

Cancer remains a significant clinical unmet need due to its prevalence, mortality and economic impact. For these reasons, cancer is a highly researched topic area and 3D bioprinting is one major technology that allows printing of 3D cancer modes in vitro in a bid to better understand cancer migratory mechanisms and subsequently, how to arrest cancer cell invasion and proliferation.

Advances in 3D bioprinting cancer migration span across various cancer pathologies, bioprinting techniques and migratory materials, all to a high standard. One such example is in the adoption of drop-on-demand 3D bioprinting to generate a versatile migratory model with relevance to breast, lung and ovarian cancer [24]. This high-throughput system utilised hydrogels of defined stiffness and peptide (RGD) content to assess the effect of different microenvironments on cancer cell migration to define underlying biological mechanisms and to identify novel therapeutic options. It was established that, whilst non-invasive breast cancer cells (MCF7) appeared to grow similarly in hydrogels of different stiffness and peptide content, the invasive breast cancer variant MDA-MB-231 was only highly proliferative in RGD-presenting hydrogels (independent of stiffness). With regards to migratory behaviour, non-invasive MCF7 cells formed multiple spheroids in Matrigel with a more proliferative capacity, whereas MDA-MB-231 cells were highly invasive exhibiting a stellate, protrusion morphology over 7 days, further verified by E-Cadherin staining of epithelial-like cells and vimentin staining of mesenchymal cells, respectively. Finally, a ROCK inhibitor (rho-associated protein kinase inhibitor often used in tumour metastasis inhibition) was seen to have a profound effect on MDA-MB-231 migration, markedly reducing cell migration distance and speed compared with non-treated controls.

Using robotic dispensing by pneumatic extrusion, 3D bioprinted co-cultures of neuroblastoma (IMR5) spheroids and endothelial (HUVEC) cells were evaluated and cell interactions were reported to contribute to solid tumour aggressive behaviour, denoted by an increase in migratory index in co-cultures versus spheroid culture alone [25]. With GelMA as the representative tumour ECM and Carbopol as a supportive embedding medium, perfusable vascular channels were fashioned to recapitulate the dynamic tumour microenvironment. When co-cultured, neuroblastoma/HUVEC cell–cell interaction was enhanced with significant neuroblastoma cell migration towards the GelMA tissue and HUVEC infiltration towards the hypoxic regions of the tumour, suggesting interactions between tumour cells and vasculature.

Opting for a laser direct-write approach, other researchers have focused on fabricating biomimetic constructs of breast cancer cell invasion into fabricated adipose tissue mimics and assessing the physiological contribution of obesity to breast cancer risk [26]. Breast cancer cells (MCF-7 or MDA-MB-231) were embedded in alginate-collagen beads, which were further embedded in an adipose-derived stem cell (ASC)-containing matrix obtained from post-menopausal women of healthy or overweight BMI. Following observations over 14 days, breast cancer cells co-cultured with high BMI adipocytes demonstrated migration and proliferation across the microbeads into the surrounding matrix whilst breast cancer cells co-cultured with healthy BMI adipocytes remained confined to the alginate-collagen microbeads—particularly with MCF-7 cells. Additionally, cancer cells were positive for epithelial to mesenchymal transition (EMT) markers E-cadherin, β-catenin and tenascin C, indicative of increased cellular migration.

4.2 Skin

As skin is a dynamic and complex structure, skin models must also be suitably complex and 3D bioprinting is the favourable method for attaining such structures. During the natural wound healing phases following trauma to the skin, fibroblasts are known to migrate and remodel the underlying dermis whilst endothelial cells traverse the matrix forming new vasculature and keratinocytes close over at the wound surface during a series of intricate and orchestrated events. There have been several examples of 3D bioprinting skin tissue, with the vast majority of attempts representing the epidermal and dermal portions of skin, whilst more elaborate models also incorporate a hypodermal or subcutaneous layer. 3D bioprinted skin migratory systems are often fabricated using robotic extrusion of bioinks based on the thickness required to generate clinically relevant models and other bioprinting techniques are less capable of doing so. Models that have been developed in order to investigate 3D bioprinting skin and subsequently analyse cellular behaviour such as migration include the production of a multi-compartment skin analogue complete with epidermis, dual compartment dermis (papillary and reticular dermis) and hypodermis using suspended layer additive manufacturing (SLAM). Biopolymers including pectin and collagen containing a tri-culture of keratinocytes, human dermal fibroblasts and adipose mesenchymal stem cells were spatially allocated to dermal compartments (Fig. 6) using robotic extrusion 3D bioprinting followed by implantation into a simulated ex vivo porcine wound model. Integration between the implant and surrounding porcine tissue was documented with migration of associated cells between the implant and porcine matrices [3].

A highly interesting approach to bioprinting skin, and one that is becoming increasingly commonplace in research towards translation to the clinic is in situ bioprinting directly onto patient wounds. In combination with integrated wound imaging technology, research into the precise delivery of autologous or allogenic dermal fibroblasts and epidermal keratinocytes directly to the injured area has been investigated towards replacement of the skin's layered structure [27]. Bioprinted cells of these types within a fibrin-collagen hydrogel carrier have been seen to promote and accelerate wound closure and epithelialisation by enabling proliferation and migration within the wound environment following deposition using an inkjet-based bioprinter. This work demonstrates potential applications to not only treat full-thickness wounds but also towards utilisation within burns, ulcers and even deep tissue injuries.

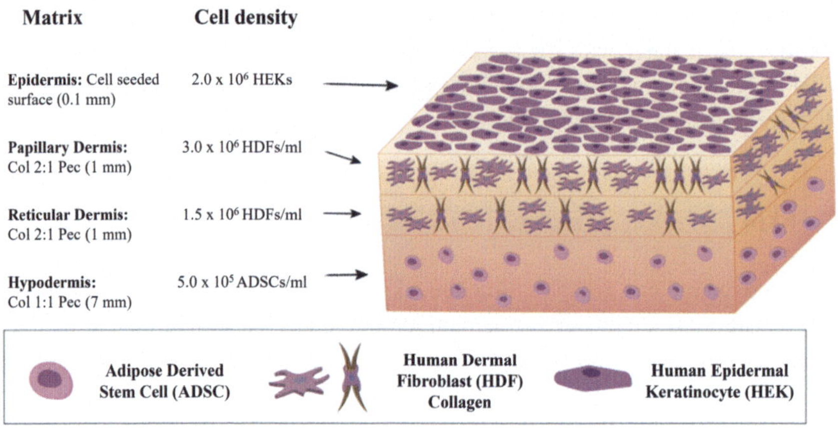

Fig. 6 Illustration of a tri-layer skin model featuring a hypodermis, dual compartment dermis and epidermis. Adipocytes differentiated from adipose-derived mesenchymal stem cells (ADSC), human dermal fibroblasts (HDF) unattached and attached to collagen and human epidermal keratinocytes (HEK) are specifically localised within the layered scaffold and encapsulated within a graded collagen (Col) / pectin (Pec) matrix blend in an arrangement that is synonymous to native tissue

4.3 Muscle

During embryonic myogenesis, skeletal myogenic progenitors migrate to generate the anlagen of limbs, diaphragm and tongue by means of directional cell migration. Following post-natal injury to skeletal muscle, however, regeneration of muscle fibres is achieved through recruitment of muscle satellite stem cells. Intramuscular or intravascular transplant cell therapy via allogenic or autologous cell injection may be adopted to aid in muscle tissue repair, although patients suffering from degenerative diseases affecting multiple large muscles may not always achieve functional restoration [28]. The predominant reason for this failure is due to the lack of migration of donor cells, and therefore much time is being committed to researching the mechanisms of myogenic stem / progenitor cell migration.

One such study was performed to assess myoblast migration by micropatterning GelMA using robotic microextrusion [29]. C2C13 mouse myoblasts were seeded onto bioprinted GelMA in parallel and perpendicular lines and fingerprint patterns to assess directed collective cell migration. Cell migration was observed as a function of distance travelled, migration velocity and also directionality. In extension, myotubes were generated and demonstrated contractility upon electrical stimulation. The physiological utility of this technology was further demonstrated through the formation of radial and circular muscle fibres analogous to the human iris that are able to functionally contract and control pupil aperture in response to external stimuli. True-to-size bioprinted discs were fashioned to represent the iris and the orientation angle distribution of bioengineered radial muscle fibres was highly aligned to those within in vivo environments.

Towards determining the feasibility of treating critical-sized human skeletal muscle injuries following the establishment of subcutaneously implanted mouse muscle in rats, research has focused on generating human-scale skeletal muscle constructs using muscle progenitor cells isolated from human muscle tissue biopsies [30]. These cells were suspended in fibrinogen, gelatin, hyaluronic acid and glycerol to form the bioink, followed by printing using pneumatic extrusion and defined culture medium was subsequently used to induce myofibre formation. The resultant printed muscle tissue mimic had characteristics of native muscle, composed of multi-layered myofibres following cell migration and assembly into spatially organised bundles. These were then successfully implanted within in vivo rat muscle defects where subsequent vascularisation and muscle function restoration were observed. Here it has been proven that awareness of native tissue architecture can be used in 3D bioprinting sophisticated tissue replicas that allow monitoring and manipulation of cell migration towards a multitude of different applications.

Take-Home Message

- 3D bioprinting allows the careful deposition of cells, biomaterials and chemical cues in graduated layers for the fabrication of 3D tissue mimics.
- The major steps towards 3D bioprinting tissue models include model design, print design, model fabrication, model propagation and application, although these processes can be tailored depending on the 3D part being generated.
- Laser-induced forward transfer, inkjet printing and robotic dispensing are the predominant strategies of 3D bioprinting, and the selection of a particular technology depends on the required resolution of the construct, fabrication speed, biopolymer viscosity and gelation speed and also associated costs of the equipment.
- 3D bioprinting has been utilised as a technique for establishing 3D migratory microenvironments, with examples of cancer, skin and muscle replicas, towards studying biological pathways in health and disease, the use of drugs or other therapeutic strategies and the assessment of construct suitability for implantation.
- 3D bioprinting continues to show promise in future applications of manufacturing 3D migratory environments and the limitations that once stood in the way of these progressions are gradually being overcome.

Acknowledgements Some of the materials in Figure 5 were obtained from Servier Medical Art (http://smart.servier.com/), licensed under a Creative Common Attribution 3.0 Generic License (https://creativecommons.org/licenses/by/3.0/).

References

1. Papaioannou T, Manolesou D, Dimakakos E, Tsoucalas G, Vavuranakis M, Tousoulis D. 3D bioprinting methods and techniques: applications on artificial blood vessel fabrication. Acta Cardiol Sin. 2019;35(3):284–9.
2. Sbirkov Y, Molander D, Milet C, Bodurov I, Atanasov B, Penkov R, et al. A colorectal cancer 3D bioprinting workflow as a platform for disease modeling and chemotherapeutic screening. Front Bioeng Biotechnol. 2021;9
3. Moakes R, Senior J, Robinson T, Chipara M, Atansov A, Naylor A, et al. A suspended layer additive manufacturing approach to the bioprinting of tri-layered skin equivalents. APL Bioeng. 2021;5(4):046103.
4. Pai R, Anupama Sekar J, Ajit S, Velayudhan S, Kasoju N, Anil Kumar PR. Three-dimensional bioprinting of tissues and organs. In: Mohanan PV, editor. Biomedical product and materials evaluation. London: Elsevier; 2022. p. 135–50.
5. Park K, Okano T, Ottenbrite R, Peppas N. Biomedical applications of hydrogels handbook. 1st ed. New York: Springer; 2010.
6. Li H, Tan C, Li L. Review of 3D printable hydrogels and constructs. Mater Design. 2018;159:20–38.
7. O'Brien F. Biomaterials & scaffolds for tissue engineering. Mater Today. 2011;14(3):88–95.
8. Nam S, Ricles L, Suggs L, Emelianov S. Imaging strategies for tissue engineering applications. Tissue Eng B Rev. 2014;21(1):88–102.
9. Ozbolat I, Khoda A, Marchany M, Gardella J, Koc B. Hybrid tissue scaffolds for controlled release applications. Virtual Phys Prototyp. 2012;7(1):37–47.
10. Kai C, Jacob G, Mei T. Interface between CAD and rapid prototyping systems. Part 1: A study of existing interfaces. Int J Adv Manuf Technol. 1997;13(8):566–70.
11. Rajapakse C, Peng M, Lee D, Shinde R, Khalid A, Hong A et al. Rapid 3D bioprinting from medical images: an application to bone scaffolding. In: Medical imaging 2018: imaging informatics for healthcare, research, and applications, 10579. 2018.
12. Gungor-Ozkerim P, Inci I, Zhang Y, Khademhosseini A, Dokmeci M. Bioinks for 3D bioprinting: an overview. Biomater Sci. 2018;6(5):915–46.
13. Peltola S, Melchels F, Grijpma D, Kellomäki M. A review of rapid prototyping techniques for tissue engineering purposes. Ann Med. 2009;40(4):268–80.
14. Smith A, Senior J. Alginate hydrogels with tuneable properties. Adv Biochem Eng Biotechnol. 2021;178:37–61.
15. Senior J, Cooke M, Grover L, Smith A. Fabrication of complex hydrogel structures using suspended layer additive manufacturing (SLAM). Adv Funct Mater. 2019;29(49):1904845.
16. Scott M, Nguyen V, Levi B, James A. Current methods of adipogenic differentiation of mesenchymal stem cells. Stem Cells Dev. 2011;20(10):1793–804.
17. Langenbach F, Handschel J. Effects of dexamethasone, ascorbic acid and β-glycerophosphate on the osteogenic differentiation of stem cells in vitro. Stem Cell Res Ther. 2013;4(5)
18. Benhardt H, Cosgriff-Hernandez E. The role of mechanical loading in ligament tissue engineering. Tissue Eng B Rev. 2009;15(4):467–75.
19. Guillemot F, Souquet A, Catros S, Guillotin B. Laser-assisted cell printing: principle, physical parameters versus cell fate and perspectives in tissue engineering. Nanomedicine. 2010;5(3):507–15.
20. Malda J, Visser J, Melchels F, Jüngst T, Hennink W, Dhert W, et al. 25th anniversary article: engineering hydrogels for biofabrication. Adv Mater. 2013;25(36):5011–28.
21. Kim J, Choi J, Kim B, Chan Choi Y, Cho Y. Piezoelectric inkjet printing of polymers: Stem cell patterning on polymer substrates. Polymer. 2010;51(10):2147–54.

22. Agarwal S, Saha S, Balla V, Pal A, Barui A, Bodhak S. Current developments in 3D bioprinting for tissue and organ regeneration—a review. Front Mech Eng. 2020;6
23. Murphy S, Atala A. 3D bioprinting of tissues and organs. Nat Biotechnol. 2014;32(8):773–85.
24. Jung M, Skhinas J, Du E, Tolentino M, Utama R, Engel M, et al. A high-throughput 3D bioprinted cancer cell migration and invasion model with versatile and broad biological applicability. Biorxiv. 2021;20 https://doi.org/10.1101/2021.12.28.474387.
25. Ning L, Shim J, Tomov M, Liu R, Mehta R, Mingee A, et al. A 3D bioprinted in vitro model of neuroblastoma recapitulates dynamic tumor-endothelial cell interactions contributing to solid tumor aggressive behavior. Adv Sci. 2022;9(23):2200244.
26. Vinson B, Phamduy T, Shipman J, Riggs B, Strong A, Sklare S, et al. Laser direct-write based fabrication of a spatially-defined, biomimetic construct as a potential model for breast cancer cell invasion into adipose tissue. Biofabrication. 2017;9(2):025013.
27. Albanna M, Binder KW, Murphy SV, et al. *In situ* bioprinting of autologous skin cells accelerates wound healing of extensive excisional full-thickness wounds. Sci Rep. 2019;9:1856.
28. Choi S, Ferrari G, Tedesco F. Cellular dynamics of myogenic cell migration: molecular mechanisms and implications for skeletal muscle cell therapies. EMBO Mol Med. 2020;12(12)
29. Du W, Hong S, Scapin G, Goulard M, Shah D. Directed collective cell migration using three-dimensional bioprinted micropatterns on thermoresponsive surfaces for myotube formation. ACS Biomater Sci Eng. 2019;5(8):3935–43.
30. Kim J, Seol YJ, Ko I, et al. 3D bioprinted human skeletal muscle constructs for muscle function restoration. Sci Rep. 2018;8:12307.

Using High-Resolution Imaging to Investigate Cell Migration In Vitro

Michelle Peckham

What You Will Learn in This Chapter

A wide array of fluorescence microscopy approaches is used to image cultured cells in vitro. This type of approach is widely used for imaging the behaviour of specific proteins within cells to understand their dynamic behaviour and location. Within the last 20 years, multiple different forms of fluorescent microscopies have been developed to achieve this. With the vast array of different approaches available, it can be difficult to know where to start. This chapter provides a simple introduction to high-resolution fluorescence microscopy approaches to image cells in vitro for beginners, how to choose the right approach and how to optimise your imaging.

1 Background to Fluorescence Microscopy: Fluorescent Proteins and Super-Resolution Imaging

A widely used approach to locating proteins in cells is to express the cDNA for the protein fused to a fluorescent protein (FP) such as eGFP (enhanced green fluorescent protein) in cells and perform live cell imaging. A more sophisticated approach is to use CRISPR or a related approach to label the endogenous protein [1]. This has the advantage that

M. Peckham (✉)
School of Molecular and Cellular Biology, Faculty of Biological Sciences, University of Leeds, Leeds, UK
e-mail: m.peckham@leeds.ac.uk

A. Brüning-Richardson, S. Knipp (eds.), *Cell Migration in Development, Health and Disease*, Learning Materials in Biosciences,
https://doi.org/10.1007/978-3-031-64532-7_9

the fluorescently labelled protein of interest should be expressed at endogenous levels. However, this is a more challenging technique. In addition, small tags (Halo, SNAP) that can then be labelled with cell-permeable dyes and other approaches are used [2]. Other forms of FPs are continually being developed [3], each with specific advantages and properties.

Transient expression of eGFP fused to the protein of interest is a straightforward way to initially assess if the fluorescently tagged protein localises correctly in live cells. Its localisation should be similar to that observed when using antibodies to determine the localisation of the protein of interest. It also enables the researcher to quickly test if fusing eGFP on the N- or C-terminus works best, or if one or the other can compromise the localisation of the protein. For example, α-actinin in which eGFP is fused to the C-terminus localises correctly, but α-actinin in which eGFP is fused to the N-terminus does not [4]. Although many FP variants have been described since GFP was first used as a reporter many years ago [5], it still performs really well, and we routinely use the enhanced version of GFP (eGFP; [6]) for initial assessments of FP fusions in cells. If starting from scratch, then we recommend trying StayGold [7].

Whilst eGFP is useful for many experiments, we recommend using FPbase [8], to choose suitable FPs for specific experiments, especially when combining multiple FPs. The excitation and emission spectra need to be carefully checked to make sure that users have the appropriate light sources (lasers) to excite the FP at the correct wavelength, and appropriate filters to image the emitted fluorescence. It is important to check the overlap of the emission spectra when using multiple fluorophores, to make sure they are well separated. FPbase is useful in being able to select the brightest fluorophores, as it provides values for the theoretical brightness (linked to the quantum efficiency (QE) for each FP). However, it is always best to test specific FPs in your specific experimental system and compare them to eGFP. Finally, FPbase also indicates if a specific FP tends to dimerise, which typically should be avoided when using FPs to tag proteins of interest.

Resolution and Super Resolution The resolution of a widefield fluorescence microscope is approximately half the wavelength of the emitted light being imaged (e.g. ~250 nm for a wavelength of exciting light of 488 nm, the wavelength used to excite eGFP (Fig. 1a, b)). What this means, practically, is that two fluorescent objects in an image that are less than 250 nm apart, cannot be resolved as two separate objects in the final image, but will be observed as one single blurry fluorescent object. A wide range of novel microscopical approaches have been developed to overcome this problem with resolution and in some cases, FPs have played a role in developing some of these.

The main reason why widefield light microscopes are limited to a resolution of ~250 nm is due to diffraction (bending) of light as it passes through the sample and objective lens. Simply put, small objects diffract the light to wider angles than larger objects (Fig. 1c). The result is that a small point object in a sample is imaged as a blurry spot. In XY, this blurry spot is also referred to as the Airy disc (Fig. 1d). In XYZ, the blurry spot is also referred to as the point spread function. The size of the numerical aperture (NA) of the

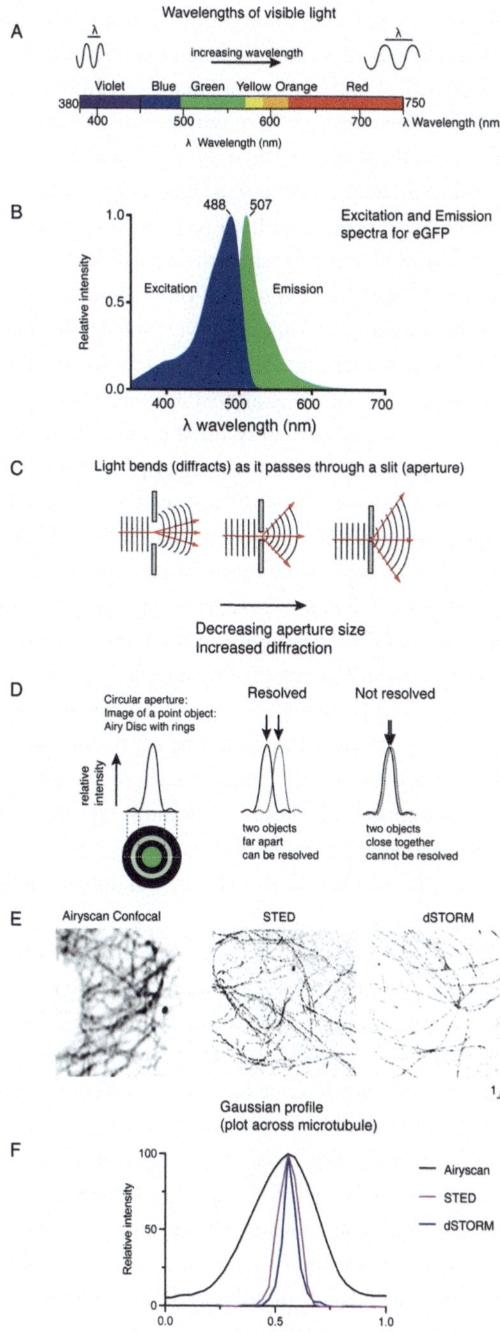

Fig. 1 Resolution and super resolution. (**a**) Wavelengths of visible light from short (~350 nm) to long (~750 nm). (**b**) The wavelengths of light used to excite eGFP (blue: peak at 488 nm) and the light emitted from excited eGFP (green: peak at 507 nm). (**c**) As light waves pass through a slit they bend (diffract). The narrower the slit, the more they diffract. (**d**) In microscopes apertures are circular rather than slits. The appearance of the final diffracted spot of light is shown (Airy disc), with its Gaussian profile (from line drawn across the centre of the disc). When two spots are far apart, the two spots are resolvable, but when they are close together they are not (as illustrated here by the two profiles for each spot). (**e, f**) Example images from Airyscan confocal, STED and dSTORM of tubulin in microtubules, labelled using an antibody and an appropriate fluorophore (on the secondary antibody), together with an example profile for a region of microtubule using each of the three techniques. The improved resolution is linked to the narrower profile

lens limits how large an angle of light it can collect from any object; the larger the NA, the greater the amount of light it can collect. However, the highest NA for most objective lenses is 1.4. (Incidentally, the image resolution for any lens with an NA of 1.4 is the same, even when one lens has a magnification of ×40 and the other of ×100). Diffraction also depends on the wavelength of the light, with longer wavelengths (towards the red end of the spectrum) diffracting light to a larger angle than shorter wavelengths, and on the refractive index of the imaging medium (e.g. oil immersion vs water or air immersion lenses). The physics behind image formation was solved in 1873 by Ernst Abbe. Using his equation ($d = \lambda/2$ N.A.), the best resolution is approximately half the wavelength of the light used to image the sample. This effectively means that two objects closer than ~250 nm cannot be resolved as two separate spots (Fig. 1d). Incidentally, this is the resolution in XY (lateral). Resolution in Z (axial) is almost 3 times lower.

The easiest and probably most accessible approach for super-resolution imaging is to use an enhanced form of microscopy provided on some Zeiss confocal microscopes, called 'Airyscan'. Confocal microscopy uses a point source of light (laser scanning) to excite the sample, and the emitted fluorescent light has to pass through a pinhole before it is detected. This approach rejects the out-of-focus fluorescence resulting in sharper images. However, it also results in a small increase in lateral resolution [9]. Reducing the size of the pinhole would increase this resolution further, but the resulting image would be very dim, as most of the light would be rejected by the pinhole, and not be recorded on the detector. Airyscan overcomes this by using a 32-element array, in which each element has an effective pinhole of 0.2. The 32 images recorded by each element are combined and processed to generate a final image with ~1.7 fold increase in lateral resolution, and near normal intensity. As this is a commercial solution, commonly included with a confocal system, this approach is widely available, and straightforward to implement. Moreover, it works with all standard dyes and fluorophores, as used in standard widefield or confocal imaging, as illustrated here for tubulin (Fig. 1e, f).

An alternative approach to Airyscan, with similar improvement in resolution is structured illumination (SIM). Many imaging facilities are now equipped with some sort of SIM microscope. In simple terms, SIM uses a fine pattern of light to illuminate the sample [10]. This interacts with the fine pattern within the sample itself (of a similar frequency) generating an interference pattern that effectively reduces the spread of the diffracted light by half, thus doubling resolution. To be able to arithmetically separate the real image from the pattern, typically, 9 images need to be obtained, in which the pattern is rotated between each one. These images are then processed via Fourier transformation and then back into real space to generate an image of the original sample at twice the resolution. The original SIM microscope used a physical grating, but new and commercial versions now use spatial light modulators that generate patterns more quickly. This approach has been further developed to generate a lattice pattern rather than stripes in the Zeiss Elyra SIM and a newer implementation (SIM^2).

Our laboratory has an iSIM microscope, which works in a slightly different way [11, 12]. The sample is illuminated by a pattern of light that is generated by a lens array, in which a laser is rapidly scanned over each of the multiple lenses simultaneously. The emitted light passes through a pinhole array, and then through a second lens array that contracts each spot by ×2 before being detected by a scientific CMOS camera. The final obtainable resolution (after a subsequent deconvolution step) is ~ ×2-fold, similar to that of Airyscan. A commercial version of this microscope has also been developed (Visitek). The key advantage of this approach is that the improvement in resolution is achieved through the optical design, and thus it does not require any image processing. Further advances for iSIM have included improved deconvolution using machine learning approaches [13].

Two further main types of super-resolution include STED (stimulated emission depletion microscopy; [14, 15]) and single molecule localisation microscopies (SMLM). In simple terms, STED is a modified form of confocal microscopy. In addition to illuminating the sample with a scanning point source of light, a second 'depletion' laser, which is donut shaped is also scanned over the sample at the same time. This switches off the fluorescence of the fluorophores within the donut, effectively shrinking the size of the spot that can emit fluorescence. The improvement in resolution depends on the square root of the intensity of the depletion laser, which can be varied. Typically, this is around 50 nm in XY, on a commercial microscope (STEDYCON, supplied by Abberior). This resolution improvement is entirely optical, and does not require any image processing, but the final images do benefit from deconvolution. However, unless the microscope also has a depletion laser for the axial direction, resolution in Z is no better than confocal. In addition, as this approach depends on the interaction of the depletion laser with the dye, specific dyes need to be used for the best performance in STED (see https://abberior.rocks/dyes-labels/) and eGFP constructs do not work very well. In the example shown here (Fig. 1e, f), a 'star-red' dye was used. More typically, constructs with Halo-Tags are used, with specific STED dyes designed for live cell imaging using STED (https://abberior.rocks/dyes-labels/abberior-live/).

Single-molecule localisation microscopies include PALM (photoactivatable light microscopy) and dSTORM (direct Stochastic Optical Reconstruction Microscopy). These techniques can provide a resolution of ~20 nm. PALM was first demonstrated using photoactivatable GFP (paGFP; [16]). paGFP is not fluorescent until it is excited by short wavelength (405 nm) light. By controlling the levels of illumination, only a small subset of molecules become fluorescent at any one time, and these tend to be well separated spatially (e.g. separated by more than 250 nm). Thus, these molecules can be resolved, and the position of each molecule is identified accurately. Eventually, the molecules bleach, and small numbers of new fluorescent molecules appear. By imaging small numbers of fluorescent molecules that appear randomly from frame to frame, over many, many frames (~5000–10,000), the positions of all the molecules are then identified and used to reconstruct the final image.

For PALM to work well, the rate at which the FP interconverts between the dark and fluorescent states must be low compared to the rate of conversion initiated by light activation. In addition, the fluorescence of the dark (inactive) FP must be low compared to that of the active (fluorescent) molecule, a property known as the contrast ratio [17]. paGFP turns out not to be the best FP for PALM as it has a relatively low contrast ratio, whilst that for pamCherry is 400× better [18]. The drawback of both these FPs is that transiently transfected cells are not fluorescent, and thus cannot be identified until they are illuminated with 405 nm wavelength of light.

Currently, the most commonly used FP used in PALM is derived from EosFP [19, 20]. EosFP is a type of fluorophore that photoconverts from a green to a red fluorescent state, when illuminated with 405 nm wavelength of light, and it has a high contrast ratio [18]. Transfected cells can thus be identified before imaging them using PALM, as they fluoresce when excited with 488 nm wavelength light. EosFP has since been engineered to be monomeric [21] and mEos3.2 is now commonly used. mEosFP is also useful in analysing protein dynamics, in a technique analogous to FRAP (fluorescence recovery after photobleaching). For example, we have used this FP to analyse the turnover of α-actinin-2 (ACTN2) in the Z-discs of cardiomyocytes [22].

STORM, originally developed by Xiowei Zhuang at about the same time as PALM [23], used antibodies labelled with two fluorescent dyes, in which one of the two dyes is bleached, a small number of the bleached molecules is switched back on by exciting the second dye, these are well separated spatially, and their positions accurately identified. It was later discovered that adding β-mercaptoethanol (which encourages some dyes to enter a dark 'triplet' state) and oxygen scavengers (which slow the re-formation of the ground state) to the imaging buffer, keeps most of the dye molecules in a dark state. A small number of dye molecules can then re-enter the ground state when illuminated with short wavelength light (405 nm) and can then be excited and fluoresce following illumination with the appropriate wavelength of light for that dye, and positions identified accurately [24, 25]. This approach is known as dSTORM (directSTORM) and is the most commonly used approach for SMLM-based super resolution for fixed cells. However, the best dye for dSTORM is Alexa 647 (as shown in the example here: Fig. 1e, f) [26].

All of the SMLM techniques are challenging to use in live cell imaging as they all require several thousand frames to be collected to reconstruct the final image. Using antibodies for visualising proteins within living cells is clearly challenging, as it relies on a good method to deliver the labelled antibodies, and a benign effect of the antibodies once inside the cells. Moreover, the challenge is to be able to track single fluorophores, and yet for the fluorophores to be well enough spatially separated to reconstruct a final image at high resolution. However, photoactivatable or switchable fluorescent proteins have been used in live cell imaging, a technique named sptPALM. sptPALM was first described using EosFP tagged molecules, in which small subsets of molecules were switched from green to red fluorescence, and the red fluorescent molecules were tracked and their positions identified [27]. However, this does require illuminating cells with 405 nm wavelength light, to induce the switch from green to red, which tends to be more cytotoxic for cells than longer wavelengths of light. Despite this, sptPALM is likely to become more popular in the future, with improvements in illumination technologies, and the greater need for imaging in live cells.

Finally, whilst not strictly a super-resolution imaging approach, light sheet imaging, and specifically lattice light sheet imaging is starting to become more widely used. In widefield, confocal and super-resolution imaging, only a single XY plane is imaged at any one time, but the entire sample in the field of view is illuminated. Fluorophores above and below the plane of imaging are also excited and fluoresce but are not imaged. This can lead to photobleaching and phototoxicity, which ideally should be avoided in live cell imaging. In light sheet (selective plane illumination) microscopy, the sample is illuminated by a thin sheet of light, which is the same as the plane that is being imaged (focal plane) [28]. Fluorophores outside this plane of light are not excited, thus, there is no out-of-focus fluorescence, which improves axial resolution, and photobleaching and toxicity effects are reduced.

There are many forms of light sheet microscopes, and these mostly differ in the width of the light sheet that they are able to generate. Ideally, the width of the light sheet should be much narrower than the sample being imaged. Moreover, the lateral width of the light sheet also determines the axial resolution. Lattice light sheet (LLS) generates an ultrathin light sheet that is suited to imaging single cells, spheroids, small organoids and small organisms such as *Drosophila* embryos [29]. The lattice light sheet uses a specific form of illumination to generate a narrow light sheet, down to ~500 nm in width (and 12 μm long), much smaller than the dimensions of a single cell (around 20–50 μm in width). In LLS, the light sheet is rapidly scanned across the cell (or field of view) to generate the image and an entire cell volume can be scanned in less than a second. The resolution obtainable is about 290 nm in XY and as little as 450 nm in Z, and thus the resolution in XY is more similar to that in Z compared to confocal. The original version of LLS uses dipping lens and an upright microscope configuration (objective is above the sample), to avoid optical aberrations introduced due to a change in refractive index between water and glass. A commercial version of LLS developed by Zeiss is now available, which uses an inverted configuration, and overcomes the optical aberrations through the use of an Alvarez plate. Experimentally, light sheet is also being developed for imaging single molecules in living cells [30].

Overall, there is a wide array of different imaging approaches to choose from. Which microscope you do eventually choose is going to depend on if it is easily available, if it is equipped for live cell imaging, and if it is able to image at the resolution and speeds that you need for your samples? Typically, these advanced microscopes are found in facilities and looked after by well-trained staff who can advise and help you with your imaging. For more broader reading, two recent reviews summarise a variety of super-resolution and other imaging approaches in more detail [31, 32] and can help you to decide which type of imaging you need for your specific experiments.

2 Preparation and Transfection of Cells for Live Cell Fluorescent Imaging

Most cultured cells can be used in live cell fluorescent imaging, for example using eGFP-tagged proteins. To test out a new eGFP-tagged protein, we typically use a transient transfection method, and easy-to-transfect cultured cell lines (e.g. HEK-293 (Human embryonic kidney 293) or COS-7 (African green monkey kidney fibroblast-like)). However, whilst easy to transfect, the cytoskeleton and general morphology of these cells can appear less well-developed than other cell lines, and we find they can be less suited for specific experiments designed to interrogate cell migration and the cytoskeleton. Therefore, we choose other cell types depending on the experiment we want to undertake. For example, B-16-f10 cells (a mouse melanoma cell line) are useful for research into actin dynamics in the lamellipodium [33], muscle myoblasts key for research into myoblast fusion [34, 35] and PC-3 (human prostate cancer) cells are useful for investigations into how myosin isoforms specifically affect cytoskeletal organisation and cell migration [36]. All of these cell lines and many others are available from ATCC (American Tissue Type Collection).

DNA We occasionally use DNA expression plasmids purified by miniprep for a quick test, but transfection rates are lower. More routinely, we use maxiprepped DNA, purified using a Qiagen Plasmid kit, following the manufacturer's instructions. The purified DNA is sterile as a result of the ethanol or isopropanol final washes. We measure the concentration using a nanodrop at a wavelength of 260 nm. Measuring the 260/280 ratio can be helpful in checking purity. We store the purified DNA in sterile TE (Tris-EDTA) or distilled water at −20 °C, aiming for a final concentration of ~1 mg/ml.

Glass Culture Surfaces and Methods for Cleaning For live cell imaging, we typically use glass-bottomed dishes. Plastic-bottomed dishes are highly autofluorescent, are poor optically and should be avoided. Typically, we use glass-bottomed dishes from IBIDI (Cat.No:81218-200) or equivalent, which have a #1.5 glass coverslip on the bottom of the dish (170 μm thick). This thickness is important for obtaining the best images, as most microscope objectives are optimised for this thickness. We occasionally use glass-bottomed multiwell dishes as an alternative. The dishes may need pretreating to enable the cells to stick to the surface with poly-L-lysine, laminin or other attachment factor. This needs to be tested for each cell type used.

Alternatively, for fixing and staining transfected cells, we use #1.5 glass coverslips, either 13 mm in diameter or square (20 × 20 mm). These complimentary experiments on fixed cells help us interpret protein localisation in live cells. In this case, it is critical to make sure that the coverslips are washed first. Coverslips straight out of the box are dusty, dust is fluorescent and will impact on image quality (Sterile glass-bottomed dishes are usually clean enough to use straight away). There are many different ways that the coverslips can be washed. The simplest is to wash them in hot water with a small amount of liquid detergent (washing up liquid). We put the whole pack (~100 coverslips) into a 50-ml falcon tube filled with hot water and a small squirt of detergent, continuously invert the tube for 1–2 min, pour away the detergent mix, wash ×5 (5 complete changes) in hot water, then ×5 in cold water, ×5 in distilled water, and then store the coverslips in 70% ethanol. Alternatively, the coverslips can be washed in a similar way, using 1 M HCL instead of detergent, followed by ×10 washes in distilled water (complete volume changes) before storing in 70% ethanol.

For more challenging super-resolution applications, specifically dSTORM of fixed cells, we use an extensive acid wash. We use #1.5 glass square (20 × 20 mm) coverslips, which we place in a coverslip minirack (e.g. C-14784 from Thermofisher: acid resistant). We wear gloves to prevent contact with the caustic chemicals used with the skin. In the fume hood, we first combine 125 ml distilled water, 25 ml ammonium hydroxide and 25 ml hydrogen peroxide, measured out using a 100-ml measuring cylinder, in a 250-ml glass beaker. The beaker is then placed on a hot stirrer set to 80 °C, with a stirrer in the bottom. The coverslips in their rack are placed into the beaker such that the coverslips are submerged and left in the cleaning solution for at least 3 h, whilst stirring. At the end of this incubation period, the rack and coverslips are transferred into a fresh 250 ml beaker with at least 150 ml of clean distilled water, and then washed a further 9 times, with fresh 150 ml of distilled water each time. The rack is then transferred into a beaker of methanol, or 70% ethanol. At this point, single coverslips can be removed using tweezers, and sterilised by briefly passing the coverslip through the tip of a Bunsen flame, before placing it into a clean dry staining rack, in a suitable container. Once this is done for all the coverslips, they can be covered and stored dry. However, care is needed here as it is very easy for the coverslips to crack in the heat. Cells are grown on these coverslips, pre-coated with attachment factors as necessary. They are then fixed, stained, and inverted onto microscope slides with a single glass cavity which we fill with dSTORM imaging buffer and seal, ready for imaging.

Transfection of Cells The day before transfection, the cells are passaged, and plated out at the required cell density on imaging dishes or coverslips, coated as required if necessary. To do this, first the growth medium is removed from the cells, and sufficient TrypLE to cover the growing surface (e.g. ~1 ml for a 25-cm^2 flask) is added, the flask is placed back in the incubator (37 °C) for 1–2 min, then observed using a phase contrast microscope (Nikon, inverted, ×10 objective) to check for cell detachment. Next, fresh growth medium (e.g. DMEM, 10% serum, pen/strep, for HEK293 cells: 10 ml for a 25-cm^2 flask) is added to the flask to inactivate the TrypLE, the medium containing the detached cells is removed, added to a 15-ml sterile falcon tube, and centrifuged for 5 min at 1000 rpm to pellet the cells. Almost all of the growth medium is aspirated, the cell pellet is loosened by gently flicking the bottom of the tube and then resuspended in 1 ml of growth medium. A small (~10 μl) aliquot of cell suspension is removed and used to count the cell density using a Neubauer haemocytometer. The cell concentration is adjusted to ~5 × 10^5 cells/ml, by adding additional growth medium as required. 400 μl (~20,000 cells) is added to the glass dish, which has a growth area of 3.14 cm^2. This results in ~60,000 cells/cm^2 (or 1×10^6 μm^2), ~50–80% confluent. The cells are allowed to attach for 5–10 min, then further growth medium is added to a final volume of ~2 ml. Alternatively, we add the cells in small 50 μl drops across the dish, allow the cells to attach for 5–10 min, and then add 2 ml of growth medium. This can be a useful way to obtain a variable cell density across the dish.

To transfect the cells, we typically use Fugene 6, but there are many alternatives, and we do occasionally use nucleofection (electroporation). Ideally, different transfection methods should be tested for each new cell line. All of the following steps are performed in a tissue culture hood under sterile conditions. FuGENE 6 is stored in the fridge, until use, and then allowed to come up to room temperature for about 15 min prior to transfection. We normally use 3:1 of reagent to DNA (6 μl fugene, 2 μg DNA). Briefly, 6 μl of FuGENE 6 is directly pipetted into ~92 μl serum-free medium (DMEM, without antibiotics) in a 1.5-ml sterile Eppendorf tube and mixed by flicking the tube briefly. The mixture is incubated for 5 min, and then 2 μg of DNA plasmid (in 2 μl) is added, such that the final volume is 100 μl (initial volume of DMEM needs to be adjusted for volume of DNA to be added). The solution is mixed by briefly flicking the tube and left to incubate for a minimum of 15 min at room temperature. This mixture is then added dropwise to the cells, and the dish is placed back in the incubator. The cells are then ready to image 24–48 h later.

In addition to transfection, there are many small-molecule fluorescent probes that can be used in imaging, which can just be added to live cells and then the cells can be imaged after a short incubation period (~1 h). We have used SPY-DNA, SPY-F-actin (fluorescent latrunculin) and SPY-tubulin (fluorescent taxol) [37]. These are now all commercially available from Spirochrome.com and all seem to work reasonably well. There are also fluorescent probes for organelles such as mitochondria (Mito-Dye (Biotium) and MitoTracker (Thermofisher) and lysosomes as well as membrane mitochondrial and endoplasmic reticulum membrane tension reporters (also commercially available from Spirochrome). As with everything, these should all be tested in your system to see how well they work for you. Companies are usually willing to provide test samples for free, to allow you to do this.

Spheroids Occasionally, we make small spheroids using transfected cells. We typically use permanently transfected cells, but transiently transfected cells can also be used. To generate permanently transfected cells, the method of transfection is the same as described above, but the cells are plated into a well of a 6-well culture plate for transfection. The day after transfection, we passage the cells into a fresh well and add fresh growth medium supplemented with G418 (geneticin: an aminoglycoside antibiotic). The majority of eGFP (or other) expression plasmids contain a resistance gene that allows cells to survive in the presence of G418. We determine the concentration of G418 required for each cell line beforehand, such that >90% of untransfected cells are dead within 5 days. After 7–10 days of selection, the cells can be kept as a mixed line of eGFP-construct expressing cells, or fluorescent cells can be isolated by FACS (flow-assisted cytometric sorting), plated out at ~1 cell per well in a 96-well dish and allowed to grow up to obtain unique clones. Alternatively, at the point of first adding G418, the cells can be counted and plated at low density (~1 × 10^4 cells/ml in 20 ml, on a 25-cm^2 dish) for clonal selection, and then individual clones selected 7–10 days later.

To make spheroids, we use two main approaches. In the first, we use non-adherent dishes (e.g. Nuclon Sphera multiwell dish). Cells are trypsinised and counted as before, and diluted to 5 × 10^3 cells/ml. The appropriate volume is then added to each well of the non-adherent plate (e.g. 200 μl for a 96-well, 500 μl for a 24-well plate and so on). Spheroids are ready to image about 3–4 days later. Alternatively, we use a simple hanging drop approach. About 10 ml of sterile PBS is added to the bottom of a 100-mm diameter plastic culture dish, cells are diluted to ~1–2 × 10^3 cells/ml and multiple drops (50 μl each) of the cell suspension (~800 cells) are pipetted onto the upturned lid, such that the drops are well spaced. The lid is then carefully inverted and placed on top of the culture dish. The cells form spheroids within ~2 days.

3 Preparation and Transfection of Cells for Live Cell Imaging

24–48 hours after transfection, we inspect the cells on a widefield inverted fluorescent microscope to check that a reasonable number (>20%) of cells are expressing the eGFP (or other FP) construct of choice. We then routinely add 10–20 mM sterile HEPES buffer (from a 1 M stock) to the medium prior to imaging. We do this as most cell media contain sodium bicarbonate, which is both required for healthy cell growth, and provides pH buffering in combination with the CO_2 in the incubator to maintain the correct pH. However, depending on the microscope available, it is not always possible to maintain a 5% CO_2 environment. HEPES works reasonably well for a few hours to maintain pH instead.

Using the standard growth medium for live cell imaging is good for cell health, but standard medium contains phenol-red (as a pH indicator) and serum, which are both fluorescent. If the eGFP signal is weak, then we have used commercial imaging solutions that are phenol-red and serum free. These include Live cell imaging solution (Fisher Scientific) and Fluorobrite DMEM (Thermofisher Scientific). Live cell imaging solution contains 140 mM NaCl, 2.5 mM KCl, 1.8 mM $CaCl_2$, 1.0 mM $MgCl_2$ and 20 mM HEPES, but does not contain glucose or serum, and thus is only suitable for short (up to ~4 h) of live cell imaging. Fluorobrite is effectively phenol-red free DMEM and will support longer-term culture when supplemented with serum and glutamine (or glutamax), and additionally needs the addition of HEPES for non-CO_2 environments.

To image spheroids, we use low melting point agarose (suitable for cell culture, e.g. from Sigma Aldrich) to immobilise them. We make up a 1% solution (0.5 g of agarose is added to 50 ml of water in a sterile 50 ml Falcon tube), heat this in a microwave until the agarose dissolves, and cool to ~40 °C in a water bath or equivalent. A single spheroid is then carefully transferred from suspension culture (drops or plates), into 1 well of an 8 well glass-bottomed dish in ~100 μl of growth medium, supplemented with HEPES as required, 100–200 μl of the warm agarose is added, and the dish transferred back to the incubator. The ratio of agarose to growth medium will determine how 'solid' the final gel is once the agarose has been set. We do not use Matrigel for imaging, as it scatters too much light, which degrades image quality.

4 Confocal and Airyscan Imaging

We routinely use a Zeiss LSM 880 confocal equipped with fast Airyscan for live cell imaging, using the ×40 N.A. 1.4 oil immersion objective. The microscope is equipped with a heated stage, which we switch on and allow to heat up to 37 °C for 30–60 min before imaging. This helps to avoid thermal drift, and thus samples going out of focus, once imaging starts. The sample is placed on the stage and first examined using the eyepieces, before switching to confocal mode. Even if the plan is to use Airyscan, we normally take an initial image using the confocal mode, exciting eGFP with the 488-nm wavelength laser, Airy set to 1, and adjusting laser intensity and gain to obtain an initial image with good signal to noise (high signal from the eGFP, low noise from the background). We then switch to Airyscan, select the SR (super-resolution mode) and again adjust laser intensity and gain as required. We then select the start and end points in z to collect a z-stack. If we initially need to characterise protein organisation throughout the entire depth of the entire cell, then we will collect a z-stack from the top to the bottom of the entire cell (Fig. 2a). This helps us determine where, in z, we wish to collect further data, and which region of the cell we might want to crop into (e.g. the leading edge of the cell). Subsequently, we will use a single plane or short z-stack, for a small region of interest (Fig. 2b).

Selecting the size and position of the z-stack and the timing interval is sample dependent in live cell imaging. The first time we image any new construct we will test out different sizes of z-stack and timing intervals. For fast-moving objects within the sample, we need to image at the highest frame rate possible. In some cases, we can predict this beforehand. For example, kinesin moves along microtubules at a speed of ~0.5–1 μm/s. To be able to reliably track single fluorescent kinesin spots from frame to frame, using a microscope with a resolution of ~0.3 μm, we can calculate that we need a minimum frame rate of two frames per second to be able to accurately identify the same spot from frame to frame and measure the speed at which it moved [38]. In other cases, without prior knowledge, we try different frame rates to determine which one works best for the behaviour we wish to quantify.

For example, we can use the Airyscan to image a large Z-stack of a cell expressing the eGFP-tagged protein of interest, such as eGFP-tagged Myo1E (Fig. 3). Although the Z-stack of almost the entire cell provides a good overview of its morphology and indicates that eGFP-Myo1E is mostly located in the cellular cortex and in filopodial structures, imaging the entire Z-stack over the entire cell for a time-lapse sequence, would be slow

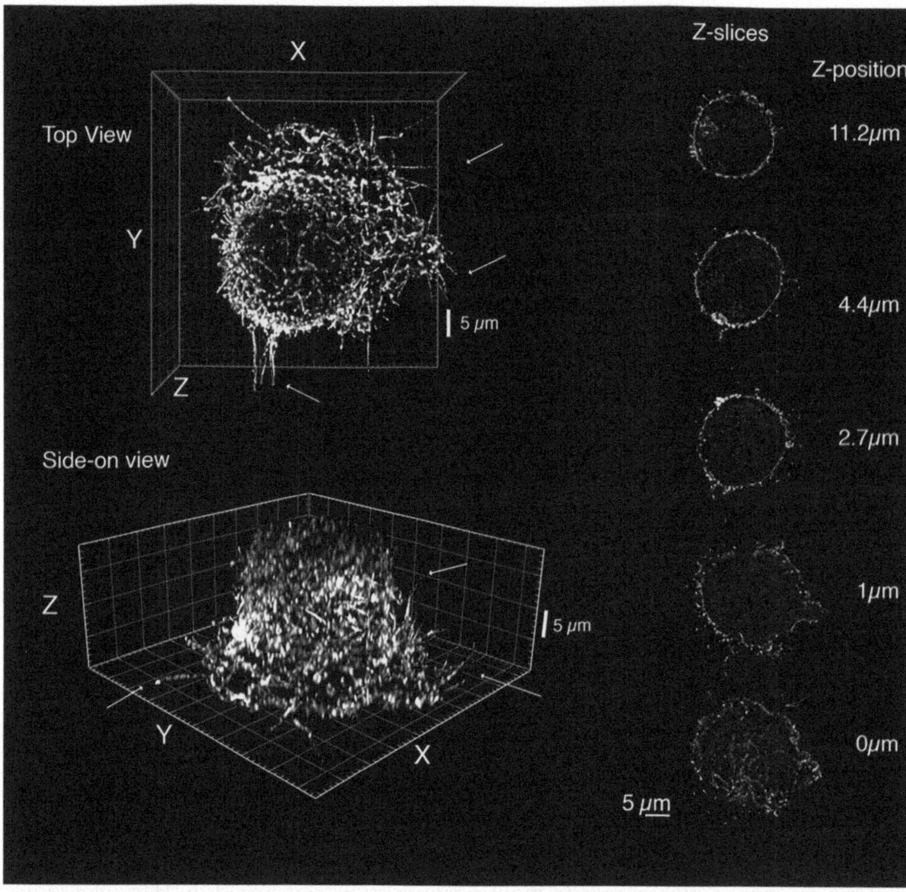

Fig. 2 Airyscan Imaging of a live PC3 cell expressing eGFP-Myo1E. A Z-stack (900 (38.32 μm) × 900 pixels in XY and 88 slices (steps) 0.173 μm apart: 15.22 μm in Z) using a zoom of 5.5, took 52 s. A top view (all the individual images in Z projected into a single image, looking down from the top) and a side on view shows the overall shape of the cell. eGFP-Myo1E positive filopodia (arrowed) are observed. Individual slices are shown on the right-hand side at different depths from the basal region of the cell (0 μm) towards the top (11.2 μm)

(about 1 Z-stack per minute), would not capture fast events, and is likely to be phototoxic. However, the Z-stack does show that many filopodia are located towards the basal region of the cell, and at the cell periphery, enabling us to make a reasonable judgement about where it is most important to image if we need to quantify the behaviour of the filopodia, for example.

For this example, we chose to reduce the size of the Z-stack to 3.29 μm (19 × 0.173 μm steps), and we cropped in on a smaller region near the edge of the cell that is rich in protrusions (XY:444 (18.9 μm) × 444 pixels) by increasing the zoom to ×10.9. This enabled us to capture a Z-stack at a much faster rate (One Z-stack in <1 s). This is still

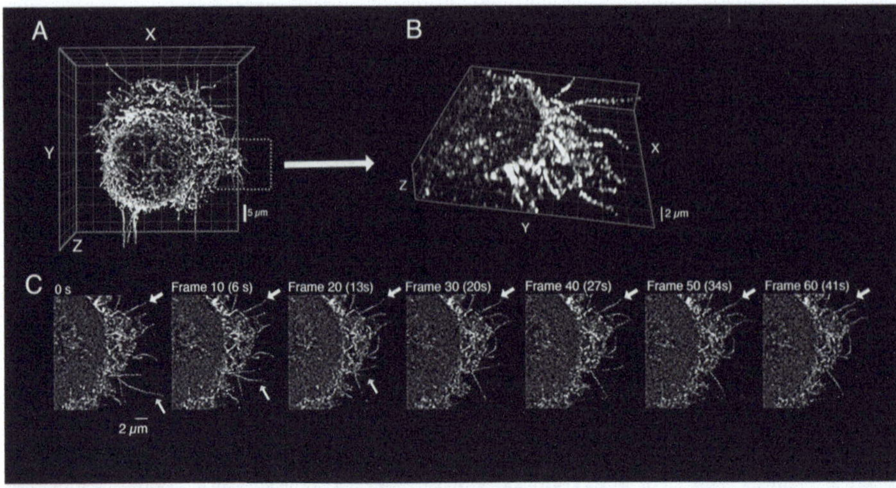

Fig. 3 Fast Airyscan imaging of a region of interest (ROI) for a PC3-cell expressing eGFP-Myo1E. From the original Z-stack of the whole cell, we chose a ROI (indicated by the dashed square in **a**) and imaged this (ROI: dimensions are 18.9 × 18.9 × 3.29 μm: XYZ). (**b**) Shows a side-view projection image of the first frame. (**c**) Shows individual maximum intensity projections (MIPs: all the images are combined into a single image, based on maximum intensity) for selected time points. Thick arrow indicates a filopodium that remains through most of the images. Thinner arrow (below) indicates a shrinking filopodium, only present up to frame 20 (13 s)

relatively slow, and if speed were an issue, we could crop into a smaller region, and reduce the Z-stack still further, to a single frame if necessary, and/or reduce the resolution of the image. The key decision is how fast do we need to image, and can we just image a small region of the cell, to quantify the behaviour we are interested in. Here, the image capture rate is fast enough to observe the overall behaviour of the eGFP-Myo1E filopodia (growing, shrinking, overall numbers). However, if we needed to measure the dynamic movement of eGFP-Myo1E within the filopodia, we would likely need to increase the imaging rate by >5 times.

The Airyscan can be too slow for imaging rapid events, especially when a larger field of view is needed. In this case, we have switched to iSIM to image rapidly moving objects such as cytoplasmic granules containing virus protein in live cells [39]. The iSIM can capture images at up to 100 frames/s for larger fields of view (90 μm × 80 μm; 1745 × 1550 pixels) and the minimum Z-step size is 0.05 μm. To capture a Z-stack of 3.29 μm using a similar size of Z-step (0.173 μm) is thus 5× faster than using the Airyscan, but with a much larger field of view. Scanning over a smaller error enables faster image capture rates. The final resolution of iSIM is similar to that obtained by Airyscan. However, the homebuilt iSIM is only equipped with two lasers (488 nm and 561 nm), and the rate of photobleaching on the iSIM is faster than it is on the Airyscan. Thus, this microscope is best for high-resolution imaging over short time periods, and specifically for rapidly moving objects.

Finally, as more and more experiments are starting to use 3D cellular models such as spheroids and organoids, we have also used confocal imaging to image these types of samples. However, here we are more interested in imaging several cells in the spheroid at once, and using a relatively fast frame rate, such that we can assess dynamic behaviour in a small region of the spheroid, and in this case, we imaged in a single plane only. This type of imaging is possible using the confocal and does show differences between different types of knockdown cells (Fig. 4). However, it would be more challenging to image larger Z-stacks at speed using this approach, whilst avoiding toxicity.

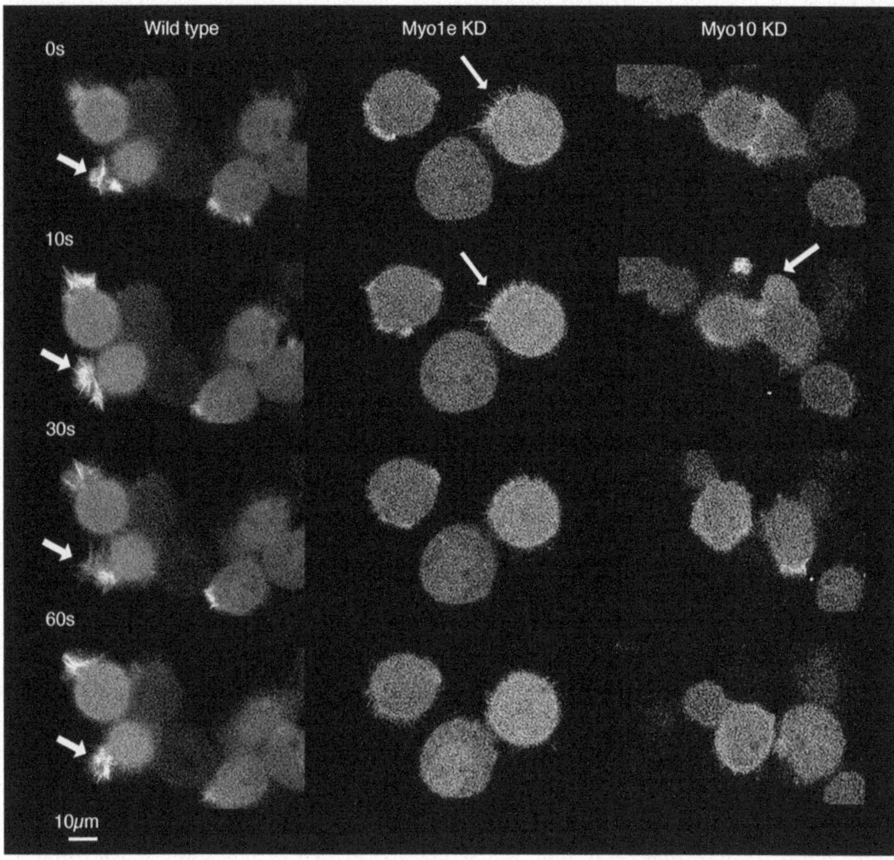

Fig. 4 Confocal imaging of PC3 cell spheroids, expressing eGFP-lifeact. eGFP-lifeact labels F-actin in cells [40]. A single plane and selected time points from 0 to 60 s are shown. The images compared wild-type PC3 cells, with those in which Myo1e of Myo10 has been knocked down (KD) using siRNA. The lack of well-focussed, eGFP-lifeact rich (and thus actin rich—arrowed for wild type) protrusions are evident in both types of knockdown cells. In Myo1e KD cells, fine protrusions are still present (arrowed), but in Myo10 KD cells, the cells only seem competent to form blebs (arrowed)

Lattice Light Sheet

An alternative approach for gentle imaging of a large volume of cells, or 3D imaging, is to use the lattice light sheet microscope. The sample preparation for this is exactly the same as for live cell imaging on the confocal, and glass-bottomed dishes (#1.5) are required. The lattice light sheet is equipped with three lasers for excitation at 488 nm, 561 nm and 640 nm (Imaging in the 'blue' range is usually more photodamaging for cells). It uses a water immersion lens for imaging (48×, NA of 1.0). We typically use the 1000 nm × 30 μm light sheet for imaging cells, spheroids and small organoids.

In a lattice light sheet, the light sheet is scanned across the sample at an angle [29] at a rapid rate at intervals of 0.2 μm (e.g. an image consisting of 10 planes would cover 2 μm). As with any microscopy, the speed of imaging the sample depends on how large a field of view you need to image, and this is adjustable in the software. However, a single time point (complete scan) using the 1000 nm × 30 μm light sheet across a large field of view (Fig. 5) that encompasses multiple cells in a small spheroid only took 4 s. Cropping and imaging a smaller ROI would increase the image rate markedly. However, in this case, the larger field of view enables us to rapidly image many cells together in 3D, and visualise their interactions with each other (e.g. Fig. 5c, arrows).

Processing and Analysis

Finally, images are data. As with any data, they need to be processed and analysed. To process the data, we commonly deconvolve all the Airyscan and STED images of fixed cells, to provide a small (~1.4 fold) increase in resolution. We use Huygens Professional software or the Zeiss confocal software (Zen Blue) to do this. We do not usually do this for live cell images but do choose to do so if we need to see fine detail more easily. LLS images can also be deconvolved, and this needs to be done before deskewing (and the ability to do this is included in the Zen Blue software).

We analyse images in numerous ways, depending on what we want to quantify. We either use commercial software (Arrivis, Imaris, Huygens) to do this, or FiJi (Image J bundle). The ImageJ/Fiji website provides a downloadable version of ImageJ (https://imagej.net/software/fiji/). It is open-source software [41] and many people do generate and upload useful plugins for the software that you can find here (https://imagej.net/plugins/). We have also developed our own software (e.g. an ImageJ plugin to analyse spheroids [42]. ICY also provides free useful software for analysing images and this integrates with both ImageJ and Matlab (https://icy.bioimageanalysis.org/). In addition, we sometimes use Cell Profiler (https://cellprofiler.org/), now in version 4 [43]. All of these are worth trying to find the analysis best suited for your specific question. If you like coding (especially using Python) and need to develop segmentation of objects such as nuclei, organelles and cell bodies, then there is also Napari (https://napari.org/stable/) and Cellpose (https://cellpose.readthedocs.io/en/latest/index.html) [44]. We have also written Python-based software to analyse patterns in super-resolution images [45].

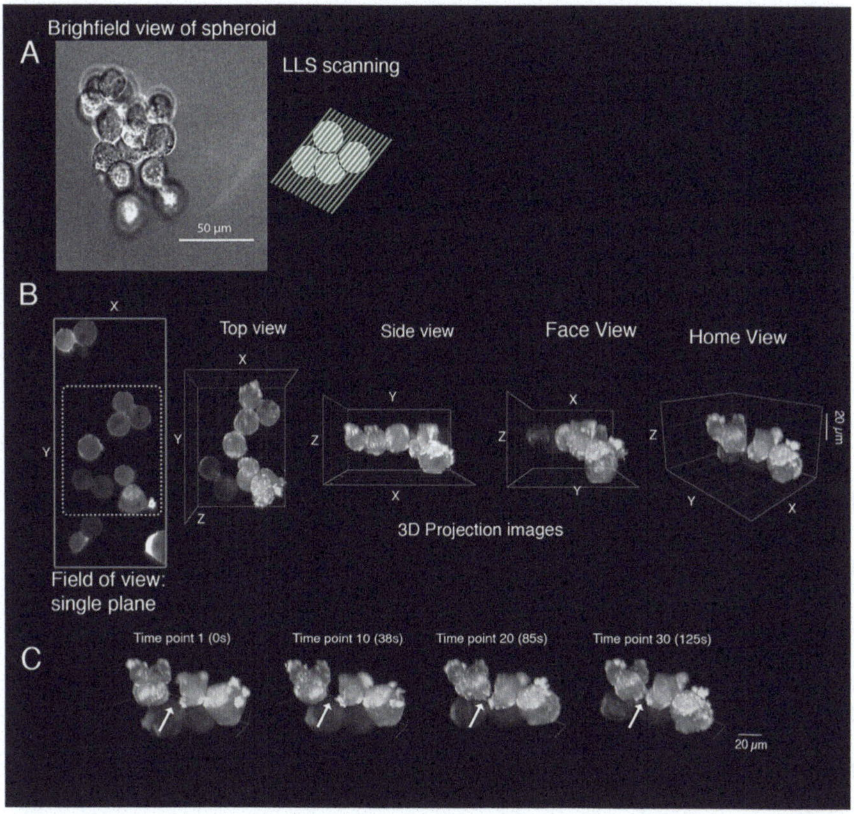

Fig. 5 Lattice light sheet imaging of a small spheroid (cells express eGFP-Lifeact). (**a**) Shows a brightfield image of a spheroid. The adjacent diagram shows how the light sheet (green lines) is scanned across the spheroid. For the example shown in (**b**, **c**), 380 slices in the Z direction were obtained (54.95 μm) per time point, X was 14.83 μm, Y was 317.82 μm. and each scan took ~4 s. The image was then deskewed to generate the final images. (**b**) Shows a single plane of the resulting deskewed image (for eGFP-Lifeact expressing cells), and 3D projection images viewing the spheroid from different angles (Top, side and face and home views). (**c**) The side view at different time points: the arrow shows how two cells move together and interact within the spheroid

Take Home Message

- Live cell imaging is important to understand the dynamics of cell behaviour, how the cytoskeleton reorganises with time, how organelles move around and so on.

(continued)

- Many new approaches to live cell imaging have been developed over the past few years, and the resolution of the resultant images is improving.
- Super-resolution imaging techniques are challenging to apply to live cell imaging, but Airyscan, SIM (or iSIM) and light sheet are relatively straightforward to use.
- You should think of images as the 'data' and make sure you design your experiments appropriately, so that you can analyse what you need (e.g. choose the appropriate resolution, imaging speed and region of interest that you need to answer your question and choose the appropriate image analysis that will help you analyse and quantify your data).

References

1. Bukhari H, Muller T. Endogenous fluorescence tagging by CRISPR. Trends Cell Biol. 2019;29:912–28.
2. Liu J, Cui Z. Fluorescent Labeling of proteins of interest in live cells: beyond fluorescent proteins. Bioconjug Chem. 2020;31:1587–95.
3. Cranfill PJ, Sell BR, Baird MA, Allen JR, Lavagnino Z, de Gruiter HM, Kremers GJ, Davidson MW, Ustione A, Piston DW. Quantitative assessment of fluorescent proteins. Nat Methods. 2016;13:557–62.
4. Dabiri GA, Turnacioglu KK, Sanger JM, Sanger JW. Myofibrillogenesis visualized in living embryonic cardiomyocytes. Proc Natl Acad Sci USA. 1997;94:9493–8.
5. Chalfie M, Tu Y, Euskirchen G, Ward WW, Prasher DC. Green fluorescent protein as a marker for gene expression. Science. 1994;263:802–5.
6. Yang TT, Cheng L, Kain SR. Optimized codon usage and chromophore mutations provide enhanced sensitivity with the green fluorescent protein. Nucleic Acids Res. 1996;24:4592–3.
7. Hirano M, Ando R, Shimozono S, Sugiyama M, Takeda N, Kurokawa H, et al. A highly photostable and bright green fluorescent protein. Nat Biotechnol. 2022;40(7):1132–42.
8. Lambert TJ. FPbase: a community-editable fluorescent protein database. Nat Methods. 2019;16:277–8.
9. Wilson T. Resolution and optical sectioning in the confocal microscope. J Microsc. 2011;244:113–21.
10. Gustafsson MG. Surpassing the lateral resolution limit by a factor of two using structured illumination microscopy. J Microsc. 2000;198:82–7.
11. Curd A, Cleasby A, Makowska K, York A, Shroff H, Peckham M. Construction of an instant structured illumination microscope. Methods. 2015;88:37–47.
12. York AG, Chandris P, Nogare DD, Head J, Wawrzusin P, Fischer RS, Chitnis A, Shroff H. Instant super-resolution imaging in live cells and embryos via analog image processing. Nat Methods. 2013;10:1122–6.
13. Guo M, Li Y, Su Y, Lambert T, Nogare DD, Moyle MW, Duncan LH, Ikegami R, Santella A, Rey-Suarez I, et al. Rapid image deconvolution and multiview fusion for optical microscopy. Nat Biotechnol. 2020;38:1337–46.

14. Hell SW, Wichmann J. Breaking the diffraction resolution limit by stimulated emission: stimulated-emission-depletion fluorescence microscopy. Opt Lett. 1994;19:780–2.
15. Klar TA, Hell SW. Subdiffraction resolution in far-field fluorescence microscopy. Opt Lett. 1999;24:954–6.
16. Betzig E, Patterson GH, Sougrat R, Lindwasser OW, Olenych S, Bonifacino JS, Davidson MW, Lippincott-Schwartz J, Hess HF. Imaging intracellular fluorescent proteins at nanometer resolution. Science. 2006;313:1642–5.
17. Hess ST, Girirajan TP, Mason MD. Ultra-high resolution imaging by fluorescence photoactivation localization microscopy. Biophys J. 2006;91:4258–72.
18. Allen JR, Ross ST, Davidson MW. Sample preparation for single molecule localization microscopy. Phys Chem Chem Phys. 2013;15:18771–83.
19. McKinney SA, Murphy CS, Hazelwood KL, Davidson MW, Looger LL. A bright and photostable photoconvertible fluorescent protein. Nat Methods. 2009;6:131–3.
20. Wiedenmann J, Ivanchenko S, Oswald F, Schmitt F, Rocker C, Salih A, Spindler KD, Nienhaus GU. EosFP, a fluorescent marker protein with UV-inducible green-to-red fluorescence conversion. Proc Natl Acad Sci USA. 2004;101:15905–10.
21. Zhang M, Chang H, Zhang Y, Yu J, Wu L, Ji W, Chen J, Liu B, Lu J, Liu Y, et al. Rational design of true monomeric and bright photoactivatable fluorescent proteins. Nat Methods. 2012;9:727–9.
22. Haywood NJ, Wolny M, Rogers B, Trinh CH, Shuping Y, Edwards TA, Peckham M. Hypertrophic cardiomyopathy mutations in the calponin-homology domain of ACTN2 affect actin binding and cardiomyocyte Z-disc incorporation. Biochem J. 2016;473:2485–93.
23. Rust MJ, Bates M, Zhuang X. Sub-diffraction-limit imaging by stochastic optical reconstruction microscopy (STORM). Nat Methods. 2006;3:793–5.
24. Heilemann M, van de Linde S, Mukherjee A, Sauer M. Super-resolution imaging with small organic fluorophores. Angew Chem Int Ed Engl. 2009;48:6903–8.
25. Heilemann M, van de Linde S, Schuttpelz M, Kasper R, Seefeldt B, Mukherjee A, Tinnefeld P, Sauer M. Subdiffraction-resolution fluorescence imaging with conventional fluorescent probes. Angew Chem Int Ed Engl. 2008;47:6172–6.
26. Dempsey GT, Vaughan JC, Chen KH, Bates M, Zhuang X. Evaluation of fluorophores for optimal performance in localization-based super-resolution imaging. Nat Methods. 2011;8:1027–36.
27. Manley S, Gillette JM, Patterson GH, Shroff H, Hess HF, Betzig E, Lippincott-Schwartz J. High-density mapping of single-molecule trajectories with photoactivated localization microscopy. Nat Methods. 2008;5:155–7.
28. Huisken J, Swoger J, Del Bene F, Wittbrodt J, Stelzer EH. Optical sectioning deep inside live embryos by selective plane illumination microscopy. Science. 2004;305:1007–9.
29. Chen BC, Legant WR, Wang K, Shao L, Milkie DE, Davidson MW, Janetopoulos C, Wu XS, Hammer JA 3rd, Liu Z, et al. Lattice light-sheet microscopy: imaging molecules to embryos at high spatiotemporal resolution. Science. 2014;346:1257998.
30. Ponjavic A, McColl J, Carr AR, Santos AM, Kulenkampff K, Lippert A, Davis SJ, Klenerman D, Lee SF. Single-molecule light-sheet imaging of suspended T cells. Biophys J. 2018;114:2200–11.
31. Jacquemet G, Carisey AF, Hamidi H, Henriques R, Leterrier C. The cell biologist's guide to super-resolution microscopy. J Cell Sci. 2020;133(11):jcs240713.
32. Schermelleh L, Ferrand A, Huser T, Eggeling C, Sauer M, Biehlmaier O, Drummen GPC. Super-resolution microscopy demystified. Nat Cell Biol. 2019;21:72–84.
33. Lopata A, Hughes R, Tiede C, Heissler SM, Sellers JR, Knight PJ, Tomlinson D, Peckham M. Affimer proteins for F-actin: novel affinity reagents that label F-actin in live and fixed cells. Sci Rep. 2018;8:6572.

34 Swailes NT, Colegrave M, Knight PJ, Peckham M. Non-muscle myosins 2A and 2B drive changes in cell morphology that occur as myoblasts align and fuse. J Cell Sci. 2006;119:3561–70.
35. Swailes NT, Knight PJ, Peckham M. Actin filament organization in aligned prefusion myoblasts. J Anat. 2004;205:381–91.
36. Makowska KA, Hughes RE, White KJ, Wells CM, Peckham M. Specific myosins control actin organization, cell morphology, and migration in prostate cancer cells. Cell Rep. 2015;13:2118–25.
37. Lukinavicius G, Reymond L, D'Este E, Masharina A, Gottfert F, Ta H, Guther A, Fournier M, Rizzo S, Waldmann H, et al. Fluorogenic probes for live-cell imaging of the cytoskeleton. Nat Methods. 2014;11:731–3.
38. Dunn S, Morrison EE, Liverpool TB, Molina-Paris C, Cross RA, Alonso MC, Peckham M. Differential trafficking of Kif5c on tyrosinated and detyrosinated microtubules in live cells. J Cell Sci. 2008;121:1085–95.
39. Remenyi R, Gao Y, Hughes RE, Curd A, Zothner C, Peckham M, Merits A, Harris M. Persistent replication of a Chikungunya virus replicon in human cells is associated with presence of stable cytoplasmic granules containing nonstructural protein 3. J Virol. 2018;92(16):e00477–18.
40. Riedl J, Crevenna AH, Kessenbrock K, Yu JH, Neukirchen D, Bista M, Bradke F, Jenne D, Holak TA, Werb Z, et al. Lifeact: a versatile marker to visualize F-actin. Nat Methods. 2008;5:605–7.
41. Schindelin J, Arganda-Carreras I, Frise E, Kaynig V, Longair M, Pietzsch T, Preibisch S, Rueden C, Saalfeld S, Schmid B, et al. Fiji: an open-source platform for biological-image analysis. Nat Methods. 2012;9:676–82.
42. Ketchen S, Rohwedder A, Knipp S, Esteves F, Struve N, Peckham M, Ladbury JE, Curd A, Short SC, Bruning-Richardson A. A novel workflow for three-dimensional analysis of tumour cell migration. Interface Focus. 2020;10:20190070.
43. Stirling DR, Swain-Bowden MJ, Lucas AM, Carpenter AE, Cimini BA, Goodman A. CellProfiler 4: improvements in speed, utility and usability. BMC Bioinformatics. 2021;22:433.
44. Stringer C, Wang T, Michaelos M, Pachitariu M. Cellpose: a generalist algorithm for cellular segmentation. Nat Methods. 2021;18:100–6.
45. Curd AP, Leng J, Hughes RE, Cleasby AJ, Rogers B, Trinh CH, Baird MA, Takagi Y, Tiede C, Sieben C, et al. Nanoscale pattern extraction from relative positions of sparse 3D localizations. Nano Lett. 2021;21:1213–20.

Further Reading

Useful Reviews

Grimm JB, Lavis LD. Caveat fluorophore: an insiders' guide to small-molecule fluorescent labels. Nat Methods. 2022;19:149–58.

Jacquemet G, Carisey AF, Hamidi H, Henriques R, Leterrier C. The cell biologist's guide to super-resolution microscopy. J Cell Sci. 2020;133(11):jcs240713.

Lemon WC, McDole K. Live-cell imaging in the era of too many microscopes. Curr Opin Cell Biol. 2020;66:34–42.

Nikon Microscopy. For many useful articles on imaging, explaining the basics. https://www.microscopyu.com/

Online Microscopy Resources. https://www.rms.org.uk/library/online-microscopy-resources.html. And courses and workshops run by the Royal Microscopical Society. https://www.rms.org.uk/

On-line training in microscopy. https://myscope.training/

Reiche MA, Aaron JS, Boehm U, DeSantis MC, Hobson CM, Khuon S, Lee RM, Chew TL. When light meets biology - how the specimen affects quantitative microscopy. J Cell Sci. 2022;135(6):jcs259656.

Schermelleh L, Ferrand A, Huser T, Eggeling C, Sauer M, Biehlmaier O, Drummen GPC. Super-resolution microscopy demystified. Nat Cell Biol. 2019;21:72–84.

Computational Modelling of Cell Migration

Roman Bauer

What You Will Learn in This Chapter

Here, you will learn about the utilization of computational modelling and simulation as well-established tools for research, including the study of biological systems. To this end, you will read about the main computational approaches as well as their advantages and disadvantages.

Due to the importance of cell migration in many biological phenomena and the involved complexities, numerous computational models of cell migration at the single-cell and collective-cell levels exist. In this chapter, an overview of this interdisciplinary field is given, and well-established computational techniques and software tools are explained. Moreover, the main research domains where cell migration has been computationally approached are described, along with a review of example studies from the literature.

Overall, this chapter will provide you with the knowledge to understand the fundamentals of computational modelling when applied to cell migration, give a high-level overview of relevant works, and discuss potential avenues for future research.

R. Bauer (✉)
University of Surrey, Guildford, UK
e-mail: r.bauer@surrey.ac.uk

A. Brüning-Richardson, S. Knipp (eds.), *Cell Migration in Development, Health and Disease*, Learning Materials in Biosciences,
https://doi.org/10.1007/978-3-031-64532-7_10

1 Introduction

The migration of cells is a core process in numerous biological systems and is essential to many biological processes [1, 2]. Indeed, the complexity of such biological systems requires underlying generative mechanisms to coordinate and streamline movements. Hence, cell migration is a key process in numerous contexts, such as embryonic development, the functioning of the immune system, wound healing, and cancer. Moreover, cell migration plays crucial roles in both health and disease, and elucidating its underlying processes is necessary for understanding various biological systems and associated pathologies.

Cell migration is a highly complex process. Like any other biological process, it is a multiscale phenomenon that comprises networks of various biological and physical processes at different spatial and temporal levels. In this context, protrusions of the cell membrane are necessary to probe the local physical environment. Depending on the sensed inputs, different biological pathways are triggered, and the associated cell behavior of motility is produced.

Since changes in cellular location usually lead to different inputs in terms of cell sensation (e.g., molecular or mechanical sensations), a closed-loop action-and-reaction process is created. Due to the emerging complexities and highly nonlinear dynamics of this type of process, it is often problematic to intuitively understand and predict many phenomena of cell migration.

Given the difficulties in comprehending many complex biological systems, computational modelling has gained significant attention as a valuable tool for biomedical research. This is because computational modelling allows us to extend and improve the capabilities of our usually vague mental models. It allows us to formulate quantitative hypotheses that can be tested in silico as well as experimentally in the wet lab. Moreover, computational modelling can be used to describe and formulate hypotheses about locations or systems that are otherwise very difficult to observe, whether for practical or ethical reasons (e.g., the development of the mammalian brain or the healing of severe skin wounds).

Due to its attractive characteristics, computational modelling has become, in addition to the classical experimental and purely theoretical approaches, a well-established component of the scientific method. Along these lines, computational methods have helped gain insights into various scientific questions that otherwise would have remained unanswered. For instance, bioinformatics techniques have revolutionized genomics and are being used for the treatment of diseases such as cystic fibrosis and cancer [3, 4].

Given that cell migration is a key biological behavior abundant in biological systems, it constitutes a crucial factor in the endeavor to understand how these systems organize themselves and evolve. It is therefore not surprising that it has become a core component of many computational models. In this chapter, processes of cell motility will be described, and relevant literature will be reviewed. We will focus on cell migration in the following five contexts: single-cell migration, biological development, tissue engineering, wound healing, and cancer growth. A selected set of relevant scientific publications will be reviewed, and the underlying approaches explained.

2 The Biology of Cell Migration

2.1 Types of Cell Migration

Cell migration is a highly heterogeneous process that significantly varies depending on the context [5]. The most researched types of actin-dependent cell movement are arguably *chemotaxis* and *haptotaxis*. *Chemotaxis* occurs through the sensing of extracellular cues, such as growth factors, chemical gradients, and extracellular matrix (ECM) components. Eukaryotic cells and insect cells generally employ lamellipodia/filopodia as their mechanism of action. *Haptotaxis* denotes cellular locomotion directed in response to concentration gradients of nondiffusible adhesive molecules within the ECM [6].

In contrast to *chemotaxis* and *haptotaxis*, *durotaxis* does not require any extracellular matrix molecules to guide movement. Instead, it is the rigidity of the substrate that plays the key role in cell movement. The study of Lo et al. [7] shows that individual cell movement can be instigated by mere physical interactions between the cells and the underlying substrate.

Other types of cell motility include *thermotaxis* [8], *electrotaxis* (also known as *galvanotaxis*) [9], *topotaxis* [10], *phototaxis* [11], and *geotaxis* [12]. However, these are relatively poorly understood, and computational work building on relevant experimental data is rare.

Overall, the machinery for cell motility functions very differently depending on the taxis. This is because diverse signaling networks are induced depending on what drives cell motility, e.g., gradients of chemokines (*chemotaxis*), ECM density (*haptotaxis*), substrate rigidity (*durotaxis*), etc. Using computational modelling, the associated biological and physical mechanisms can be incorporated at various levels of detail.

2.2 Constituents of Cell Motility

Cell migration requires membrane protrusions at the cell front [13]. There are different ways these protrusions are realized: eukaryotic cell motility uses three prevalent forms of protrusion. The first employs polymerization of the cytoskeletal protein actin and is called actin-dependent cell migration [14]. The second form, known as "amoeboid migration," involves moving using so-called blebs [15]. Finally, the third form functions by rotation of thin helical appendages, i.e., flagellar-dependent swimming.

Actin-dependent cell movement is a broad type of migration, and amoeboid migration functions via protrusions similar to those found in actin-dependent movement but differ in their composition and mechanism of motility. The third form is mainly relevant for cells such as bacteria migrating in liquid environments. Arguably, most computational modelling has focused on the first and second forms of cell motility. In the following, a brief overview is given of the main constituents of cell motility that are considered in computational models.

The actin-myosin system: Actin-dependent cell migration of eukaryotic cells is implemented through the so-called actin-myosin system. The actin-myosin system comprises two different components at the front and rear ends of the cell. In particular, actin filaments are produced at the front via a cellular process called "treadmilling" [16]. The rear component comprises myosin that binds to the bundled actin filaments at the front and generates a contractile stress. Overall, the actin-myosin system acts as a motor that drives motility.

Lamellipodia: These are wide and flat membrane protrusions at the leading edge of cells [17]. The leading edge of cells invokes the assembly of new actin filaments from existing ones and thus largely controls the rate of actin polymerization that drives lamellipodial protrusion. Notably, lamellipodia-based migration is not universal and represents only one of many mechanisms of cell motility in three-dimensional (3D) environments [18].

Filopodia: Filopodia are actin-rich, short, and finger-like protrusions at the leading edge that act to sense the local microenvironment. They can act alone or in combination with other types of protrusion. Filopodia can sense chemical and mechanical properties of the local environment but are not responsible for the motility of the cell body.

Blebbing: Most research on cell migration has focused on motility mechanisms involving the lamellipodia. However, a different type of cell migration (often called amoeboid cell migration) is based on blebs [15, 19]. Like lamellipodia, blebs are protrusions of the cell membrane that play an active role in the movement of many cell types. However, blebs are not propelled by the actin-myosin system but through hydrostatic pressure and cytoplasmic flow.

Overall, the motility of cells is a heterogeneous process that can leverage different mechanisms and may be used by cells in different situations. This heterogeneity entails that the understanding of cell migration requires consideration of the physical implementation of motility on a small-scale, detailed level. Alternatively, it requires simplifications to study behavioral patterns of cells on a larger, collective scale.

2.3 Guidance of Cell Movement and Interactions with Local Environment

It is well-established that many different physical factors can influence cellular movement. To this end, cells have a plethora of ways of sensing their local physical neighborhood. A number of questions can be asked regarding the exact mechanisms for sensing these extracellular factors and conveying instructions to the machinery that generates the forces necessary for motility [20].

An abundant feature of biological systems, especially during development, is the presence of substances in extracellular space that guide the migration of cells. Extracellular guidance cues can invoke signaling cascades in the migrating cell, leading to the polymerization and coordinated extension of protrusions that then collectively form small focal complexes that exert a mechanical force, leading to the forward movement of the cell body.

There are several different possibilities for how the spatial distribution of extracellular cues is established. One possibility is through the cells' differential expression of genetic markers [21]. Such differentiated cellular composition can give rise to the spatially patterned secretion of extracellular or membrane-bound substances. Notably, there are often multiple substances interacting in a well-organized fashion. A possible mechanism for such inter-substance interactions is reaction-diffusion systems, as originally proposed by Alan Turing in 1952 [22]. A relevant computational model that produces spatial and biologically realistic patterns via cellular production and secretion of interacting morphogens is described in [23].

As previously described, in addition to extracellular substances, physical/material properties of the substrate and ECM can also influence cellular movements, i.e., they can act as guidance cues. In this context, the study of Zhao et al. [24] presents a computational model that probes the effects of biochemical and mechanical cues on cell migration. They find that biochemical cues allow for better directionality and persistence, while mechanical cues are better suited for the coordination of collective cell migration.

The type of computational techniques used to model these directed movements is of high importance to adequately account for the characteristics of these guidance cues. There is no one-size-fits-all computational approach to model the guidance of cell movements, so it is up to the modeler to choose and appropriately design the model depending on the context, available computational resources, and research aims.

3 Types of Computational Models

There are a wide range of computational models that have been used to formulate and assess hypotheses of cell migration. The following review of relevant computational techniques should not be understood as a basic overview of the main lines of previous research.

3.1 Continuum Models and the Finite Element Method

One main computational approach to model biological systems is the so-called continuum model, which plays a crucial role in modern engineering technology applications. These models essentially formulate the system as a spatial continuum of interlaced segments. The motions across this continuous scaffold are then governed by specific differential equations that describe the interactions of variables between neighbors under specified boundary conditions, such as Newton's law of motion or the principle of conservation of mass. Notably, these interactions can be based on various mechanisms, e.g., through mechanical forces or biochemical reactions.

The most prevalent of these continuum models is the finite element method (FEM). In FEM, systems are discretized into large numbers of interacting subregions (i.e., the elements) that collectively form a grid. The types of problems for which FEM can be used are diverse and include various problems such as structural analysis problems [25], heat transfer problems [26], or electromagnetic systems [27]. In the context of cell migration, FEM is a particularly well-suited model for tissue mechanics and biophysical aspects of cell motility.

From a computational perspective, FEM is well-suited for highly parallelized and effective simulations. This is because the behavior of the whole domain can be modelled as the summation of the adjacent elements, which can be mathematically captured by an element matrix. Matrices can be combined into a single matrix that represents the whole structure and formulates a linear system of equations to obtain a set of unknowns in the model [28]. Such matrix operations can, due to the usually limited number of mutually interacting elements, be computed relatively effectively with large numbers of processors.

Another powerful continuum modelling technique is phase-field modelling. One of its main advantages is that it allows for the natural modelling of the cell membrane as an interface, enabling the formulation of the interplay between the cells and the extracellular environment [29]. Moreover, it has additional advantages in terms of computational efficacy. It initially found applications in domains different from cell migration but has more recently been taken up to model cell motion.

Numerical simulations based on continuum modelling are very well-suited for detailed simulations of cell morphology during migration, as they allow for the modelling of cell mechanics and membrane deformation, which play crucial roles in cell motility. Indeed, the most detailed computational models of cell motility arguably employ the continuum approach, as it allows for the natural formulation and simulation of mathematical descriptions of membrane shape and dynamics.

On the other side, continuum models tend to oversimplify by not supporting the modelling of the multiscale nature of cellular system dynamics. In other words, the highly complex networks of genetic, signaling, membrane-bound, and various physical interactions cannot be integrated into a unified and coherent continuum model. This reduces their explanatory power and renders them particularly prone to issues when extrapolating to novel situations, i.e., situations that are different from those covered by the available experimental data.

3.2 Discrete/Agent-Based Modelling

In discrete models, elements act autonomously and independently in a shared physical environment. Different subclasses of this modelling approach exist. In some cases, the elements are located on a regularly spaced grid in 2D or 3D, while more advanced software locations can take continuous values that are only limited by the resolution of the spatial coordinates. However, discrete and agent-based modelling (ABM) are fundamentally different from continuum modelling in that the simulated elements are inherently separated from the space within which they reside.

Due to its adherence to the principles of autonomy and local information exchange, ABM has become a highly used modelling approach in many scientific fields. One of its main advantages compared to continuum modelling is its adequacy for biological systems, because cells can be modelled as the fundamental elements. ABM is a multiscale approach and supports the modelling of cell behaviors as controlled by gene regulatory dynamics and cellular pathways. Moreover, ABM allows for the accounting of physical processes such as intercellular mechanics and the diffusion of substances in an integrative, multiscale computational framework.

One of the disadvantages of ABM is that it can quickly become very demanding from a computational perspective. Hence, the usage of highly optimized code is becoming increasingly important in order to simulate realistic systems [30]. The recently published BioDynaMo software [31] represents a high-performance software that can simulate large numbers, i.e., in the order of billions of interacting agents in an efficient manner.

3.3 Hybrid Computational Modelling

In addition to the FEM and ABM approaches, combining these two can often be used to leverage their strengths while mitigating their weaknesses. For instance, ABM can be used to easily model various behaviors of cells, including cell migration, proliferation, growth, differentiation, death, and cellular secretion of substances while accounting for biomechanical dynamics. By leveraging the computationally less intense continuum modelling approach, a hybrid approach can scale to comparably large system sizes.

Indeed, the combination of continuum and agent-based models has become an attractive approach. The work of [32] makes use of FEM interfaced with the modelling of individual cells, as performed with ABM, to simulate cell dynamics in epithelial layers. They demonstrate their model in two different scenarios of collective cell migration. The study of [33] also employs such a hybrid approach to model the growth dynamics of realistically sized glioblastoma (GBM), including cell motility during progression. The authors confirm experimental findings, including different distributions of cancer and host cells.

4 Computational Models of Cell Migration

As described above, the field of computational biology has gained significant momentum. To a large extent, the increased availability of computational resources, both in hardware and software, has facilitated the uptake of computational methods. It is therefore not surprising that the number of relevant publications has been consistently increasing since the early 1990s (Fig. 1).

4.1 Software for the Modelling of Cell Migration

There are multiple software and software packages to model cell migration in biological tissues. Table 1 shows a selected list of some of the most prevalent ones, which are actively maintained and used in different biomedically relevant contexts.

It is noteworthy that numerous computational frameworks and software have been employed to model cell migration. Naturally, these come with their own features, so it is important to understand the advantages and disadvantages of each software before starting a model implementation. A selected list of such software and relevant features is shown in Table 1.

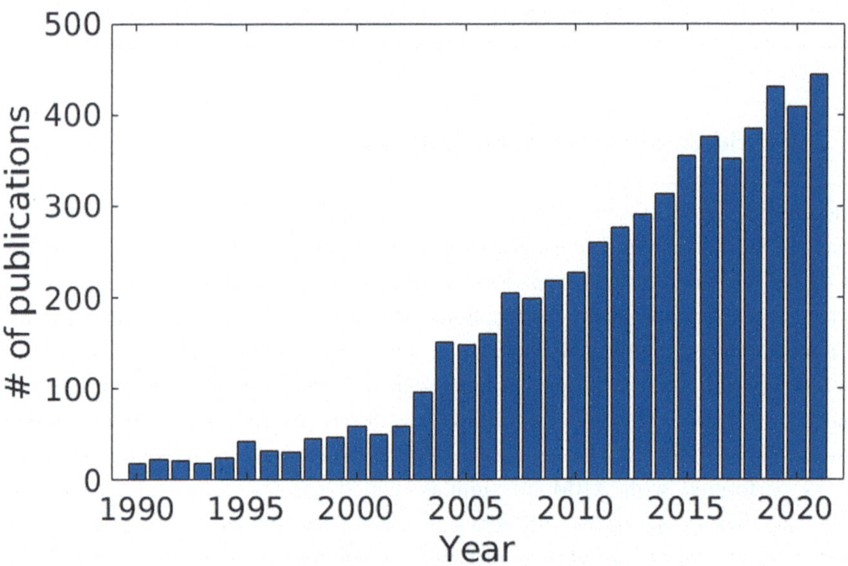

Fig. 1 Scientific publications on computational modelling of cell migration. A search for the terms "computational model" and "cell migration" on PubMed shows that the number of scientific works on the overlap between these topics has been steadily increasing since the 1990s. This increase has been enabled by the availability of better and more powerful hardware, software, and the increasing importance of computational models in biology

Table 1 Software for the modelling of cell migration. Numerous software options exist that allow modelling and simulating the migration of cells. Here, a non-exhaustive list of some of the most used and/or recent software is provided, along with relevant features and publications

Software name	Modelling approach	Based on	Parallelized?	License	References
BioDynaMo	ABM	C++	Yes	Free	[31]
COMSOL	Continuum	Java	Yes	Commercial	[34]
Cx3D	ABM	Java	No	Free	[35]
MecaGen	ABM	C++	Free	Free	[36]
NetLogo	ABM	Scala & Java	No	Free	[37, 38]
PhysiCell	ABM	C++	Free	Free	[39]

4.2 Applications

Here, several scientific domains where computer models of cell migration have been applied are described:

4.2.1 Single-Cell Migration Mechanics

Cell migration has been modelled and simulated for single cells or entire populations of cells. A numerical model that can capture the detailed dynamics of individual cells during migration and their morphological changes will help to better understand the internal control processes as well as the observed behaviors in different contexts. Marée et al. [40] present a 2D model that includes intracellular signaling and morphological dynamics. Cao et al. [41] also propose a 2D model that integrates biochemical signaling, cell mechanics, and deformations. They demonstrate that it can account for changing modes of cell motility and confirm their model predictions with experimental data. The study of [42] presents a hybrid agent-based/FEM model that includes numerous cellular characteristics, namely, membrane protrusions, membrane tensions, hydrostatic pressure, and matrix adhesion. Notably, this work also combines amoeboid migration with actin-dependent migration.

Numerous studies investigate the types of cell motility. For instance, Tan and Chaim [43] computationally model two different types of sensing during chemotaxis. They model the cell membrane as a 1D ring and investigate the impact of different parameters such as noise and cell speed. The study of [44] presents a 2D FEM-based model for cell migration in a chemotactic field. The authors show that a set of reaction-diffusion equations can be used to reproduce how a single cell moves along an extracellular substance gradient while periodically changing its morphology in accordance with well-established dynamics. Furthermore, the study provides a prediction on the mechanism by which cells can increase their persistence and polarity by generating their own local gradients of chemoattractants.

More recently, 3D cell migration based on lamellipodia in arbitrarily shaped and topographically structured surroundings has been computationally modelled [45]. Another FEM model of cell migration, as presented in [46], has the capability to account for mechanical and chemical guidance. Heck et al. [47] present a hybrid model that captures the dynamics of actin protrusion in interaction with extracellular matrix properties. The study of [48] presents a continuum model where migration of a single cell is simulated in 3D, taking into account interactions between cell mechanics and cellular signaling. The results show agreement with their experimental data in terms of the switching between different modes of migration. The study of [49] presents a 3D phase-field model that can be used to simulate chemotaxis in an individual cell, also taking into account rigid obstacles and fibers. Figure 2 shows a sequence of images obtained from this model.

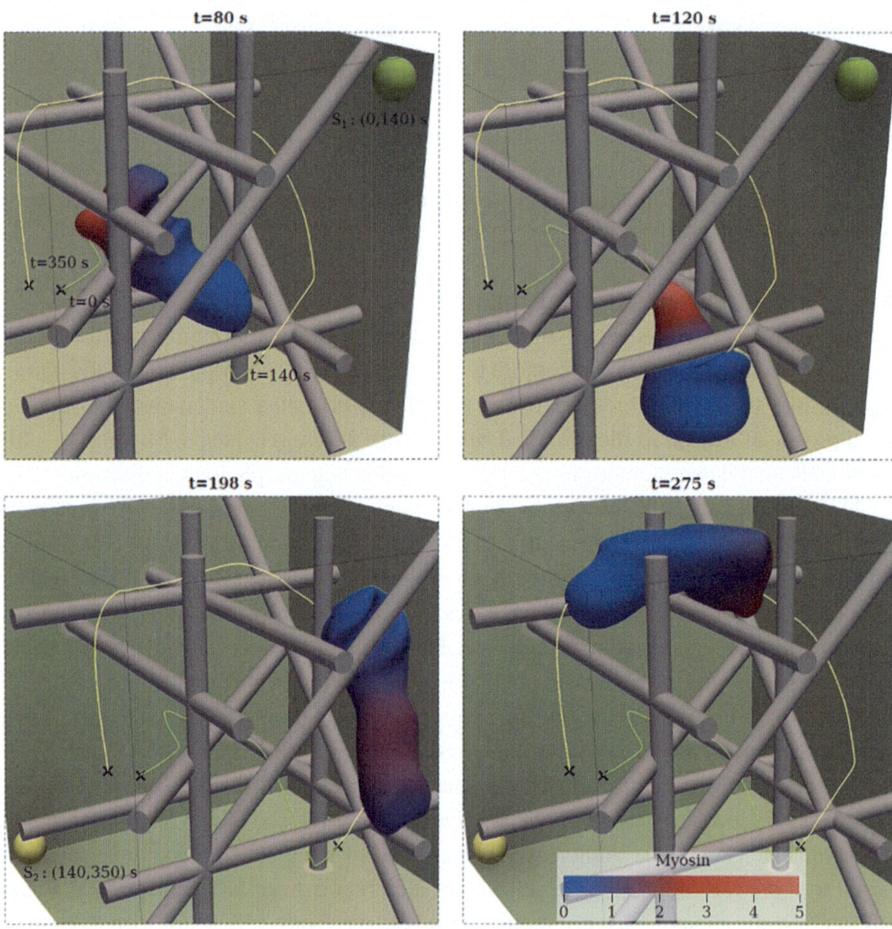

Fig. 2 Sequence of images showing 3D migration in a fibrous environment. The cell path is shown as a yellow line, while the chemoattractant distributions are indicated on three walls of the domain at $t = 80$, 198, and 275 s. Sources of the chemical substance providing guidance are shown as spheres S_1 (green sphere) and S_2 (yellow sphere). The coloring of the cell membrane indicates concentrations of intracellular cell signaling variables in this computational model. Figure from [49]

Computational models have also been used to study interactions between migratory cells. Reference [50] models single-cell migration in a 2D ECM. They then employ their model to reproduce multicellular streaming, where cells migrate along the tracks and tubes of individual leader cells. This is a particularly notable work because it demonstrates a major advantage of computational modelling by providing a mechanistic explanation for a nontrivial phenomenon that emerges from the collective action of simpler behaviors.

4.2.2 Development

One of the most fundamental research questions in biology is how complex multicellular structures organize themselves from a single precursor cell. However, experimental work on this topic is often problematic due to the difficulties in reproducing in vitro the conditions of natural development in vivo. Hence, mathematical and numerical methods provide a highly attractive, complementary approach to studying biological development. Indeed, computational modelling has become prevalent in the discipline of developmental biology [52]; numerous computational studies exist that focus on specific developmental phenomena. Here, a few example domains are described:

Embryogenesis

Cell migration is a major component of embryonic development, where cells follow genetically encoded rules and use information available in their local environment to steer their motility. Through this process, cells reach their final locations in a self-organized manner, generating the biological structure in a robust and consistent way. The early stages of embryogenesis constitute a critical period of biological development. The eminent developmental biologist Lewis Wolpert notably stated that "It is not birth, marriage, or death, but gastrulation which is truly the most important time in your life," referring to one of the crucial stages of early embryonic development.

An ABM is presented by [53], who employ NetLogo to model self-organization and cell migration in the developing zebrafish. Another recent computer model of early embryogenesis is presented in the study of [36]. The authors use the agent-based software MecaGen to simulate the first phase of epiboly. To this end, they model and simulate cell behavior as controlled by an abstract gene regulatory network (GRN) and the associated mechanical interactions.

Neural Development

Neural development constitutes a crucial component of biological development and overlaps with embryogenesis. However, neural development continues to progress beyond the embryonic stage and, in the case of humans, does so for approximately two decades. Only then do the networks of the brain responsible for higher-order cognition, self-control, and planning achieve their fully mature state.

A key component of neural development is the migration of precursor cells, which move to their final locations to establish the anatomical structure of the brain [54]. Understanding this process is crucial not only to gain insights into brain organization and function but also

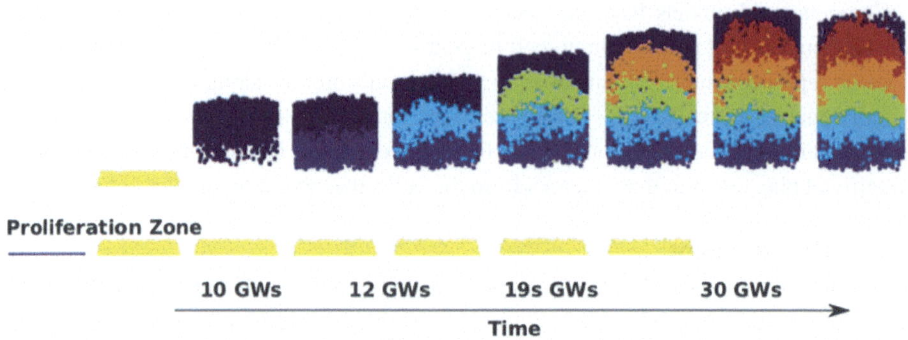

Fig. 3 Simulated cell migration during layer formation in the mammalian brain. The figure, extracted from [51], uses ABM to model and simulate cell migration during early brain development. Early neuronal precursors differentiate into different neuron types (indicated by different colors), proliferate, and migrate to their target layers to take up their final positions. The model recapitulates realistic numbers of neurons and is consistent with several experimental studies. The shown simulation is 3D, but the lateral view is provided for clearer visualization

to address neurodevelopmental disorders. Along those lines, the study of [51] employs ABM to simulate cell differentiation, cell-cell interactions and migration during cortical development in 3D, shining light on observations of the healthy brain as well as certain characteristics in autistic spectrum disorder and polymicrogyria (Fig. 3).

Most computational models of brain development have focused on elucidating the formation of fundamental brain anatomical phenomena. One striking aspect of (human) brain morphology is its folded structure. Although it is currently not entirely understood if and what the functional implications of the gyri and sulci are, significant computational work has been conducted to account for the structural development of brain folding. Neuronal migration constitutes a core developmental process in cortex folding, and divergent radial migration has been associated with cortical folding [55]. The majority of computational models of cell migration in neural development employ the FEM approach [56]. For instance, previous studies have investigated many biomechanical aspects of brain folding, such as cranial constraints [57], differential growth between cortical layers [58], the white matter neuroarchitecture [59, 60], and tissue stiffness [61].

The direct representation of axonal bundles and individual cell behavior, as could be implemented in ABM, constitutes an exciting prospect for future agent-based models of brain development. Notably, the studies of Zubler et al. [62] of Bauer et al. [51] use ABM to simulate cortical layer formation during brain development. In contrast to the previously mentioned studies on brain development, here autonomous cell migration based on genetically encoded behaviors is simulated. Moreover, this approach also accounts for various interactions between cells, including mechanical forces and the secretion or detection of diffusible substances.

4.2.3 Tissue Engineering

A biomedically very promising research domain is the synthetic growth of biological tissues. It is now possible to grow various tissues that recapitulate natural growth in a laboratory setting. This allows for a better understanding of processes that would otherwise be very problematic to observe in vivo, such as human brain development.

In addition to the predominantly scientifically motivated line of research, engineered tissues bear the promise of being used for transplantation purposes, allowing for the regeneration or improvement of diseased or injured tissues. Moreover, they can be used to test drug candidates and conduct toxicity testing. Although more research is needed to apply such engineered tissues in these clinical and medical contexts, the field has advanced significantly in recent years. It has become possible to use human stem cells and synthetically grow them into highly complex systems in vitro, such as retinal organoids and articular, bone, or cardiovascular tissues.

The engineering of biological tissues usually involves cell migration because stem cells differentiate and subsequently migrate in order for the tissue to organize itself. This migratory process has been computationally modelled in several tissue engineering scenarios, such as cardiovascular tissue [63, 64], bone tissue [65], or muscle constructs [66]. Notably, computational modelling is particularly well-suited for tissue engineering applications because of the highly interdisciplinary nature of this field, combining techniques from experimental biology, medicine, engineering, computer science, and mathematical modelling.

4.2.4 Immune System

The immune system is crucial for the normal physiology of the body. It is a highly complex system that continuously searches for and destroys pathogens, adapts, learns, and evolves. In a simplified manner, the immune system can be understood as comprising multiple interacting components of highly specialized cells.

Dendritic cells and T cells are responsible for detecting pathogens in their surrounding environment. Although many of the mechanisms that allow T cells to accomplish this task are not entirely understood, research has generated significant insights into the kinetic and molecular mechanisms regulating T-cell migration. Notably, certain diseases of the immune system have been linked to aberrant migration capabilities of T cells [67]. Hence, elucidating the mechanisms of migration of T cells and other immune cells has strong implications for the treatment of many diseases.

The fact that the immune system functions via communication among different cells and the movement of immune cells to remove microorganisms and pathogens renders it a particularly well-suited system for computational modelling. For instance, [68] model the cell motility of neutrophils under the influence of multiple chemoattractants and propose a winner-take-all mechanism that enables cells to migrate more efficiently in ambiguous environments [68]. This is an attractive hypothesis, especially because it aligns well with a similar mechanism proposed by Sakumura et al. [69] for growth cone navigation.

In contrast to brain development, simulations of the immune system tend to employ ABM more often. One of the most widely used frameworks in the context of the immune system is NetLogo [37]. Reference [70] review various computational models of the immune system, with several incorporating the migration of cells. However, due to limitations of NetLogo in terms of flexibility, scalability, and capability to simulate realistic physical environments, several other methods have been employed to model the migration of cells of the immune system. For instance, the study of Niculescu et al. [71] simulated T cells migrating in a packed epidermal environment. Notably, due to the important involvement of inflammation in wound healing, this study is relevant both to the immune system and to wound healing.

4.2.5 Wound Healing

During wound healing, specific cells are attracted by growth factors secreted at the wound. These factors allow them to migrate, proliferate, and produce extracellular matrix in a well-orchestrated manner, closing the wound and avoiding infection. A better understanding of these processes and their interactions would be of significant value in speeding up or improving wound healing. For instance, such insights could help design new types of wound dressings [72].

Wound healing is tightly linked to the previously discussed topic of the immune system. Indeed, inflammation can be a major complication in wound healing. Moreover, the study of [73] presents an ABM of inflammation and wound healing. However, the migration of cells in space is not modelled. In contrast, the study of Boon et al. [74] presents an ABM for wound contraction, which plays an important role during wound healing. The authors use the model to predict several behaviors that are in accordance with well-established phases of wound healing. The model is not quantitatively assessed based on experimental data, but it is the first cell-based model that links the immune response system to wound contraction. It is also of biomedical relevance because it can serve as a baseline model to probe the impacts of various wound healing parameters.

The work of [75] uses a sophisticated hybrid modelling approach, combining ABM and FEM, to simulate wound healing in 2D. They simulate various mechanical properties and demonstrate that the resolution of the coupled models has a high impact on these properties. This highlights the importance of the chosen parametrization of the employed frameworks and that thorough validation with experimental information is a key component of computational modelling. Relevant to this work is also the study of [76], where wound healing in early chick embryos was experimentally as well as computationally (with FEM) modelled. Based on this interdisciplinary approach, the authors derive a biologically plausible mechanism for the remarkable ability of the embryo to heal quickly without leaving scars.

4.2.6 Cancer

Cancer is defined as the uncontrolled proliferation of cells, ultimately leading to the death of the host. However, cellular invasion and the creation of metastases are crucial for high-grade tumors. Hence, a better understanding of how cancer progresses requires cell migration to be considered.

Cancer is a highly complex process that comprises numerous intracellular, intercellular, and extracellular mechanisms. Notably, cancer cells are heterogeneous; some are necrotic, some are proliferative, and others are invasive (and hence migratory). ABM allows modelling of not only different types of cancer cells but also interactions with the tumor microenvironment and the immune system, which play important roles during cancer progression. Computational analysis and modelling have become widely used in oncology [77].

Both experimental and computational work have focused on cell migration in cancer [78]. Given the importance of cell invasion in the progression of cancer, most computational models of cancer that involve spatial information incorporate behaviors of cell motility. As described earlier (Sect. 2.1), there exists a plethora of modes of cell migration. However, due to the need for more experimental data on in vivo cancer progression, computational models of cancer cells usually focus on a few tactics, such as the combination of chemotaxis and haptotaxis [6].

5 Future Outlook

Computational models have provided many insights into the workings of cell migration. Naturally, to validate these models, it is important that they are assessed based on a large variety of datasets, particularly with data from different sources and in many different contexts. Indeed, also thanks to the increased availability of computational frameworks, theoretical concepts, and computer software, it has become possible to formulate and assess different hypotheses, compare them with one another, and validate them with various experimental data.

Notably, advances in the efficient use of computer hardware [30, 31] allow for models of increasing detail, as well as the inclusion of many spatial and temporal scales. Due to ongoing technological advances, computational simulations are steadily becoming more powerful, enabling models of larger system sizes and higher relevance to real-world applications. For instance, applications in tissue engineering and tissue regeneration have the potential to revolutionize biomedicine. However, the main applications described above (single-cell migration, biological development, tissue engineering, immune system, wound healing, and cancer) are currently still predominantly in the realm of experimental biology.

There are increasingly more opportunities to leverage computational modelling, simulation, and optimization to accelerate research. Such models and simulations can identify experiments that are likely to fail, making otherwise costly experiments more efficient. Hence, in interaction with experimental research, computational models can significantly reduce the demand for research resources, facilitate the generation of novel and more realistic hypotheses, and ultimately help us gain insights into cell migration.

It is noteworthy that many domains of bioinformatics have experienced tremendous growth, particularly due to the creation of publicly available databases and datasets. For instance, the UK Biobank initiative has created a vast amount of information that is currently transforming the study of various biomedical topics on a population level [79, 80]. One promising avenue for future research is to make use of similar platforms to enable computational modelling of cell migration in different contexts. The incorporation of datasets from different labs and continued refinement and interaction with experimentalists will help provide computational models with higher accuracy and better explanatory power.

In addition to the availability of datasets, it is important that computational tools and software for analysis, modelling, and simulation are improved, refined, and made accessible to a wide research community. This is a highly challenging task because the development and maintenance of such software are expensive and demanding in terms of computational and human resources. Hence, it is important that future research software development acknowledges the need for sustainable, long-term efforts [81]. To this end, it is recommended to collaborate with software developers in similar research communities, adapt or extend existing solutions rather than reinvent the wheel, and generate as well as adopt common standard formats. Moreover, the use of open-source licenses that facilitate collaboration and academic uptake should be encouraged.

There are many opportunities for future research on cell migration using computational modelling. Firstly, it can be used to further improve our understanding of many biological phenomena. Secondly, it has high relevance for clinical applications, as highlighted by the cases of tissue engineering, the immune system, wound healing, and cancer described above. Better and more efficient ways to influence the invasion of cancer cells or speed up the wound healing process would be of significant value. Thirdly, the computational component also aligns well with the emerging field of precision medicine [3, 82]. It is anticipated that future computational models will become increasingly accurate and have significantly more predictive power, which are crucial conditions for applications in clinical settings.

Take-Home Message

- Two main computational approaches to model cell migration exist, each with its advantages and disadvantages.
- Various open-source and commercial software for computationally modelling cell migration are available.
- Computational models of cell migration can be differentiated into models of detailed, individual-level cell motility or collective cell migration.

(continued)

- Computational modelling of cell migration has gained significant momentum in studying questions related to single-cell motility, biological development, tissue engineering, the immune system, wound healing, and cancer progression.

References

1. Ridley AJ, Schwartz MA, Burridge K, Firtel RA, Ginsberg MH, Borisy G, et al. Cell migration: integrating signals from front to back. Science. 2003;302(5651):1704–9.
2. Friedl P, Gilmour D. Collective cell migration in morphogenesis, regeneration and cancer. Nat Rev Mol Cell Biol. 2009;10(7):445–57.
3. Ashley EA. Towards precision medicine. Nat Rev Genet 2016 179 [Internet]. 2016 Aug 16 [cited 2022 Jul 12];17(9):507–22. https://www.nature.com/articles/nrg.2016.86
4. Li X, Dowling EK, Yan G, Dereli Z, Bozorgui B, Imanirad P, et al. Precision combination therapies based on recurrent oncogenic coalterations. Cancer Discov [Internet]. 2022 [cited 2022 Jul 8];12:1542–59. http://aacrjournals.org/cancerdiscovery/article-pdf/12/6/1542/3153102/1542.pdf
5. Friedl P, Wolf K. Plasticity of cell migration: a multiscale tuning model. J Cell Biol. 2010;188:11–9.
6. Tao Y, Winkler M. A chemotaxis-haptotaxis model: the roles of nonlinear diffusion and logistic source. SIAM J Math Anal [Internet] 2011 [cited 2022 Jul 11];43(2):685–704. https://epubs.siam.org/terms-privacy
7. Lo CM, Wang HB, Dembo M, Wang YL. Cell movement is guided by the rigidity of the substrate. Biophys J. 2000;79(1):144–52.
8. Higazi AAR, Kniss D, Manuppello J, Barnathan ES, Cines DB. Thermotaxis of human trophoblastic cells. Placenta [Internet] 1996 [cited 2022 Jul 8];17(8):683–687. https://pubmed.ncbi.nlm.nih.gov/8916219/
9. Li Y, Xu T, Zou H, Chen X, Sun D, Yang M. Cell migration microfluidics for electrotaxis-based heterogeneity study of lung cancer cells. Biosens Bioelectron. 2017;89:837–45.
10. Park JS, Kim DH, Levchenko A. Topotaxis: a new mechanism of directed cell migration in topographic ECM gradients. Biophys J. 2018;114(6):1257–63.
11. Jekely G. Evolution of phototaxis. Philos Trans R Soc B Biol Sci. 2009;364(1531):2795–808.
12. Lowe B. The role of Ca2+ in deflection-induced excitation of motile, mechanoresponsive balancer cilia in the ctenophore statocyst. J Exp Biol. 1997;200(11):1593–606.
13. Krause M, Gautreau A. Integrin steering cell migration: lamellipodium dynamics and the regulation of directional persistence. Nat Publ Gr [Internet]. 2014 [cited 2022 Jul 11];15:577. www.nature.com/reviews/molcellbio
14. Fritz-Laylin LK. The evolution of animal cell motility. Curr Biol [Internet]. 2020;30(10):R477–82. https://doi.org/10.1016/j.cub.2020.03.026.
15. Ikenouchi J, Aoki K. A clockwork bleb: cytoskeleton, calcium, and cytoplasmic fluidity. FEBS J. 2021;289(24):7907–17.
16. Shao D, Levine H, Rappel WJ. Coupling actin flow, adhesion, and morphology in a computational cell motility model. Proc Natl Acad Sci USA. 2012;109(18):6851–6.
17. Johnson HE, King SJ, Asokan SB, Rotty JD, Bear JE, Haugh JM. F-actin bundles direct the initiation and orientation of lamellipodia through adhesion-based signaling. J Cell Biol. 2015;208(4):443–55.

18. Petrie RJ, Yamada KM. At the leading edge of three-dimensional cell migration. J Cell Sci. 2012;125:5917–26.
19. Paluch EK, Raz E. The role and regulation of blebs in cell migration. Curr Opin Cell Biol. 2013;25:582–90.
20. SenGupta S, Parent CA, Bear JE. The principles of directed cell migration. Nat Rev Mol Cell Biol. 2021;22:529–47.
21. Telley L, Agirman G, Prados J, Amberg N, Fièvre S, Oberst P, et al. Temporal patterning of apical progenitors and their daughter neurons in the developing neocortex. Science. 2019;364(6440):eaav2522.
22. Turing AM. The chemical basis of morphogenesis. Philos Trans R Soc B Biol Sci. 1952;237:37–72. http://hopf.chem.brandeis.edu/members_content/yanglingfa/pattern/Turing/TheChemical Basis of Morphogenesis.pdf
23. Majumdar R, Sixt M, Parent CA. New paradigms in the establishment and maintenance of gradients during directed cell migration. Curr Opin Cell Biol. 2014;30(1):33–40. https://doi.org/10.1016/j.ceb.2014.05.010.
24. Zhao J, Cao Y, DiPietro LA, Liang J. Dynamic cellular finite-element method for modelling large-scale cell migration and proliferation under the control of mechanical and biochemical cues: a study of re-epithelialization. J R Soc Interface. 2017;14(129)
25. Hartmann F, Katz C. Structural analysis with finite elements. New York: Springer; 2004.
26. Reddy JN, Gartling DK. The finite element method in heat transfer and fluid dynamics. 3rd ed. CRC Press; 2010.
27. Bastos JPA, Sadowski N. Electromagnetic modeling by finite element methods. Routledge; 2003.
28. Pakzad M. Parallel implementation of the finite element method on shared memory multiprocessors. PhD Thesis. Newcastle University; 1995.
29. Moure A, Gomez H. Phase-field modeling of individual and collective cell migration. Arch Comput Methods Eng. 2021;28(2):311.
30. Gonzalez-de-Aledo P, Vladimirov A, Manca M, Baugh J, Asai R, Kaiser M, et al. An optimization approach for agent-based computational models of biological development. Adv Eng Softw [Internet]. 2018 Jul 1 [cited 2022 Mar 16];121:262–75. https://doi.org/10.1016/j.advengsoft.2018.03.010
31. Breitwieser L, Hesam A, de Montigny J, Vavourakis V, Iosif A, Jennings J, et al. BioDynaMo: a modular platform for high-performance agent-based simulation. Bioinformatics. 2022;38(2):453–60.
32. González-Valverde I, García-Aznar JM. Mechanical modeling of collective cell migration: an agent-based and continuum material approach. Comput Methods Appl Mech Eng. 2018;337:246–62.
33. de Montigny J, Iosif A, Breitwieser L, Manca M, Bauer R, Vavourakis V. An in silico hybrid continuum-/agent-based procedure to modelling cancer development: interrogating the interplay amongst glioma invasion, vascularity and necrosis. Methods [Internet]. 2021;185(January 2020):94–104. https://doi.org/10.1016/j.ymeth.2020.01.006.
34. Jafari A, Broumand P, Vahab M, Khalili N. An eXtended finite element method implementation in COMSOL multiphysics: solid mechanics. Finite Elem Anal Des [Internet]. 2022;202(5):103707. https://doi.org/10.1016/j.finel.2021.103707.
35. Zubler F, Douglas R. A framework for modeling the growth and development of neurons and networks. Front Comput Neurosci. 2009;3:25.
36. Delile J, Herrmann M, Peyriéras N, Doursat R. A cell-based computational model of early embryogenesis coupling mechanical behaviour and gene regulation. Nat Commun. 2017;8:1–10.

37. Wilensky U, Evanston I. NetLogo: center for connected learning and computer-based modeling. Evanston, IL: Northwest University; 1999.
38. Tisue S, Wilensky U. NetLogo: a simple environment for modeling complexity. Conf Complex Syst. 2004.
39. Ghaffarizadeh A, Heiland R, Friedman SH, Mumenthaler SM, Macklin P. PhysiCell: an open source physics-based cell simulator for 3-D multicellular systems. PLoS Comput Biol. 2018;14:1–31.
40. Marée AFM, Grieneisen VA, Edelstein-Keshet L. How cells integrate complex stimuli: the effect of feedback from phosphoinositides and cell shape on cell polarization and motility. PLoS Comput Biol. 2012;8(3):e1002402.
41. Cao Y, Ghabache E, Rappel WJ. Plasticity of cell migration resulting from mechanochemical coupling. Elife. 2019;8:1–18.
42. Tozluoğlu M, Tournier AL, Jenkins RP, Hooper S, Bates PA, Sahai E. Matrix geometry determines optimal cancer cell migration strategy and modulates response to interventions. Nat Cell Biol. 2013;15(7):751–62.
43. Tan RZ, Chiam KH. A computational model for how cells choose temporal or spatial sensing during chemotaxis. PLoS Comput Biol. 2018;14(3):e1005966.
44. MacDonald G, Mackenzie JA, Nolan M, Insall RH. A computational method for the coupled solution of reaction–diffusion equations on evolving domains and manifolds: application to a model of cell migration and chemotaxis. J Comput Phys. 2016;309:207–26.
45. Winkler B, Aranson IS, Ziebert F. Confinement and substrate topography control cell migration in a 3D computational model. Commun Phys. 2019;2(1):82.
46. Murphy L, Madzvamuse A. A moving grid finite element method applied to a mechanobiochemical model for 3D cell migration. Appl Numer Math [Internet]. 2020;158:336–59. https://doi.org/10.1016/j.apnum.2020.08.004.
47. Heck T, Vargas DA, Smeets B, Ramon H, van Liedekerke P, van Oosterwyck H. The role of actin protrusion dynamics in cell migration through a degradable viscoelastic extracellular matrix: insights from a computational model. PLoS Comput Biol [Internet]. 2020 [cited 2022 Jul 7];16(1):e1007250. https://journals.plos.org/ploscompbiol/article?id=10.1371/journal.pcbi.1007250
48. Cao Y, Ghabache E, Miao Y, Niman C, Hakozaki H, Reck-Peterson SL, et al. A minimal computational model for three-dimensional cell migration. J R Soc Interface. 2019;16(161):20190619.
49. Moure A, Gomez H. Three-dimensional simulation of obstacle-mediated chemotaxis. Biomech Model Mechanobiol. 2018;17(5):1243–68.
50. Schlüter DK, Ramis-Conde I, Chaplain MAJ. Computational modeling of single-cell migration: the leading role of extracellular matrix fibers. Biophys J [Internet]. 2012 Sep 9 [cited 2022 Jun 14];103(6):1141.
51. Bauer R, Clowry GJ, Kaiser M. Creative destruction: a basic computational model of cortical layer formation. Cereb Cortex. 2021;31(7):3237–53.
52. Sharpe J. Computer modeling in developmental biology: growing today, essential tomorrow. Development. 2017;144(23):4214–25.
53. Dalle Nogare D, Chitnis AB. NetLogo agent-based models as tools for understanding the self-organization of cell fate, morphogenesis and collective migration of the zebrafish posterior lateral line primordium. Semin Cell Dev Biol [Internet]. 2020;100(December 2019):186–98. https://doi.org/10.1016/j.semcdb.2019.12.015.
54. Rubenstein J, Rakic P. Cellular migration and formation of neuronal connections: comprehensive developmental neuroscience. Academic Press; 2013.
55. Del-Valle-anton L, Borrell V. Folding brains: from development to disease modeling. Physiol Rev. 2022;102(2):511–50.

56. Darayi M, Hoffman ME, Sayut J, Wang S, Demirci N, Consolini J, et al. Computational models of cortical folding: a review of common approaches. J Biomech. 2022;139:110851.
57. Nie J, Guo L, Li G, Faraco C, Stephen Miller L, Liu T. A computational model of cerebral cortex folding. J Theor Biol [Internet]. 2010;264(2):467–78. https://doi.org/10.1016/j.jtbi.2010.02.002.
58. Tallinen T, Chung JY, Biggins JS, Mahadevan L. Gyrification from constrained cortical expansion. Proc Natl Acad Sci USA. 2014;111(35):12667–72.
59. Bayly P V., Okamoto RJ, Xu G, Shi Y, Taber LA. A cortical folding model incorporating stress-dependent growth explains gyral wavelengths and stress patterns in the developing brain. Phys Biol [Internet]. 2013 Feb [cited 2022 Jul 12];10(1):016005.
60. Garcia KE, Wang X, Kroenke CD. A model of tension-induced fiber growth predicts white matter organization during brain folding. Nat Commun. 2021;12(1):6681.
61. Razavi MJ, Zhang T, Li X, Liu T, Wang X. Role of mechanical factors in cortical folding development. Phys Rev E. 2015;92(3):032701.
62. Zubler F, Hauri A, Pfister S, Bauer R, Anderson JC, Whatley AM, et al. Simulating cortical development as a self constructing process: a novel multi-scale approach combining molecular and physical aspects. PLoS Comput Biol [Internet]. 2013 [cited 2022 Mar 14];9(8):e1003173. https://journals.plos.org/ploscompbiol/article?id=10.1371/journal.pcbi.1003173
63. Soares JS, Sacks MS. A triphasic constrained mixture model of engineered tissue formation under in vitro dynamic mechanical conditioning. Biomech Model Mechanobiol. 2016;15(2):293–316.
64. Loerakker S, Ristori T. Computational modeling for cardiovascular tissue engineering: the importance of including cell behavior in growth and remodeling algorithms. Curr Opin Biomed Eng. 2020;15:1–9.
65. Burova I, Wall I, Shipley RJ. Mathematical and computational models for bone tissue engineering in bioreactor systems. J Tissue Eng. 2019;10:2041731419827922.
66. Torii R, Velliou RI, Hodgson D, Mudera V. Modelling multi-scale cell–tissue interaction of tissue-engineered muscle constructs. J Tissue Eng. 2018;9:2041731418787141.
67. Wiedemann C, Amann B, Degroote RL, Witte T, Deeg CA. Aberrant migratory behavior of immune cells in recurrent autoimmune uveitis in horses. Front Cell Dev Biol. 2020;8(March):1–15.
68. Byrne MB, Kimura Y, Kapoor A, He Y, Mattam KS, Hasan KM, et al. Oscillatory behavior of neutrophils under opposing chemoattractant gradients supports a winner-take-all mechanism. PLoS One. 2014;9(1):e85726.
69. Sakumura Y, Tsukada Y, Yamamoto N, Ishii S. A molecular model for axon guidance based on cross talk between Rho GTPases. Biophys J. 2005;89(2):812–22.
70. Chiacchio F, Pennisi M, Russo G, Motta S, Pappalardo F. Agent-based modeling of the immune system: NetLogo, a promising framework. Biomed Res Int. 2014;2014:907171.
71. Niculescu I, Textor J, De Boer RJ. Crawling and gliding: a computational model for shape-driven cell migration. PLoS Comput Biol. 2015;11:e1004280.
72. Dong R, Guo B. Smart wound dressings for wound healing. Nano Today. 2021;41:101290.
73. Mi Q, Rivière B, Clermont G, Steed DL, Vodovotz Y. Agent-based model of inflammation and wound healing: insights into diabetic foot ulcer pathology and the role of transforming growth factor-β1. Wound Repair Regen [Internet]. 2007 Sep 1 [cited 2022 Jul 14];15(5):671–82. https://onlinelibrary.wiley.com/doi/full/10.1111/j.1524-475X.2007.00271.x
74. Boon WM, Koppenol DC, Vermolen FJ. A multi-agent cell-based model for wound contraction. [cited 2022 Jul 8]. https://doi.org/10.1016/j.jbiomech.2015.11.058
75. Lee JJ, Talman L, Peirce SM, Holmes JW. Spatial scaling in multiscale models: methods for coupling agent-based and finite-element models of wound healing. Biomech Model Mechanobiol [Internet]. 2019;18(5):1297–309. https://doi.org/10.1007/s10237-019-01145-1.

76. Wyczalkowski MA, Varner VD, Taber LA. Computational and experimental study of the mechanics of embryonic wound healing. J Mech Behav Biomed Mater. 2013;28:125–46.
77. Axenie C, Bauer R, Martínez MR. The multiple dimensions of networks in cancer: a perspective. Symmetry 2021;13:1559 [Internet]. 2021 Aug 25 [cited 2022 Jul 12];13(9):1559. https://www.mdpi.com/2073-8994/13/9/1559/htm
78. Spatarelu CP, Zhang H, Nguyen DT, Han X, Liu R, Guo Q, et al. Biomechanics of collective cell migration in cancer progression: experimental and computational methods. ACS Biomater Sci Eng. 2019;5(8):3766–87.
79. Sudlow C, Gallacher J, Allen N, Beral V, Burton P, Danesh J, et al. UK Biobank: an open access resource for identifying the causes of a wide range of complex diseases of middle and old age. PLoS Med. 2015;12(3):e1001779.
80. Ollier W, Sprosen T, Peakman T. UK biobank: from concept to reality. Pharmacogenomics. 2005;6(6):639–46.
81. Venters CC, Capilla R, Betz S, Penzenstadler B, Crick T, Crouch S, et al. Software sustainability: research and practice from a software architecture viewpoint. J Syst Softw. 2018;138(1):174–88.
82. Letai A. Functional precision cancer medicine-moving beyond pure genomics. Nat Med [Internet]. 2017;23(9):1028–35. https://doi.org/10.1038/nm.4389.

Software Applications for the Analysis of Cell Migration

Arndt Rohwedder

What You Will Learn in This Chapter
This chapter aims to provide an overview of software available for analysing raw data from migratory studies. It discusses significant parameters for evaluating migratory events and serves as a guide to selecting the most suitable software based on the acquisition technique used in the experiment. Special focus is given to open-source solutions. Finally, internet links are provided to help find useful analytical software.

1 Introduction

The development of software for the analysis and quantification of cell migration events is a very fast-growing field. Any list of available software will be outdated even while this chapter is written. Therefore, only a few established examples will be quoted in the following text.

Any computational analysis of cell migration events relies on the fact that there are recorded differences between two time points. This can be continuous time series recorded in 2D or 3D on an imaging system but also impedance measurements. Other analysis

A. Rohwedder (✉)
Johannes Kepler University Linz, Linz, Austria
e-mail: arndt.rohwedder@jku.at

A. Brüning-Richardson, S. Knipp (eds.), *Cell Migration in Development, Health and Disease*, Learning Materials in Biosciences,
https://doi.org/10.1007/978-3-031-64532-7_11

methods may rely on starting conditions compared with endpoint conditions, a method that could be considered a condensed time series.

Measurements may rely on changes in location or morphological changes, or both over time. Under certain conditions, even single time points might contain migratory information, provided sufficient background information on the experimental subject is available. In light of this variety of methods, the analytical task has to be split between pure quantification, which has to be adjusted to the method of recording, and data analysis. A huge amount of software exists to cope with both tasks, either in combination or as separated elements.

Taking into account that in all living systems, a broad variation of cellular responses exists, a considerable number of experimental data/replicates are necessary. With the advent of high-throughput recording systems, powerful computational analysis systems are necessary to cope with the huge amount of data. Available storage has become less of a problem in recent years, but the amount of data generated in always shorter intervals and higher dimensions has also increased. Therefore, analytical software more and more has to be integrated into the interpretation of the pure results. Integration of data from different acquisition methods, allowing a comparison of experimental setups, still provides a challenge.

2 General Considerations

Three layers of software for the analysis of cell migration can be considered:

- Acquisition: Software necessary to control the acquisition instrument and manage data storage. Usually provided by the manufacturer of the instrument and requires more specialist knowledge. Nevertheless, some open-source software available can be tailored to purpose.
- Quantification: Software to extract cell migration-relevant parameters from the acquired data. The software is subject to extensive development and exists in a broad range of specialisations. Both commercial and open-source software are available.
- Data analysis: Software for handling quantified data to extract biologically relevant information. Several statistical packages, commercial as well as free, are available. High-throughput measurements increasingly require the application of statistical software and, also increasingly, machine learning applications.

For the purpose of analysing cell migration, two approaches are possible: the most common and time-resolved techniques involve the recording and subsequent analysis of time series.

Although other technologies, like impedance measurements [1, 2], are available, the vast majority of data for analysing cell migration come from imaging applications. Quantification of experimental data in non-imaging techniques is very different from imaging-based methods and is mostly closely hardware connected. This limits the user's choice to the subsequent data interpretation/analysis.

Even when limiting the scope to imaging-related techniques, a multitude of options apply, resulting in a broad range of potential interpretations and problems of translating results.

Imaging in general relates to the acquisition and analysis of multidimensional data stored in different formats. A single image here represents an at least 2D array of values; thus, an image contains as much information as pixels have been acquired.

Alternative to entire series of images, single scans can be used to analyse the shape and orientation of cells or cell aggregates, respectively. This requires some assumptions regarding the shape a cell predominantly adopts when in motion.

The challenge here is to isolate the relevant information for cell migration purposes. This is where the software comes in.

When considering cell migration analysis, it is important to identify the relevant parameters. And there are a few: not only the speed but also directionality, frequency of migration events, overall length of migration, chain migration, mode of migration (e.g. nucleokinesis or amoeboid) and probably more. The number of parameters becomes even more numerous if the third dimension and/or a complex environment is included.

The task then will be to decide which parameters are relevant for the experimental conditions addressed. This is a highly critical part of the analysis. At this point, there is a high risk of the introduction of bias by the experimenter or neglect of potentially relevant parameters. Any analysis software, on the one hand, should extract as many features of the measurements as possible and, on the other hand, support the user in choosing the most relevant data. Hardly any software combines all those features, requiring users to combine several classes of software.

Numerous relevant imaging techniques exist, bearing the potential of providing information on cell migratory behaviour. Image data can come from instruments as diverse as phase contrast, confocal microscopy or magnetic resonance imaging. Associated data, therefore, come in a multitude of dimensionalities and, consequently, resolutions in time and space. Nevertheless, there is an urgent need for a unifying approach, improving the translation of analysis results between systems applied.

The process starts with the acquisition technique applied. Most available imaging systems come with dedicated software provided by the manufacturer, although increasingly open systems are being developed.

Some manufacturers have already integrated analysis suites into their acquisition software to provide at least some basic image analysis capabilities. Some of the most prominent manufacturers are as follows:

Zeiss uses the Zen software for a multitude of imaging systems. It comes with several basic image analysis tools, like image enhancement and/or segmentation, and can be extended by purchasable modules. The LAS-X software from Leica provides equivalent options, and a purpose-specific version can be purchased. The Olympus cellSens software focuses on acquisition, but the company provides separate analysis tools. Nikon provides the NIS-Elements software for their microscopes.

On the open-source side, there is the μManager, probably the most prominent. It focuses on acquisition and does not provide remarkable analysis tools. However, as an open-source system, it provides the opportunity for the user to modify the code for specialised purposes.

Several parameters describe the migratory behaviour of cells, but no acquisition technique is able to access all of them. Often, a single parameter is not able to uncover differences between migrating cells. Table 1 lists a number of potential migration-relevant parameters and which technique has the potential to measure them. The higher the number of measured parameters, the harder the choice to find the most representative ones. When considering software/workflow development, availability and practicability are important elements. However elegant the usage of certain programming environments might be, the main aim is to provide working and reliable software for the quantification and analysis of imaging data. Over recent years, the interpreted language Python has become a powerful tool for this purpose. It not only is easy to learn and understand but also provides comfortable specialised imaging-related tools that can be integrated easily. Python is handy because it is very widespread, cheap and well-maintained. It also provides, with only a few additional activities, the option to turn a script into stand-alone software to be distributed on computers without previous Python installation.

User bias or preference for a certain parameter might lose out on relevant information. Therefore, in the final step of analysing cell migration, statistical data analysis must include evaluation of the parameter's significance, e.g. by using principal component analysis (PCA).

3 Time-Resolved Migration Assays

Due to the high data content of such measurements, these measurements are usually limited to 2D.

In its most basic form, migration assays are performed under phase-contrast light microscopy. Here, cells are captured in a time series, usually with a digital camera device in a series of greyscale images (Fig. 1). The most famous are the wound healing assays, in which a gap in a cell monolayer is produced, and cells migrating into this gap are observed (Fig. 2). The gap is not mandatory for the analysis of migratory behaviour and, in the case of chemoattraction/chemorepulsion, is not helpful, but it provides the easiest technique to achieve high-throughput data on cell speed and orientation in two dimensions. Software for the analysis of migrating cells exists in abundance, both commercially and as open source/freeware.

Table 1 Parameters with potential migratory information

Parameter	Description	Technology
X—Extension (complete)	Maximum breadth of cells or aggregates in an image	2D and 3D imaging
X—Extension (largest)	Maximum breadth of the largest cell or aggregate in an image	2D and 3D imaging
Y—Extension (complete)	Maximum length of cells or aggregates in an image	2D and 3D imaging
Y—Extension (largest)	Maximum length of the largest cell or aggregate	2D and 3D imaging
Z—Extension	Maximum height of cells or aggregates in an image	3D imaging
X/Y extension ratio (complete)	Identification of predominant extension of the image in the plane	2D and 3D imaging
X/Y extension ratio (largest)	Identification of predominant extension of the largest cell or aggregate in the plane	2D and 3D imaging
X/Z extension ratio	Determine the "flatness" of cells or aggregates in an image	3D imaging
Apoptosis	Number/occurrence of early-stage apoptosis	Impedance
Tightness	Distance between adjacent cells, cell-cell contacts, tight junctions	Impedance
Dislocation vector	Distance moved between two time points	Impedance, 2D and 3D imaging
Cell location	Relative location of cells	Impedance, 2D and 3D imaging
Surface area	Area of the membrane surface of a cell or aggregate	Impedance, 3D imaging
Number of extensions	Number of protrusions measured from a cell body	2D (limited) and 3D imaging
Mean length of extensions	Average lengths of protrusions measured from a cell body	2D (limited) and 3D imaging
Median length of extension	Median lengths of protrusions measured from a cell body	2D (limited) and 3D imaging
Variance length of extension	Variance lengths of protrusions measured from a cell body	2D (limited) and 3D imaging
Mean size of extensions	Average surface area or volume of protrusions measured from a cell body	3D imaging
Max. size of extension	Maximum surface area or volume of protrusions measured from a cell body	3D imaging
Variance size of extension	Variance surface area or volume of protrusions measured from a cell body	3D imaging

(continued)

Table 1 (continued)

Parameter	Description	Technology
Classification of extension	Parameter-based classification of measured extension: brim, broad, narrow, small	3D imaging
Fractionation	Especially in the case of aggregates: number of isolated cells from central aggregate	Impedance, 2D and 3D imaging
Average distance to central aggregate	Aggregate measurements: average distance of isolated cells from central aggregate	Impedance, 2D and 3D imaging
Median distance to central aggregate	Aggregate measurements: median distance of isolated cells from central aggregate	Impedance, 2D and 3D imaging
Variance distance to central aggregate	Aggregate measurements: variance distance of isolated cells from central aggregate	Impedance, 2D and 3D imaging
Total distance travelled	In single-cell tracking: vectorial distance between start and end points	Impedance, 2D imaging
Net distance travelled	Difference between start and end points	Impedance, 2D and 3D imaging
Mean curvilinear speed	For time series: average of speed between time points	Impedance, 2D imaging
Mean straight-line speed	Speed between start and end points	Impedance, 2D and 3D imaging
Directionality ratio	Identification of predominant directions	Impedance, 2D imaging
Max. distance travelled	Migratory capability of cells	Impedance, 2D and 3D imaging
Change in open wound area	Area change of cell free area in wound healing assays	Impedance, 2D imaging
Speed of wound closure	Time required to close a wound in a wound healing assay	Impedance, 2D imaging
Shape of wound closure	Determination if cell migration occurs global or central	Impedance, 2D imaging

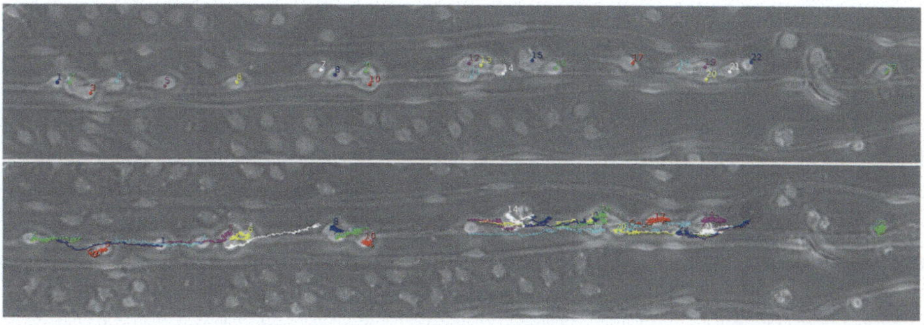

Fig. 1 Migrating Locusta neurons. Courtesy of Sabine Knipp

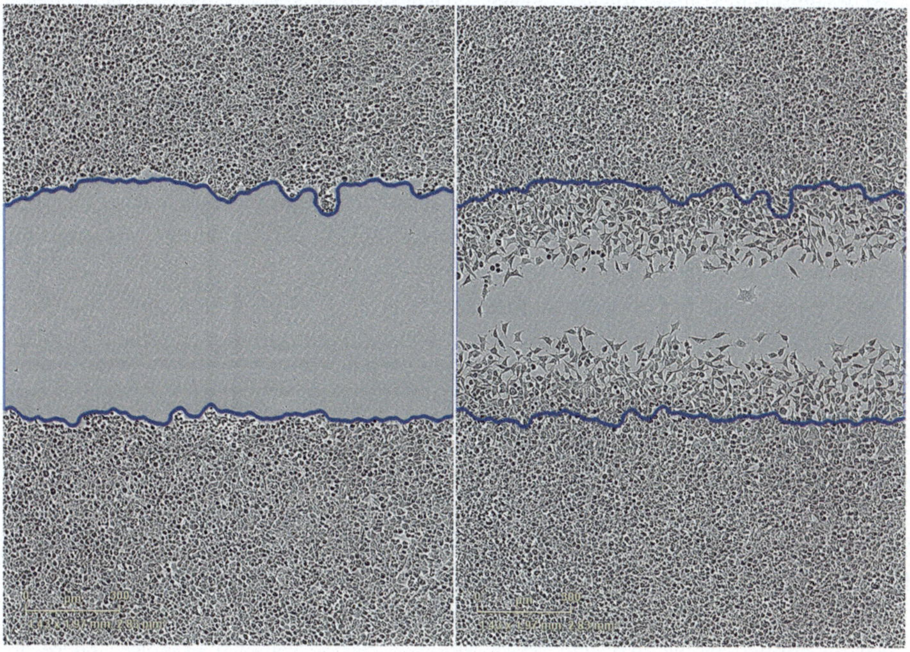

Fig. 2 Wound healing assay with HEK293 cells. Courtesy of Sabine Knipp

There are commercial solutions combining hardware and analysis software based on imaging.

4 Impedance Measurements

This technology is not as widespread as imaging-based methods but provides reliable measurements of cell shape and location. Impedance measurements exploit the conductive properties of cell membranes and, by alterations in frequencies and electrical conductivity, deduce the shape and position of individual cells or cell clusters. One of the advantages of impedance-based measurements is its label-free application, suitable for live cell measurements while avoiding phototoxicity. A good review on that technique can be found in Chiu et al., 2019. The advent of microfabrication and decentralised manufacturing enables the specialised application of this technology to be adapted to expert applications [3].

Due to the necessary specialist equipment for impedance-based measurement systems, acquisition open-source software is not available. Parameters that can be acquired using impedance technology include cell morphology, cell apoptosis, cell proliferation and cell migration [1, 4]. To access these parameters, the experimental setup has to be chosen carefully, rendering impedance measurement acquisition more complicated than imaging-based techniques. Unlike imaging-based measurements, current frequency, electrode position and current intensity have to be considered.

The basic impedance measurements have to be translated into the desired parameters. Again, there is a lack of widespread, easily accessible software to extract the parameter information. The expensive equipment, provided only by a handful of manufacturers or even laboratory-specific constructions, clearly represents a disadvantage over the much cheaper imaging-centred approaches. Despite its clearly advantageous precise measurement technology, impedance measurements will most likely remain a niche technique until they become more widespread.

Upon retrieval of the above parameters, this type of measurement faces the same analytical challenges as optical/microscopical measurement techniques. This part of the data analysis will be discussed later in this chapter.

5 Imaging Technology

When imaging technologies are applied, some general steps are usually implemented to allow quantification. These steps can be applied in hidden layers of the software used or require adjustment by the user. To separate background from cells, an intensity threshold is usually applied in the first step, often followed by the removal of noise. There are a number of different threshold algorithms, and choosing the correct algorithm can be tricky. Intensity threshold can often be adjusted by the user, although some software applies threshold estimation automatically. At this step, image resolution differences can already cause problems in the transferability of quantification results: noise in low-resolution images could potentially lead to the removal of single cells and thus mask potential migration events from a central object. In high-resolution images, on the other hand, automatic noise-reducing algorithms could result in false-positive interpretations of cells by the software. The next step is the identification of cell boundaries using algorithms like edge detection or watershed, which provide outlines of the cells. Different image quantification software use different algorithms for this purpose and demand user knowledge to varying degrees. Open-source as well as commercial analysis software exists with different degrees of specialisation. While user adjustments have minor effects on time-series migration assays, where the cell's path provides the main information, morphological measurements are strongly affected. User bias can easily interfere with acquiring comparable data.

6 Acquisition Software

A broad range of acquisition software is available for this method, usually provided by the manufacturer of the imaging hardware. Nevertheless, open-source software also exists that can be adjusted to specialised user demands. The best example here might be μManager [5], which provides support for multiple imaging hardware and integration with the free image analysis software Fiji/ImageJ [6]. Other frameworks require programming language integration and, by that, achieve a high degree of modularity, such as the Python library: Python Microscope [7]. A more generalised approach, ImSwitch, requires more expert background [8]. Time-series-based migration assays are already demanding in terms of storage space by nature, and often these methods are combined with high-throughput measurements. Adaptation of acquisition software might be demanding in this context. Commercially available software provided by manufacturers of microscopes do not necessarily offer the option of high-throughput acquisition, depending on the type of microscopic device. Often, the software can be adapted by purchasing additional modules.

7 Imaging Time-Series Migration Assays

The most widespread technique to evaluate the migratory properties of cells or cell aggregates is the wound healing assay. It has become one of the most frequently used techniques to measure cell migration in two dimensions. The technique is comparatively easy: a 2D cell culture under consideration is grown to confluency, then a cut, or "wound", is applied, and the invading cells into this cleared region are recorded (see Fig. 2 for an example).

While options for real-time measurements exist, most of the data extraction takes place on recorded time series. Deduced from the statement above, one can imagine that storage and administration of imaging data require frameworks in themselves. Some platforms, both free and commercial, exist that also provide the integration of third-party software or custom-made analysis scripts. One such example is the Open Microscopy Environment (OME), which provides access from a number of languages and a standardised file storage system while connecting databases and controlling access [9]. Different approaches can be used to organise imaging data for morphological analysis. Systems intended to combine imaging data from different sources, including morphological information, exist [10] but are intended to provide reference material rather than organisational structures for the lab. Online imaging databases like the Image Data Resource (IDR) exist to provide reference material and submit published imaging data that might be helpful for evaluating experimental results [11].

7.1 Single-Cell Migration Assays

To retrieve migratory data from imaging time series, a number of tools exist. An excellent overview can be found in Barry et al. [12] or Masuzzo et al. [13]. In its simplest form, the user will have to mark single cells in each subsequent time frame, and the data for this particular cell will be stored separately. An example of single-cell-based cell migration quantification techniques is the MTrackJ ImageJ plugin, designed to be used for a low number of cells and to be quantified in a time series [14]. This becomes impractical if the cell number increases. Automated cell detection and tracking methods are therefore more suitable for high-throughput data analysis. The MATLAB-based CellTracker provides a well-documented specialised application that is able to provide reliable tracking analysis [15]. An example using the same image analysis platform as MTrackJ is TrackMate [16]. Using automated tracking software comes with some requirements for time series to be analysed. The software uses segmentation techniques to isolate single cells and therefore needs to be able to identify cells from their background. Although this works well under optimal imaging conditions, images usually require some preparation before the analysis can be applied. Potential modifications can include contrast adjustment or background subtraction. The application of machine learning has the potential to overcome quality issues in time series and has recently been included in the ImageJ TrackMate plugin [17]. The requirement to control the reliability of automated analysis software will still remain with the user. This is true not only for open-source software but also for commercial software. The advantage of open-source software in this context is the ability of the user to track applied algorithms, an option that does not exist in commercial software. Faulty results produced by commercial software usually cannot be adjusted by the user, and the correction fully depends on the service of the software company.

8 2D Wound Healing Assay

The analysis needs to be altered when a classical wound healing assay is used to determine the migratory capabilities of a 2D cell culture. Unlike the single-cell tracking software discussed, this software relies on the detection of areas of confluent cell lawns versus open areas without cells. Analysis in wound healing assays usually measures the time required to close the gaps produced at the start of the experiment in a closed cell lawn. Open-source tools for ImageJ exist and, at least partially, enable high-throughput analysis of important parameters from this type of experiment. For instance, a basic analysis without high-throughput option is TScratch [18], while high-throughput capabilities are promised by the plugin Wound_healing_size_tool [19].

Modern commercial high-throughput imaging systems often already include a range of automatic cell migration analysis tools, e.g. the ImageXpress systems from Molecular Devices® or Incucyte from Sartorius®, to mention just two widespread machines. The underlying software for the analysis of the imaging provides results very similar to those of the open-source software but also usually "predigests" reports for users for immediate use.

Although it is potentially possible to produce 3D time series, the acquisition tends to be tricky, especially if the objects are able to migrate out of the focal plane.

9 Morphology-Based Software Tools

Interpretation of cellular morphology for the purpose of migratory events requires either prior information about the cell type under consideration or acquisition of reference data. Reference data would be the cellular situation at a certain time point to any treatment. This requires less acquisition data than time series but excludes the determination of entire migration paths. If a cellular response in three dimensions is to be considered, two time points rather than continuous time series are more likely the tool of choice.

10 2D Morphology Analysis

A multitude of image analysis tools are available for retrieving information based on 2D images. Several tools are already implemented as plugins for ImageJ/Fiji to allow a number of the above parameters to be extracted (Fig. 3; e.g. [12, 20, 21]). Often, such tools are very specialised. For example, there are morphological plugins specialised in the quantification of ultra-resolution accessible features, like spines on neurons [22]. Such specialised software can rarely be used to extract migratory information from single cells or cell groups, although it provides helpful additional information.

The problem of morphological quantification from 2D cell micrographs represents a subject that has been intensely worked on for a long time. Therefore, a number of solutions

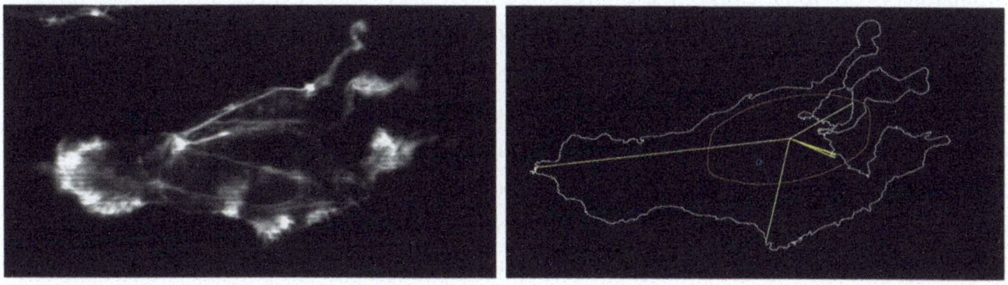

Fig. 3 Single glioblastoma cell from Ketchen et al. [20]

exist. Apart from commercially available multipurpose image analysis software like Imaris or VoloCity, to name just two, open-source software is also very diverse. Software that requires a programming language installed on the machine might be the most frequently found solution. However, there are some potential limitations. Although open-source software using MATLAB® as a base can be considered "free" in itself, it still requires the comparatively expensive background of MATLAB itself. One excellent MATLAB®-based open-source solution for morphological analysis is ShapeMetrics [23]. Translation or transfer of the relevant script into the free variant of MATLAB®, GNU Octave, might be possible but requires strong background knowledge in this programming language.

Easier to access and often already installed on many computers is the interpreted language Python. Nevertheless, due to dependencies, running a script in Python might not be straightforward. In Python, the development quickly moved towards 3D analysis, leaving a number of libraries that require some basic programming skills. One example is a quantification library for neuronal morphology [24]. This example highlights the problem of identifying suitable software. Although software for the quantification of neuronal morphology is highly popular [25], data resulting from such analysis rarely contribute significantly to information on cellular migration. Python has become more attractive for the analysis of 2D image data with the availability of machine learning algorithms that can be easily included in the workflow [26]. In terms of cell morphological analysis, machine learning has the potential to separate different cell types, but care needs to be taken when certain parameters are of interest. In cases of supervised learning, the parameters are rarely recorded. With the increasing number of high-content image acquisition systems, high-throughput and machine learning capabilities have become more and more important. Some of the above multipurpose software solutions already provide capabilities to analyse sets of images in one go but often require adjustments. Some workflows address these problems in terms of morphology quantification and subsequent analysis [27].

11 3D Morphology Analysis

3D analysis has increasingly become of interest with high numbers of 3D cell cultures, like spheroid matrix embedding. The advantage of analysing cells or cell aggregates in three dimensions is obvious. Cells, like the rest of us, not only occupy three dimensions but also move in three dimensions. When analysed in two dimensions, especially when derived from recorded image stacks, information on extensions tends to be lost or misinterpreted: an extension in one slide of the stack will lose contact with the cell body in another slide and would be interpreted as an individual entity. Additionally, the shape of a cell body could mask smaller extensions if those are located in different slides than the maximal size of the cell/aggregate body. It is therefore crucial that a 3D recorded cell/aggregate morphological analysis also respects the dimensionality. Parameters derived from 3D morphological analysis generally represent an extension of parameters recorded from 2D morphological analysis (see Table 1). Computationally, 3D image analysis requires significantly more effort. Software solutions therefore need to take this into account. More recent developments apply various technologies to handle this problem, and in theory, there should be no problem quantifying 3D imaging data on a standard PC. Many of the multipurpose software solutions mentioned above already include 3D analysis. Implementations for ImageJ/Fiji that handle 3D analysis have been developed, but it should be noted that ImageJ/Fiji were initially developed for the analysis of 2D images, and it is not easy to "convince" ImageJ/Fiji to "think" in three dimensions. This is also true for some of the more specialised software. Software dedicated to high-content 3D analysis does exist (Fig. 4; [28]), but it is still infrequent. Python has proven to be a very useful tool for handling 3D data, due to its easily installable modules. Some modules are ideally suited for data extraction and integration into statistical packages for mass data analysis (see links below).

12 Data Analysis

Although the subsequent data analysis depends on each lab's established and preferred methods, it may be of interest that this part could also be integrated into the image analysis workflow. Workflows based on MATLAB (GNU Octave), R or Python allow for a large amount of modularity, which can be highly valuable when high-content methods are applied. Libraries/modules are available in large quantities and excellent quality and usually do not require a degree in informatics to be used. Covering these would require an entire book at least in parts. I would also encourage thinking outside the field. Many quantification problems have been solved in engineering, archaeology and even security. After all, for the computer, a human is simply a central body with extensions.

13 Choices

The amount of potentially usable software is large. To decide which software to use, it is worth considering what is needed and what resources (personnel and financial) are available.

For example, if data analysis from 3D acquisition is planned, the software clearly has to support 3D image analysis. Next, how much time is needed to achieve useful data analysis with the specific software? Is it necessary to browse through a large number of manuals or even take a course on the software? Often, multipurpose analysis software is very well-equipped with a large number of quantification functions but can be confusing for beginners. Using libraries in, say, Python, would additionally require at least some basic programming skills. On the other hand, a scripting approach can be much more target-oriented. Commercial software is usually designed to handle a multitude of image quantification issues, making it multipurpose. The advantages typically include good support, regular updates and services that allow direct contact for questions or even

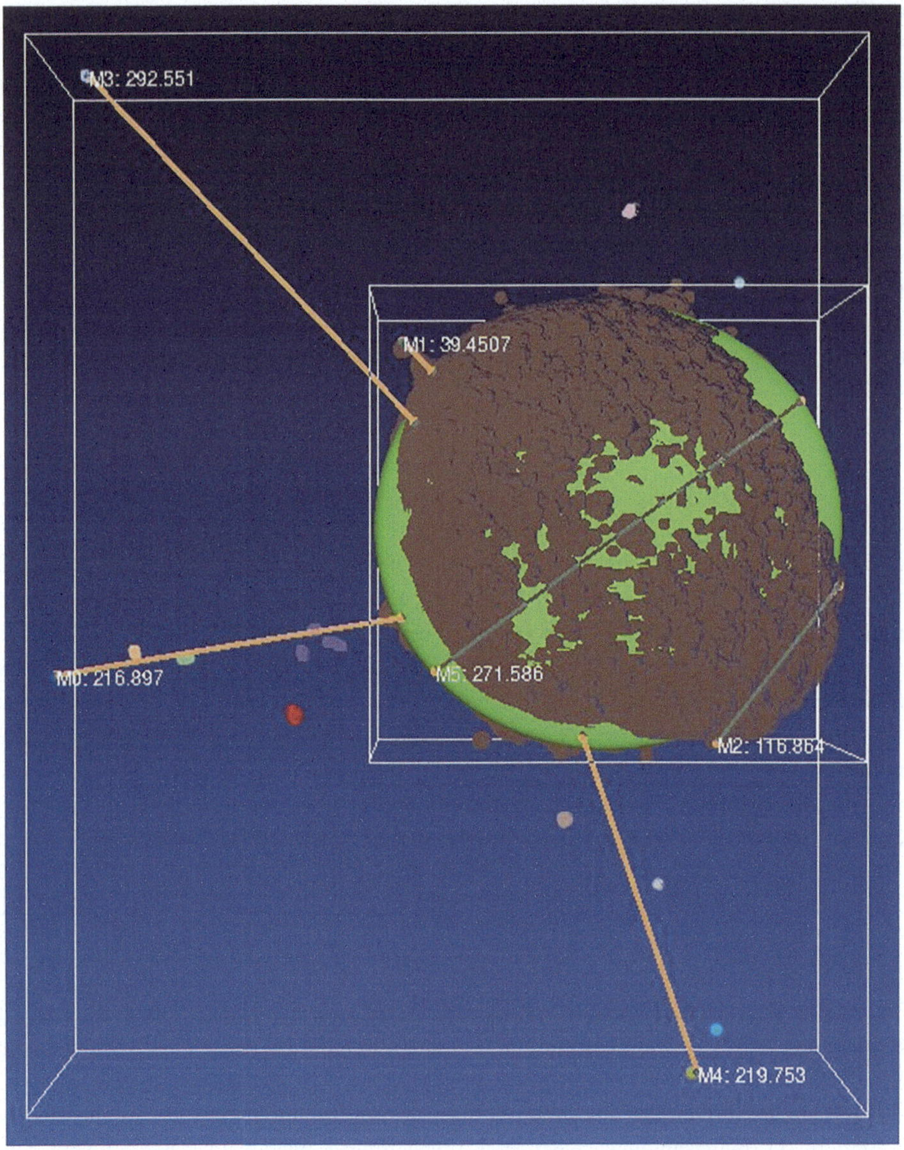

Fig. 4 Representation of a 3D spheroid, with some relevant parameters highlighted

courses. However, the drawbacks, aside from the costs, can include limited workflow control. Although publications might use a particular software, the reliability of the hidden algorithms can be questioned. With commercial software, there is no option to control correctness. If used for publicly funded research, the use of software developed with public funding should be considered.

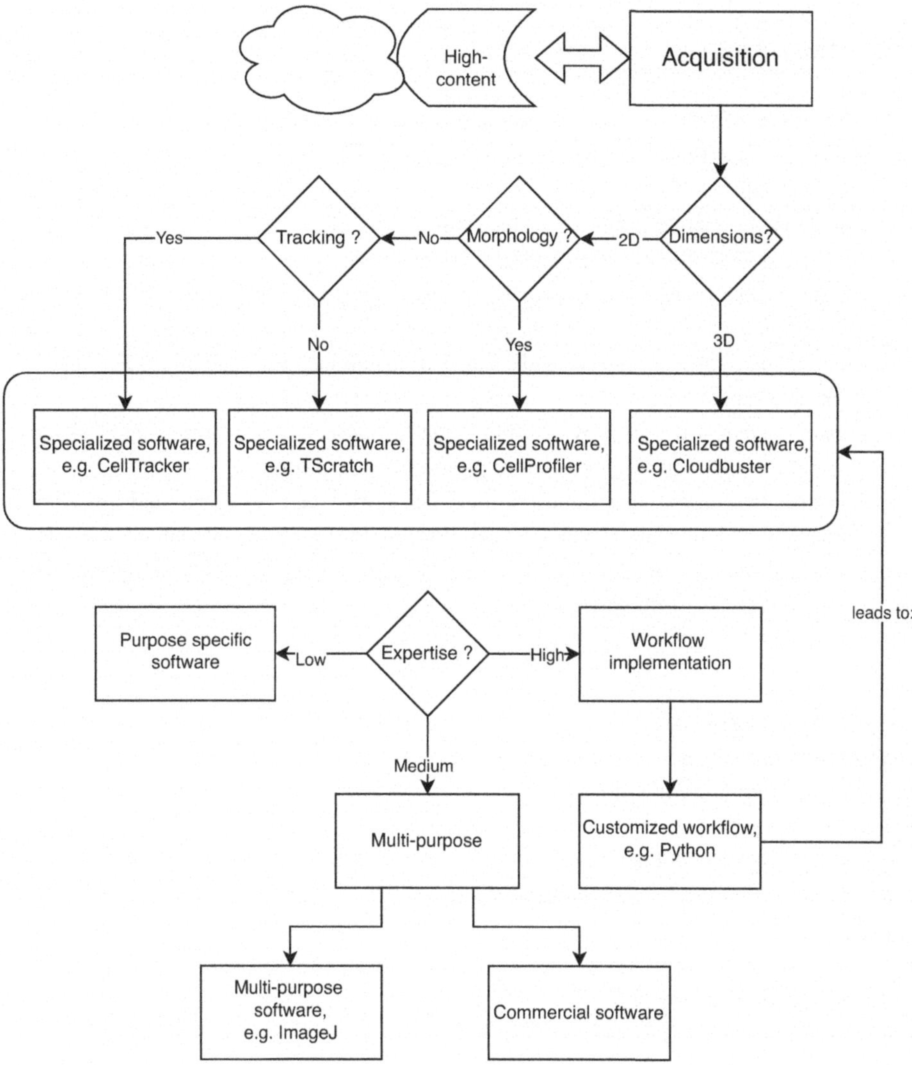

Fig. 5 Decisions to consider. The examples mentioned in the graph are representative of the software discussed in the text

The graph in Fig. 5 can help make decisions about which software to choose. The examples mentioned in the graph are intended to represent a group of software options discussed in the text.

Take-Home Message

- The choice of analysis software depends on the available expertise. The broader the range of applications for the software, the more time it takes to use it for a specific task.
- Specialised software helps to produce results quickly.
- Multipurpose software often requires higher user expertise but, in most cases, can be transformed into specialised software through the integration of plug-ins/modules.
- Open-source software provides algorithm transparency but generally lacks quick support options.
- Closed-source software provides high support but lacks transparency and risks a complete loss of data if the software product is discontinued.
- Modern interpreted programming languages are easy to learn and provide numerous modules to produce specialised workflows, ideally suited for the analytical task in question. The learning effort required is quickly compensated by the resulting useful software.

Links for Free Software

Programming languages:
https://www.python.org/downloads/
https://octave.org/index
https://cran.r-project.org/index.html
https://www.oracle.com/java/technologies/downloads/
Acquisition:
https://micro-manager.org/
https://python-microscopy.org/
https://imswitch.readthedocs.io/en/stable/
Database:
https://www.openmicroscopy.org/omero/
https://idr.openmicroscopy.org/
Image analysis:
https://imagej.nih.gov/ij/
https://imagej.net/imagej-wiki-static/Fiji
https://icy.bioimageanalysis.org/
https://qupath.github.io/
https://www.celltracker.website/index.html
https://www.cse-lab.ethz.ch/tscratchdownload/
https://github.com/AlejandraArnedo/Wound-healing-size-tool/

https://github.com/KerosuoLab/Shapemetrics
https://github.com/djpbarry/Adapt
https://github.com/swanandlab/SMorph
https://github.com/ARRohwedder/Surface_and_Direction_Detector
https://github.com/ARRohwedder/Cloudbuster
https://github.com/kukionfr/VAMPIRE_open
https://imagej.net/imagej-wiki-static/TrackMate
https://imagescience.org/meijering/software/mtrackj/
Helpful python modules:
http://www.open3d.org/
https://pythonawesome.com/a-python-library-for-common-tasks-on-3d-point-clouds/
https://scikit-image.org/
https://scikit-learn.org/stable/
https://pypi.org/project/avifilelib/

References

1. Chiu S-P, Batsaikhan B, Huang H-M, Wang J-Y. Application of electric cell-substrate impedance sensing to investigate the cytotoxic effects of andrographolide on U-87 MG glioblastoma cell migration and apoptosis. Sensors. 2019;19:1–15.
2. Xu Y, Xie X, Duan Y, Wang L, Cheng Z, Cheng J. A review of impedance measurements of whole cells. Biosens Bioelectron. 2016;77:824–36.
3. Jahnke HG, Mewes A, Zitzmann FD, Schmidt S, Azendorf R, Robitzki AA. Electrochemical live monitoring of tumor cell migration out of micro-tumors on an innovative multiwell high-dense microelectrode array. Sci Rep. 2019;9:1–10.
4. Dowling CM, Herranz Ors C, Kiely PA. Using real-time impedance-based assays to monitor the effects of fibroblast-derived media on the adhesion, proliferation, migration and invasion of colon cancer cells. Biosci Rep. 2014;34:415–27.
5. Edelstein AD, Tsuchida MA, Amodaj N, Pinkard H, Vale RD, Stuurman N. Advanced methods of microscope control using μManager software. J Biol Methods. 2014;1:e10.
6. Schindelin J, Arganda-Carrera I, Frise E, Verena K, Mark L, Tobias P, Stephan P, Curtis R, Stephan S, Benjamin S, Jean-Yves T, Daniel JW, Volker H, Kevin E, Pavel T, Albert C. Fiji - an open platform for biological image analysis. Nat Methods. 2009; https://doi.org/10.1038/nmeth.2019.Fiji.
7. Susano Pinto DM, Phillips MA, Hall N, Mateos-Langerak J, Stoychev D, Susano Pinto T, Booth MJ, Davis I, Dobbie IM. Python-microscope—a new open-source python library for the control of microscopes. J Cell Sci. 2021;134:1–7.
8. Moreno X, Al-Kadhimi S, Alvelid J, Bodén A, Testa I. ImSwitch: generalizing microscope control in Python. J Open Source Softw. 2021;6:3394.
9. Allan C, Burel J, Moore J, Blackburn C, Linkert M, Macdonald D, Moore WJ, Neves C, Patterson A, Porter M, Tarkowska A, Loranger B, Avondo J, Lianas L, Leo S, Hands K, Hay RT, Best C, Kleywegt GJ, Zanetti G, Jason R. OME remote objects (OMERO): a flexible, model-driven data management system for experimental biology. Nat Methods. 2012;9:245–53.

10. Martone ME, Tran J, Wong WW, Sargis J, Fong L, Larson S, Lamont SP, Gupta A, Ellisman MH. The Cell Centered Database project: an update on building community resources for managing and sharing 3D imaging data. J Struct Biol. 2008;161:220–31.
11. Williams E, Moore J, Li SW, Rustici G, Tarkowska A, Chessel A, Leo S, Antal B, Ferguson RK, Sarkans U, Brazma A, Carazo Salas RE, Swedlow JR. Image data resource: a bioimage data integration and publication platform. Nat Methods. 2017;14:775–81.
12. Barry DJ, Durkin CH, Abella JV, Way M. Open source software for quantification of cell migration, protrusions, and fluorescence intensities. J Cell Biol. 2015;209:163–80.
13. Masuzzo P, Van Troys M, Ampe C, Martens L. Taking aim at moving targets in computational cell migration. Trends Cell Biol. 2016;26:88–110.
14. Meijering E, Dzyubachyk O, Smal I. Methods for cell and particle tracking. Methods Enzymol. 2012;504:183–200.
15. Piccinini F, Kiss A, Horvath P. CellTracker (not only) for dummies. Bioinformatics. 2016;32:955–7.
16. Tinevez JY, Perry N, Schindelin J, Hoopes GM, Reynolds GD, Laplantine E, Bednarek SY, Shorte SL, Eliceiri KW. TrackMate: an open and extensible platform for single-particle tracking. Methods. 2017;115:80–90.
17. Ershov D, Phan M, Pylvänäinen JW, Rigaud SU, Le Blanc L, Conway JRW, Laine RF, Roy NH, Bonazzi D, Duménil G, Jacquemet G, Tinevez J. Bringing TrackMate into the era of machine-learning and deep-learning. bioRxiv. 2021; https://doi.org/10.1101/2021.09.03.458852.
18. Gebäck T, Schulz MMP, Koumoutsakos P, Detmar M. TScratch: a novel and simple software tool for automated analysis of monolayer wound healing assays. Biotechniques. 2009;46:265–74.
19. Suarez-Arnedo A, Figueroa FT, Clavijo C, Arbeláez P, Cruz JC, Muñoz-Camargo C. An image J plugin for the high throughput image analysis of in vitro scratch wound healing assays. PLoS One. 2020;15:1–14.
20. Ketchen S, Rohwedder A, Knipp S, Esteves F, Struve N, Peckham M, Ladbury JE, Curd A, Short SC, Brüning-Richardson A. A novel workflow for three-dimensional analysis of tumour cell migration. Interface Focus. 2020; https://doi.org/10.1098/rsfs.2019.0070.
21. Sethi P, Virmani G, Gupta K, Thumu SCR, Ramanan N, Marathe S. Automated morphometric analysis with SMorph software reveals plasticity induced by antidepressant therapy in hippocampal astrocytes. J Cell Sci. 2021; https://doi.org/10.1242/jcs.258430.
22. Levet F, Tønnesen J, Nägerl UV, Sibarita JB. A software tool for quantitative analysis of nanoscale spine morphology. Methods. 2020;174:49–55.
23. Takko H, Pajanoja C, Kurtzeborn K, Hsin J, Kuure S, Kerosuo L. ShapeMetrics: a userfriendly pipeline for 3D cell segmentation and spatial tissue analysis. Dev Biol. 2020;462:7–19.
24. Torben-Nielsen B. An efficient and extendable python library to analyze neuronal morphologies. Neuroinformatics. 2014;12:619–22.
25. Nguyen C, Thompson-Peer KL. Comparing automated morphology quantification software on dendrites of uninjured and injured drosophila neurons. Neuroinformatics. 2021;19:703–17.
26. Phillip JM, Han K-S, Chen W-C, Wirtz D, Wu P-H. A robust unsupervised machine-learning method to quantify the morphological heterogeneity of cells and nuclei. Nat Protoc. 2021;16:754–74.
27. Furkel J, Knoll M, Din S, Bogert NV, Seeger T, Frey N, Abdollahi A, Katus HA, Konstandin MH. C-MORE: a high-content single-cell morphology recognition methodology for liquid biopsies toward personalized cardiovascular medicine. Cell Rep Med. 2021;2:100436.
28. Rohwedder A, Knipp S, Esteves FO, Hale M, Ketchen SE, Treanor D. 'Cloudbuster': a Python-based open source application for three-dimensional reconstruction and quantification of stacked biological imaging samples. Interface Focus. 2022;12(5):20220016. https://doi.org/10.1098/rsfs.2022.0016.

Pharmacological Strategies for Targeting Cancer Cell Migration and Invasion

Roger M. Phillips

What You Will Learn in This Chapter

In this chapter, the two key themes are (i) pharmacological strategies to target or exploit cancer cell migration and invasion and (ii) in vitro models that are widely used to either unravel the complex biology of cancer cell invasion or evaluate the efficacy of potential therapeutic agents in the early stages of preclinical drug development. You will learn that not only does each of the pharmacological strategies being tested have its specific advantages and disadvantages, but there are also broader issues that challenge the whole concept of targeting cell migration and invasion. Similarly, there is no 'ideal' experimental model to test potential therapeutic agents, and the choice of which model to use depends very much on the experimental questions being addressed. In summary, targeting the complex biology underpinning cell migration and invasion has the potential to add new drugs to the pharmacopoeia, but significant challenges (which are, in fact, opportunities) need to be addressed if this potential is to be realised in the clinic.

R. M. Phillips (✉)
School of Applied Sciences, University of Huddersfield, Queensgate, Huddersfield, UK
e-mail: r.m.phillips@hud.ac.uk

A. Brüning-Richardson, S. Knipp (eds.), *Cell Migration in Development, Health and Disease*, Learning Materials in Biosciences,
https://doi.org/10.1007/978-3-031-64532-7_12

1 Introduction

As an established hallmark of cancer, the invasion and metastasis of solid cancers represents a major challenge to healthcare providers, accounting for the majority of cases of treatment failure and mortality. A dominant feature of solid cancers is local invasion and metastasis, but despite decades of research and development, the pharmacopoeia is notably deficient in cancer therapeutics that are anti-invasive or anti-metastatic. This sobering fact is in sharp contrast to the progress that has been made in developing targeted and immunological therapeutics, and the development of anti-invasive/anti-metastatic drugs remains a significant unmet clinical need. The problem may be further compounded by recent preclinical studies demonstrating that long-term treatment of tumours with targeted drugs (PARP1 and Chk1 inhibitors) may in fact promote drug resistance, migration and invasion [1]. There is therefore a need to develop anti-invasive/anti-metastatic drugs, but multiple challenges need to be overcome in order to achieve this objective [2, 3]. Two key challenges are as follows:

(i) Developing therapeutics that selectively inhibit cancer cell migration, invasion and metastasis without interfering with essential normal processes such as embryonic development, wound healing and immune processes involving the migration of leukocytes, lymphocytes, etc. to sites of action. This problem is exacerbated by the fact that non-cytotoxic anti-invasive/anti-metastatic therapeutics will need to be administered over long periods of time in order to prevent cancer cell invasion.

(ii) The pharmacological evaluation of anti-invasive/anti-metastatic drugs in both preclinical models and clinical trials needs to be carefully designed, as this approach will not necessarily lead to tumour shrinkage. Current regulatory requirements for the approval of drugs typically require evidence of tumour shrinkage by a single test agent, and this introduces many challenges, especially as the best use of anti-invasive/anti-metastatic drugs will likely be in combination with anti-proliferative or targeted therapeutics.

The biological complexity underpinning the biology of cell migration, invasion and metastasis is being unravelled, and numerous opportunities for therapeutic intervention have emerged that address these challenges. The complexity and plasticity of these processes mean, however, that there is no 'one-size-fits-all' approach to the pharmacological evaluation of these strategies, and preclinical models need to be carefully selected and tailored to each strategy being investigated. Experimental models are at the heart of all preclinical decision-making stages, and the second part of this chapter will focus on established and new experimental models that are being used to identify novel therapeutics that target cell migration and invasion. Initially, however, a short overview of pharmacological strategies that are being developed is provided to give context for the subsequent section on experimental models.

2 Overview of Pharmacological Strategies Designed to Inhibit or Exploit Cell Migration and Invasion

The metastatic process is a complex, multistep process involving cell migration, local invasion, intravasation, circulation, extravasation and metastatic colonisation in specific organ 'niches' [4]. While strategies are being developed to target each of these stages of the process [3], multiple approaches are being developed to either inhibit or exploit the underlying mechanisms driving cell migration and local invasion (Fig. 1). The process of cancer cell invasion is also complex, and cancer cells can invade collectively as sheets or clumps of cells or individually via ameboid or mesenchymal modes. In addition, plasticity is built into the system, allowing cells to adapt and change modes of invasion depending on signalling pathways, specific conditions and interactions with the extracellular matrix and other noncancerous cells in the tumour. These challenges are not unsurmountable,

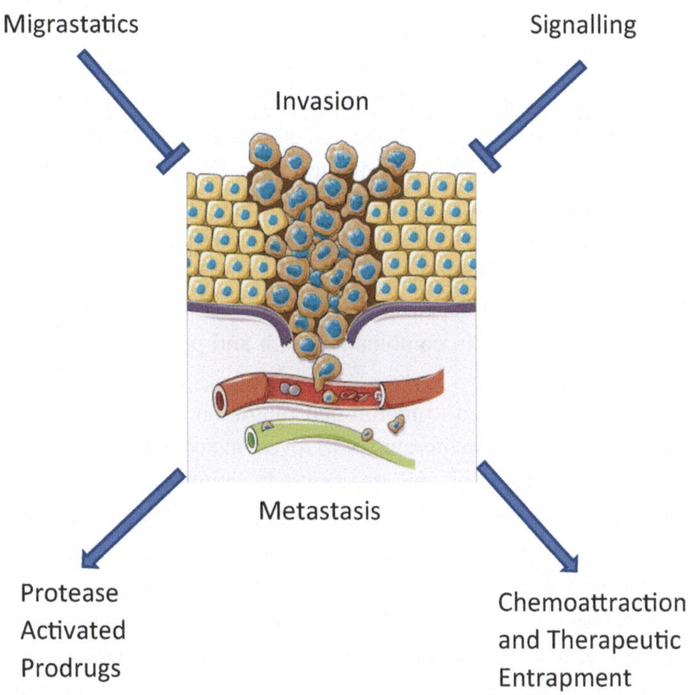

Fig. 1 Overview of therapeutic strategies being developed to target cell migration and invasion. Strategies can be broadly divided into those that inhibit key biochemical pathways and processes that drive cell migration and invasion (inhibiting specific signalling pathways or 'migrastatics' that target downstream processes common to all migrating cells) or approaches that exploit tumour cell migration biology, such as protease-activated prodrugs and the use of chemoattractants to entrap migrating cancer cells into a location where they can be treated more effectively. Image created with https://smart.servier.com/

however, and specific examples of therapeutic strategies under development are described below.

The term ‘migrastatics’ has been coined to describe drugs that interfere with all modes of cancer cell invasion, and these drugs are distinct from cytotoxic/cytostatic drugs that target cell proliferation [5]. Targeting all stages of the cell invasion process is feasible, including extracellular stimuli, ECM-cell signalling pathways, intracellular signalling pathways, small RHO-GTPases and effectors including cell contractility and actin polymerisation. While each of these approaches is feasible and attractive, it has been argued that migrastatics targeting the ultimate effectors of cell movement (actinomycin contractility and actin polymerisation) would be most effective, as these essential processes are difficult to bypass, making the emergence of resistance less likely (compared to focused targeting of upstream signalling pathways) [5]. There is considerable merit in this approach, as signalling pathways governing cell migration are highly redundant, and suppressing a single pathway could result in the emergence of resistant clones and treatment failure. A series of potential candidate migrastatic compounds that target the ultimate effector mechanisms of cell movement have been identified, details of which have been reviewed elsewhere [5]. These include myosin inhibitors, actin-destabilising (e.g. cytochalasins) and actin-stabilising (e.g. jasplakinolide) drugs, kinase inhibitors involved in the regulation of actomyosin contractility (e.g. the ROCK inhibitor Y-27632) and tropomyosin inhibitors (e.g. TR100) that disrupt the actin cytoskeleton by affecting its stability. The challenge, of course, is to find drugs that selectively inhibit cancer cell migration without inhibiting normal cell migration, especially as the administration of migrastatics is likely to be continuous and long-term. Nevertheless, multi-kinase inhibitors of ROCK pathways in particular appear promising as single agents in preclinical models, but the intention is to use these in combination with anti-proliferative drugs in order to achieve full efficacy.

Therapeutic strategies designed to inhibit cell migration are an established approach, and a more detailed analysis of the literature will reveal many more targets and approaches than described here. Inhibition of cell migration is intuitively attractive, but alternative approaches have been adopted by others in the field. Rather than inhibiting key mechanisms involved in cell migration, one alternative approach being developed is to exploit cell migration biochemistry and use it as a ‘tool’ to enhance drug delivery. For example, proteases play a key role not only in tumour cell invasion but in tumour biology more generally, and they have been an attractive target for drug development for many years. While the initial attempts to develop broad-spectrum matrix metalloproteinase (MMP) inhibitors in the 1990s did not translate into clinically useful drugs [6], more recent studies have focused on using specific proteases as biochemical ‘triggers’ to selectively release therapeutic ‘warheads’ (Fig. 2), thereby enhancing drug delivery [7, 8]. To exemplify this, the prodrug ICT2588 was developed [9], consisting of a vascular disrupting agent (azademethylchochicine) conjugated to a protease-cleavable peptide and an endcap to prevent nonspecific endopeptidase cleavage. Following cleavage of the peptide by membrane-type MMP (MT1-MMP or MMP14), tumour-selective release

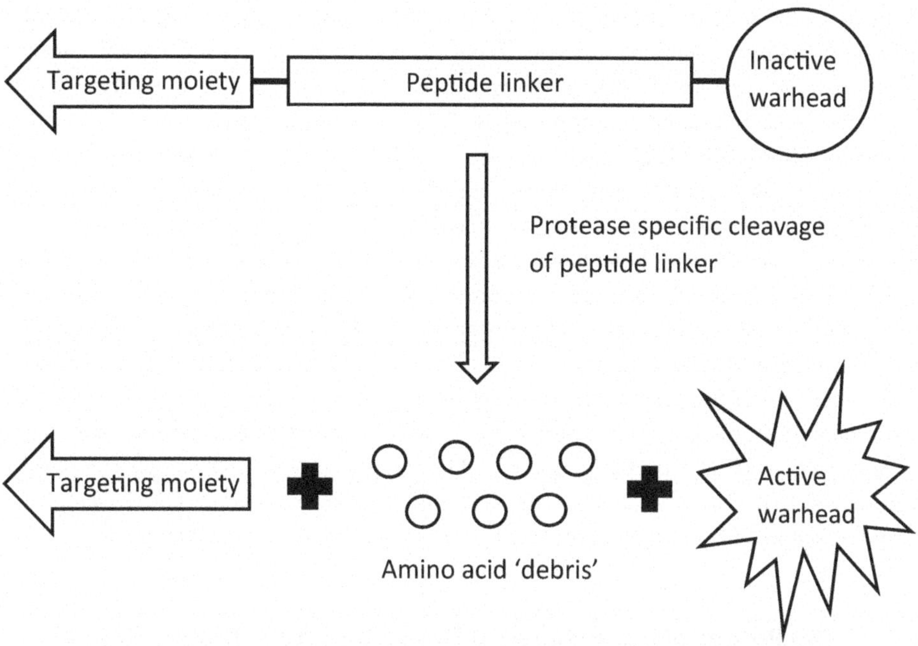

Fig. 2 Schematic outlining the concept behind the development of protease-activated prodrugs. The inactive prodrug consists of a peptide with a defined amino acid sequence linked to a warhead. At the other end of the peptide, a targeting moiety (such as a monoclonal antibody) can be added to further increase selectivity and/or efficacy. Following cleavage of the peptide linker by 'tumour-specific' proteases, the peptide is further digested by endopeptidases to release the active therapeutic warhead at the tumour site. This approach exploits one of the key biological processes that drives cell migration and invasion

of the warhead was documented alongside reduced toxicity (compared to the warhead alone), vascular collapse and haemorrhagic necrosis in HT1080 tumours in vivo [10]. The antitumour activity of ICT2588 was further significantly enhanced by co-administration with doxorubicin, which was selected as a means of reducing the 'viable rim' of cells typically observed following the administration of vasculature-disrupting agents [10]. This approach has clinical merit, as exemplified by the antibody-drug conjugate brentuximab vedotin, which uses the combined targeting strategies of an anti-CD30 antibody conjugated to a cytotoxic payload (monomethyl auristatin E) by a protease (cathepsin)-cleavable peptide. Brentuximab vedotin is approved for the treatment of CD30-positive Hodgkin lymphoma [11], and while a more comprehensive review of protease-activated prodrugs, strategies, challenges and future directions can be found elsewhere [7], the success of this approach is encouraging.

Over recent years, considerable interest has been generated in the development of loco-regional therapies for tumours such as glioblastomas (GBM). In this case, the inability to surgically remove cells that have invaded from the primary tumour into the cerebral parenchyma (because removal of a sufficiently large margin of normal tissue is impossible) results in relatively rapid recurrence of the disease, typically at the site of surgical resection. A key feature of GBM is its aggressive and invasive nature, forming multiple finger-like projections into the brain parenchyma, making complete surgical resection effectively impossible [12]. Understanding the biochemical mechanisms underpinning GBM invasion is leading to new therapeutic options [12], and one alternative approach is to exploit the role that chemoattractants play in the invasion and cell migration process. The concept here is to use chemoattractants to 'turn invading cells around' and force them to migrate to a confined area where treatment is easier (a process termed 'therapeutic entrapment'). Proof-of-concept studies using hydrogels loaded with a chemoattractant (human urotensin II) and a cytotoxic drug (doxorubicin) have demonstrated that U87MG GBM cells can be trapped in the hydrogel where they are eradicated by the action of doxorubicin [13]. In another study, biodegradable hydrogels (GlioGel) loaded with three chemoattractants (CXCL10, CCL2 and CCL11) strongly attracted F98 and U87MG GBM cells in vivo [14]. These studies provide new opportunities for treating GBM, one of the most devastating diseases, where there is a clear clinical need for new therapeutics.

3 The Potential Importance of Drug Stability in the Preclinical Development of Anti-Migratory/Anti-Invasive Drugs

As mentioned earlier, it is highly likely that drugs designed to inhibit cell migration and invasion will need to be given over a prolonged period of time (at concentrations above the minimal effective concentration); therefore, the stability of drugs in experimental test systems is an important but understudied factor. From work done in the 1970s and beyond, it is well known that many cytotoxic drugs are highly unstable in aqueous solutions, and this can lead to erroneous conclusions [15]. In addition, the stability of antineoplastic agents during preparation and storage prior to in vitro assays being conducted is also an important consideration [16]. Clearly, a comprehensive analysis of drug stability cannot be fully deployed during the very early stages of drug discovery, but in lead identification and optimisation studies, analytical [17] and/or bioassay [16]-based assessment of drug stability would be valuable for therapeutics that need to be given over long periods of time.

Take-Home Message 1 Multiple experimental strategies are being developed to either inhibit cell migration and invasion or exploit the underlying biology governing cell migration using protease-activated prodrugs and therapeutic trapping using chemoattractants (Fig. 1). These approaches require testing in experimental models, and the next section describes some of the experimental models typically used to evaluate anti-migratory/anti-invasive compounds.

4 Experimental Models for the Preclinical Evaluation of Anti-Invasive/Anti-Metastatic Drugs

A wide range of experimental models have been developed for assessing the efficacy of anti-invasive/anti-metastatic compounds in preclinical development. These can be broadly categorised into cell-free, in vitro and in vivo assays, each with its own advantages and disadvantages. There is no one experimental model that can be used universally, but as a general rule, the initial steps of the preclinical development pathway should test compounds against in vitro models that will give the compounds in question the best chance of working (to avoid rejecting potentially useful classes of compounds too early). This is followed by more advanced and challenging models as compounds progress towards the 'lead compound' stage of the developmental pipeline. A detailed overview of some experimental models that employ accurate data analysis tools (in the form of time-lapse microscopy and complex metrics analysis) to evaluate anti-invasive/anti-metastatic drugs has been published elsewhere [18, 19]. Here, these and other models are briefly reviewed, along with citations to more detailed studies for the reader to pursue, some of which are covered in another chapter in this publication.

2D 'Scratch' or 'Wound Healing' Assays In this assay, either a 'scratch' is made in a confluent monolayer of cells using a pipette tip, or a commercially available silicone culture insert is used to introduce a 'wound field' with a defined gap [20]. The ability of cells to close the 'gap' can be measured using a variety of phase contrast (Fig. 3) or scanning confocal microscopic methods (including real-time analysis using time-lapse

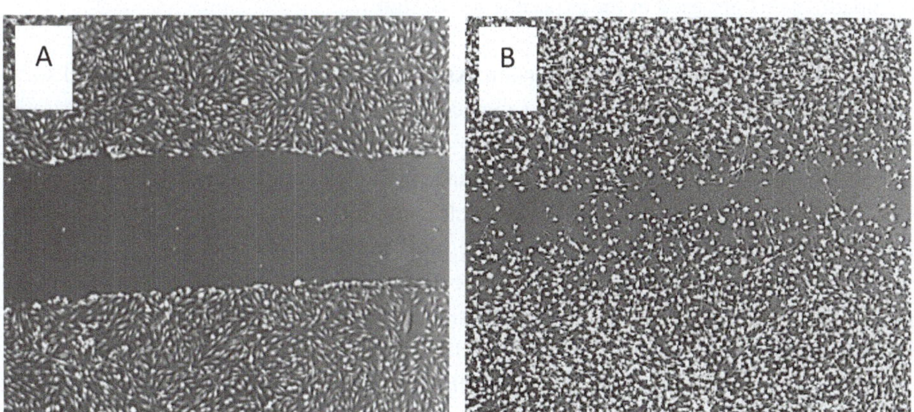

Fig. 3 Wound healing or 'scratch' assay. The assay involves making a gap in a confluent layer of adherent cells (panel **a**) using a pipette tip (hence the term 'scratch') and measuring how quickly cells close the gap at various times after making the scratch (24 h in the case of panel **b**). The other way of introducing a gap is by using specific culture inserts, but the principle of the assay is identical. These images (×10 magnification) were provided by Becky Griggs, University of Huddersfield

microscopy) and quantified using open-source software such as ImageJ. Both assays are valuable tools for assessing cell migration, but in terms of reducing the variability of wound widths, the culture insert method produced wounds with the lowest coefficient of variation [19]. The scratch assay using a pipette tip is, however, a much cheaper and faster method.

2D Individual Cell Tracking Assays In this assay, cells are seeded at a low density, the number of which varies depending on the characteristics of the cell type. Cells are then visualised using a time-lapse microscopy setup fitted with a CO_2 microscope cage incubator (set at 37 °C and 5%CO_2). A variant of this assay is the colloidal particle assay, where cells are added to plates coated with colloidal gold or quantum dots. Cells that are able to migrate phagocytose the particles, leaving behind a phagokinetic track that is free of colloidal gold or quantum dots. These tracks can be visualised microscopically and quantified. Following image capture, a variety of open-source software tools are available for measuring individual cell migration and morphological features such as membrane protrusions [21]. For example, TrackMate is free-to-download software that can be used with ImageJ. Additional software has also been developed to facilitate high-throughput analysis of cell migration, where multiple biological and pharmacological conditions can be tested in parallel [22]. Live cell time-lapse microscopy combined with individual cell tracking has many advantages over other techniques. These include the fact that other assays (such as the wound healing assay) can be influenced by cell proliferation and cell-to-cell contacts (e.g. the wound can close by a combination of cell proliferation and cell migration effects). In addition, not only can migration speed and persistence be determined, but also the migratory path of cells can be measured, particularly if chemoattractants are used in the experimental design.

Transwell Cell Migration and Invasion Assays These assays are also referred to as the Boyden chamber assay, and the setup consists of two culture chambers separated by a porous membrane. Cells are added to the top chamber, and the number of cells migrating to the lower surface of the membrane (in the lower chamber) can be determined by a variety of techniques. These include cell counting of fixed and stained cells or the use of fluorescent dyes, which eliminate the need to count cells and reduce the number of steps in the protocol. This assay can be used to study both cell migration (using uncoated membranes) and cell invasion (using membranes coated with Matrigel), and detailed protocols are described elsewhere [19]. While this is a widely used method and transwells/Boyden chambers are commercially available, one limitation of this method is that migrating cells cannot be visualised in real time.

3D Models Recent advances in the development of 3D models of tumour invasion have been reviewed elsewhere [23–25]. These models include spheroid invasion models, bioprinting, scaffolds, hydrogels, microfluidic and organotypic models that mimic the complex architecture and interactions with the microenvironment that drive tumour cell migration and invasion. Brief details of each approach, together with citations to some of the most current literature, are presented below.

Spheroid Invasion Models Seminal work on multicellular spheroids was performed in the late 1970s [26], and these models remain of considerable interest because they mimic many of the features of tumour biology, including physiological microenvironments such as oxygen and nutrient deprivation. Spheroids are aggregates of cells formed via a variety of techniques, including the hanging drop technique, cell pelleting, spinner flasks and the liquid overlay technique [27]. Establishing spheroids can, however, be challenging, as not all cell lines are capable of forming tight, regularly shaped spheroids with a defined hypoxic microenvironment, and many are simply loose aggregates of irregular size and shape that are difficult to manipulate. As an alternative, growing adherent cells on microcarrier beads has been used to study dynamic cell behaviour in vitro [28]. This method does not, however, generate a tumour microenvironment but remains a useful method to study the migratory properties of adherent cells that do not form multicellular spheroids. Multicellular spheroids are a useful model to study tumour biology and drug pharmacology in general, but in order to study cell migration and invasion, spheroids are typically embedded in various extracellular matrices (collagen, Matrigel, etc.), and the migration of cells from the surface of the spheroid can be determined manually or via time-lapse microscopy (Fig. 4). Quantification and analysis of cell migration can include maximum distance travelled from the surface of the spheroid, migration index, cell trajectories and migration velocity, for example [28]. The multicellular spheroid model is of considerable interest for the study of tumour cell biology and migration behaviour, but there are several limitations of the approach (e.g. generating spheroids of a uniform size is time-consuming, achieving uniformity in size and shape is challenging, embedding spheroids in matrices is imprecise in terms of location, etc.) that make it less suitable for high-throughput screening of potential drug candidates. Recent studies have, however, overcome some of these limitations, and these are described in detail by Shin and Anseth [24].

Microfluidic 3D Models In recent years, microfluidic models designed to study and evaluate candidate therapies have been developed. These models range in complexity, but they provide the opportunity to introduce biological (interactions between cancer cells, stromal cells, extracellular matrices, etc.), microenvironmental physiology (e.g. hypoxia and nutrient deprivation) and biomechanical (e.g. interstitial fluid pressure and shear fluid flow) features associated with the tumour microenvironment that cannot be modelled using other models. In addition to modelling the complexity of tumour biology with increasing accuracy, these models can be used to assess the issue of drug delivery and its impact on cell behaviour in poorly perfused regions of the tumour microenvironment [29, 30]. Further development of these models is envisaged, and as they become more refined, it is

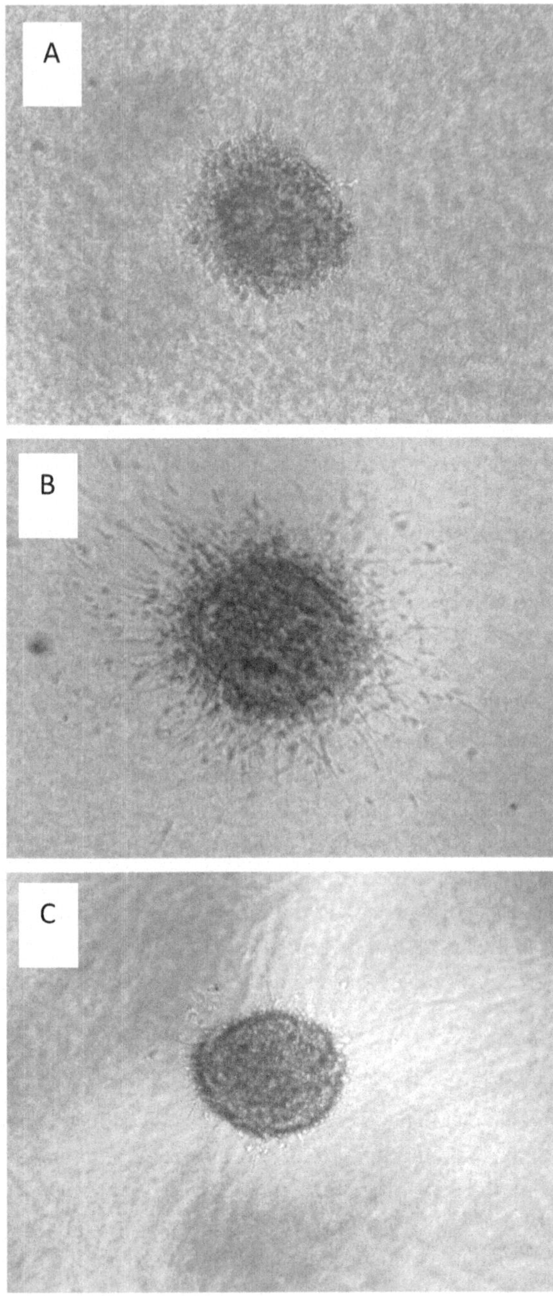

Fig. 4 Multicellular spheroid migration assay. In this example, multicellular spheroids using patient-derived GBM cells were embedded in collagen gels (panel **a**), and control spheroids were incubated at 37 °C for 24 h. Cells can be clearly seen migrating away from the surface of the spheroid (panel **b**), and cell migration was retarded by treating spheroids with the anti-migratory compound BIO-indirubin (5 μM) for 24 h (panel **c**). Cell migration and its retardation by therapeutic agents can be quantified using freely available software (e.g. ImageJ). These images (×10 magnification) were provided by Samuel Peat, University of Leeds, and Dr. Anke Brüning-Richardson, University of Huddersfield

hoped that they will play a pivotal role in the discovery of new classes of drugs targeting cancer cell invasion and migration. Full details of current developments in this field, its application to the study of cell migration and invasion and the challenges that have to be addressed have been described elsewhere [24, 31].

Bioprinting, Scaffolds, Hydrogels and Organotypic Models The tumour microenvironment is complex, and as our understanding of tumour-stroma interactions increases, experimental models that mimic these interactions are being developed. A variety of different approaches with differing degrees of complexity are being pursued, and detailed reviews of progress in this field have been presented elsewhere [23, 25]. Models include the use of porous scaffolds into which cells are seeded, attach, proliferate, differentiate and migrate, with the aim of recapitulating essential cell-cell or extracellular matrix interactions. Materials used to generate scaffolds include synthetic polymeric biomaterials (e.g. polyethylene glycol, polycaprolactone), natural biomaterials (e.g. collagen, fibrin, alginate, chitosan), ceramics (e.g. hydroxyapatite) or hydrogels (e.g. a network of crosslinked polymers with structures similar to the extracellular matrix, with Matrigel being a well-known example), all of which have advantages and disadvantages [23]. 3D bioprinting, together with cell printing, is used to create even more complex models with well-defined and reproducible architectures that facilitate the study of cell-cell, cell-matrix and multicell interactions and can also incorporate the temporal evaluation of cell responses to growth factors, drugs, etc. Finally, organotypic or organoid models of cancer aim to develop models that will enable a better prediction of therapeutic response in animals and humans than current 2D models provide. An additional objective is to use these models to reduce and replace the number of animal experiments required in preclinical drug development. Organoids or organotypic models include multicell spheroids in their simplest form but extend to stem cell and patient-derived organoids that provide models that are physiologically relevant and allow a more in-depth understanding of cell-cell and cell-matrix interactions that occur in tumours compared to models that use a single cell type. This field is rapidly growing with potential applications for personalised medicine using patient-derived organoids, tumour biology and drug discovery, details of which are extensively reviewed elsewhere [23, 25].

Take-Home Message 2 There are a wide range of in vitro experimental models that can be used to study the biology of and evaluate therapeutic strategies designed to interfere with cell migration and invasion. Each of these models has its own advantages and disadvantages, and the selection of which model to use depends on the experimental questions being asked (Fig. 5). The strengths and limitations of 2D vs 3D in vitro models have been described elsewhere [32], but essentially, the choice of model comes down to the simplicity, reproducibility and low cost of 2D models compared to 3D models that more closely mimic the complexity of tumour biology but require extensive expertise, equipment and higher cost. Significant developments in technology will yield even more

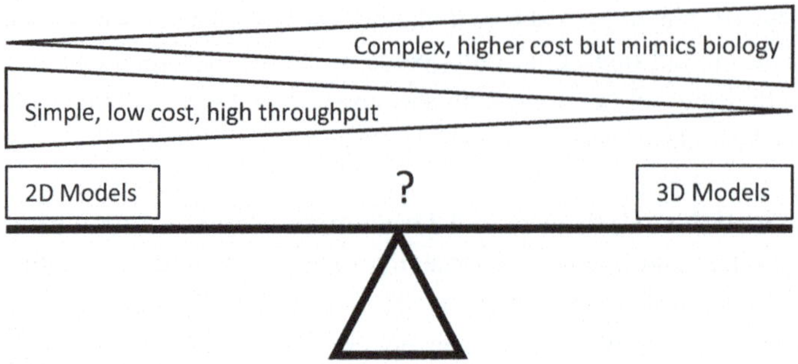

Fig. 5 Simplified overview of the range of in vitro experimental models available for studying cell migration and invasion. Essentially, the choice of model depends on the scientific questions being asked (for studying both biology and pharmacology/drug development), with options ranging from comparatively simple 2D models (e.g. the scratch assay) to complex 3D models mimicking tumour cell interactions with the cellular microenvironment, extracellular matrix and biomechanical forces (e.g. interstitial fluid pressure)

sophisticated models used in the search for a greater understanding of both tumour biology and drug development.

Final Take-Home Messages

While inhibition strategies targeting cell migration are intuitively sensible, several challenges must be addressed. First and foremost is the clinical setting in which anti-migratory/anti-invasive drugs could be most beneficial. Patients with overt metastatic disease would clearly not benefit, as the 'horse has already bolted', and the same may apply to early-stage cancer where no metastasis is clinically detectable, but neoadjuvant or adjuvant or adjuvant systemic therapy is administered on the presumption that micrometastasis is present. In cases where complete resection of the primary tumour may have occurred but the patient subsequently relapses with macrometastatic disease, disseminated tumour cells in distant organs must have already occurred prior to removal of the primary tumour. Targeting the very early stages of invasion and cell migration may therefore be ineffective at preventing or treating metastatic disease [3]. In contrast, exploiting the biology of tumour invasion through strategies such as protease-activated prodrugs [7] and therapeutic entrapment using chemoattractants [13, 14] offers potential alternative strategies that may circumvent these challenges. In the case of protease-activated prodrugs, the challenge is to ensure the selective release of therapeutic warheads within the tumour mass without damaging normal cells involved in key biological processes such as wound healing and immune cell function, where cell migration is an essential part of normal tissue biology. Progress is being made with the preclinical development

(continued)

of migrastatics, but clinical studies need to be carefully designed, as these drugs will have pharmacodynamic properties distinct from cytotoxic and targeted therapeutics. In contrast, the clinical development of protease-activated prodrugs and therapeutic entrapment strategies is likely to be conceptually easier, as classical endpoints such as tumour regression can be employed. Alongside the development of new therapeutics, preclinical models used to unravel the complexity of tumour cell interactions with the cellular and physiological tumour microenvironment continue to become more sophisticated and complex, offering real opportunities to develop and evaluate new therapeutic approaches. The selection of which experimental model to use, however, very much depends on the experimental questions being asked and the stage of preclinical drug development.

References

1. Guo N, Li MZ, Wang LM, Chen HD, Song SS, Miao ZH, et al. Repeated treatments of Capan-1 cells with PARP1 and Chk1 inhibitors promote drug resistance, migration and invasion. Cancer Biol Ther. 2022;23(1):69–82.
2. Anderson RL, Balasas T, Callaghan J, Coombes RC, Evans J, Hall JA, et al. A framework for the development of effective anti-metastatic agents. Nat Rev Clin Oncol. 2019;16(3):185–204.
3. Ganesh K, Massague J. Targeting metastatic cancer. Nat Med. 2021;27(1):34–44.
4. Fares J, Fares MY, Khachfe HH, Salhab HA, Fares Y. Molecular principles of metastasis: a hallmark of cancer revisited. Signal Transduct Target Ther. 2020;5:28.
5. Gandalovicova A, Rosel D, Fernandes M, Vesely P, Heneberg P, Cermak V, et al. Migrastatics-anti-metastatic and anti-invasion drugs: promises and challenges. Trends Cancer. 2017;3(6):391–406.
6. Coussens LM, Fingleton B, Matrisian LM. Matrix metalloproteinase inhibitors and cancer: trials and tribulations. Science. 2002;295(5564):2387–92.
7. Poreba M. Protease-activated prodrugs: strategies, challenges, and future directions. FEBS J. 2020;287(10):1936–69.
8. Vandooren J, Opdenakker G, Loadman PM, Edwards DR. Proteases in cancer drug delivery. Adv Drug Deliv Rev. 2016;97:144–55.
9. Gill JH, Loadman PM, Shnyder SD, Cooper P, Atkinson JM, Ribeiro Morais G, et al. Tumor-targeted prodrug ICT2588 demonstrates therapeutic activity against solid tumors and reduced potential for cardiovascular toxicity. Mol Pharm. 2014;11(4):1294–300.
10. Atkinson JM, Falconer RA, Edwards DR, Pennington CJ, Siller CS, Shnyder SD, et al. Development of a novel tumor-targeted vascular disrupting agent activated by membrane-type matrix metalloproteinases. Cancer Res. 2010;70(17):6902–12.
11. Scott LJ. Brentuximab vedotin: a review in CD30-positive Hodgkin lymphoma. Drugs. 2017;77(4):435–45.
12. Seker-Polat F, Pinarbasi Degirmenci N, Solaroglu I, Bagci-Onder T. Tumor cell infiltration into the brain in glioblastoma: from mechanisms to clinical perspectives. Cancers (Basel). 2022;14(2):443.

13. Kasapidou PM, de Montulle EL, Dembele KP, Mutel A, Desrues L, Gubala V, et al. Hyaluronic acid-based hydrogels loaded with chemoattractant and anticancer drug - new formulation for attracting and tackling glioma cells. Soft Matter. 2021;17(48):10846–61.
14. Dery L, Charest G, Guerin B, Akbari M, Fortin D. Chemoattraction of neoplastic glial cells with CXCL10, CCL2 and CCL11 as a paradigm for a promising therapeutic approach for primary brain tumors. Int J Mol Sci. 2021;22(22):12150.
15. Freshney RI, Paul J, Kane IM. Assay of anti-cancer drugs in tissue culture: conditions affecting their ability to incorporate 3H-leucine after drug treatment. Br J Cancer. 1975;31(1):89–99.
16. Bosanquet AG. Stability of solutions of antineoplastic agents during preparation and storage for in vitro assays. II. Assay methods, Adriamycin and the other antitumour antibiotics. Cancer Chemother Pharmacol. 1986;17(1):1–10.
17. Betteridge RF, Bosanquet AG. Automated stability determination of anticancer drugs incubated in cell culture medium. Int J Pharm. 1991;71:73–9.
18. Kramer N, Walzl A, Unger C, Rosner M, Krupitza G, Hengstschlager M, et al. In vitro cell migration and invasion assays. Mutat Res. 2013;752(1):10–24.
19. Pijuan J, Barcelo C, Moreno DF, Maiques O, Siso P, Marti RM, et al. In vitro cell migration, invasion, and adhesion assays: from cell imaging to data analysis. Front Cell Dev Biol. 2019;7:107.
20. Jonkman JE, Cathcart JA, Xu F, Bartolini ME, Amon JE, Stevens KM, et al. An introduction to the wound healing assay using live-cell microscopy. Cell Adhes Migr. 2014;8(5):440–51.
21. Barry DJ, Durkin CH, Abella JV, Way M. Open source software for quantification of cell migration, protrusions, and fluorescence intensities. J Cell Biol. 2015;209(1):163–80.
22. Masuzzo P, Huyck L, Simiczyjew A, Ampe C, Martens L, Van Troys M. An end-to-end software solution for the analysis of high-throughput single-cell migration data. Sci Rep. 2017;7:42383.
23. Rodrigues J, Heinrich MA, Teixeira LM, Prakash J. 3D in vitro model (R)evolution: unveiling tumor-stroma interactions. Trends Cancer. 2021;7(3):249–64.
24. Shin DS, Anseth KS. Recent advances in 3D models of tumor invasion. Curr Opin Biomed Eng. 2021;19:100310.
25. Hayksl MM, Nahmias C, Varon C, Martin OCB. Organotypic modeling of the tumour landscape. Front Cell Dev Biol. 2020;8:606039.
26. Sutherland RM, Durand RE. Growth and cellular characteristics of multicell spheroids. Recent Results Cancer Res. 1984;95:24–49.
27. Bialkowska K, Komorowski P, Bryszewska M, Milowska K. Spheroids as a type of three-dimensional cell cultures-examples of methods of preparation and the most important application. Int J Mol Sci. 2020;21(17):6225.
28. Liu H, Lu T, Kremers G-J, Seynhaeve ALB, ten Hagen TLM. A microcarrier-based spheroid 3D invasion assay to monitor dynamic cell movement in extracellular matrix. Biol Proc Online. 2020;22:1–12.
29. Ayuso JM, Virumbrales-Munoz M, Lacueva A, Lanuza PM, Checa-Chavarria E, Botella P, et al. Development and characterization of a microfluidic model of the tumour microenvironment. Sci Rep. 2016;6:36086.
30. Mpekris F, Angeli S, Pirentis AP, Stylianopoulos T. Stress-mediated progression of solid tumors: effect of mechanical stress on tissue oxygenation, cancer cell proliferation, and drug delivery. Biomech Model Mechanobiol. 2015;14(6):1391–402.
31. Mehta P, Rahman Z, Ten Dijke P, Boukany PE. Microfluidics meets 3D cancer cell migration. Trends Cancer. 2022;8(8):683–97.
32. Asghar W, El Assal R, Shafiee H, Pitteri S, Paulmurugan R, Demirci U. Engineering cancer microenvironments for in vitro 3-D tumor models. Mater Today (Kidlington). 2015;18(10):539–53.

Oncolytic Viruses and Cell Migration

Anke Brüning-Richardson

What You Will Learn in This Chapter

Viruses have always accompanied human life and impacted our health. Recent developments in the medical field have highlighted that it is possible to exploit the requirement of a virus to replicate within a host cell. For example, viruses can be genetically engineered to specifically target, enter and finally destroy tumour cells. These viruses are called oncolytic viruses (OVs). In addition, it has been discovered that OV infections can also affect cell migration by utilizing the host cell's microtubule cytoskeleton to reach replication sites within the cell. Consequently, finely tuned microtubule dynamics are disturbed, with follow-on effects on cell migration. This chapter will provide a general background on viruses and their life cycle, OVs, viral interaction with the microtubules and their effect on cell migration. Specific examples of these interactions are given, and migration assays that include the use of viruses are described. The newest developments in the field are summarized.

A. Brüning-Richardson (✉)
Department of Biological and Geographical Sciences, University of Huddersfield, Huddersfield, UK
e-mail: a.bruning-richardson@hud.ac.uk

A. Brüning-Richardson, S. Knipp (eds.), *Cell Migration in Development, Health and Disease*, Learning Materials in Biosciences,
https://doi.org/10.1007/978-3-031-64532-7_13

1 Viruses and Virus Replication

Viruses are believed to have evolved alongside humankind for millions of years. Evidence for this is, for example, the discovery of endogenous retroviruses inserted into the human genome [1]; when the human genome sequence was published in 2001 [2–4], it was revealed that about two-thirds of the sequence consisted of endogenous retroviruses. This intertwined relationship has been identified as one driver of human evolution, with a predicted 30% of all adaptive amino acid changes in the conserved part of the human proteome being driven by viruses [5, 6].

What are viruses? Viruses are small, infectious particles or agents consisting of a protein shell or membrane and a genetic component, which can be single- or double-stranded RNA or DNA. The genetic material contains all the information required for the virus to replicate within a host. Due to their inability to replicate without a host cell, some research has suggested that viruses are not alive but purely protein/DNA structures [7]. However, since the virus particles contain all the required building blocks for replication, it may be argued that they are living organisms. Once they have entered a host cell, they also actively de-coat and activate the replication process.

As a consequence of a virus's requirement to replicate within a host, its structure is optimized to allow attachment or binding to a specific host cell, followed by penetration of the cell membrane, then replication within and egression from the cell. Depending on the type of genetic component of the virus, DNA or RNA, replication occurs inside or outside of the host cell nucleus.

The replicative part of a virus's life cycle also includes movement within the host cell to reach the site of replication, such as viral factories located close to the host cell nucleus. The release of the newly formed virions may rely on 'hitchhiking' on microtubule (MT) tracts to reach the cell membrane and egress.

Viruses are classified based on the presence of DNA or RNA—whether this is double-stranded or single-stranded—and their particular structure, i.e. if they are enveloped or not. All viruses follow six main stages of replication:

- Attachment of the virus to the host cell
- Penetration, which can be by fusion with the host membrane or endocytosis
- Uncoating, which is the removal or degradation of the viral coat or capsid to release the genomic nucleic acid
- Replication, which may involve several stages of transcription and translation of viral proteins and genome
- Virus assembly, where the newly produced viral genome and proteins are assembled into new virions
- Virus egression or release via lysis and budding

The six stages of viral replication are shown in Fig. 1.

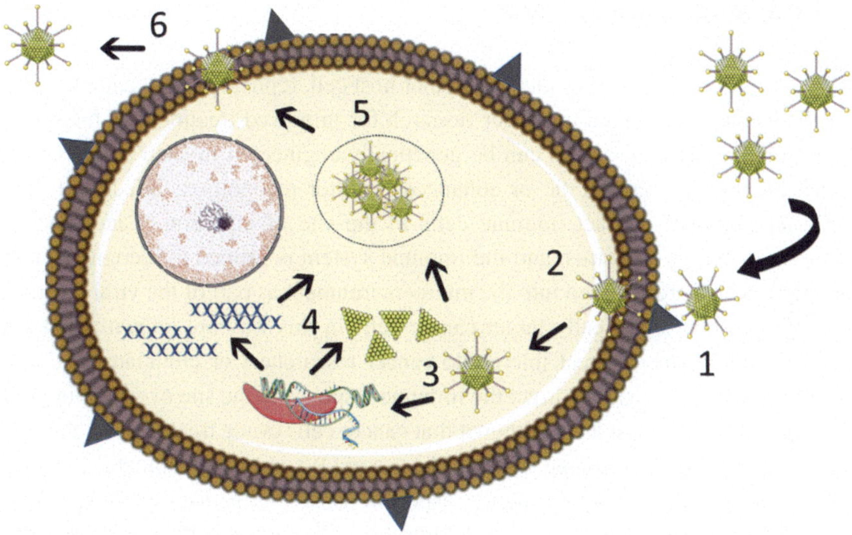

Fig. 1 Simplified life cycle of a virus characterized by six stages. (1) Attachment: A virus can interact with its host cell via recognition and attachment to a receptor molecule on the surface of the host cell. The virus has surface proteins that are part of the recognition process. (2) Viral entry: Entry is by fusion with the host cell surface or by endocytosis. (3) Uncoating: Once inside the host cell, the virus loses its protein coat either by degradation or removal. This step prepares for the replicative stage. (4) Replication: Depending on the type of virus, this step may take place in the cytoplasm or nucleus of the host cell and may involve several stages of transcription and translation of viral genome and proteins. (5) Virus assembly: The produced viral proteins and genetic components are assembled into new virions ready to be released and infect new host cells. (6) Egress: New viral particles are released from the host cell by lysis or budding

Viruses show 'tropism', which means that they have a particular host range based on the interaction of proteins located on the surface of the virus with specific receptors on the cell surface of the host cell. Many intracellular interactions of viruses with the host cell are known, involving the actin and MT cytoskeleton. All parts of the virus' replicative life cycle are facilitated by the interaction of viruses with these subcellular structures; they allow the movement of viruses to subcellular sites of replication, promote viral assembly and enable efficient cell-to-cell spread [8]. This ability of viruses to be mobile within a host cell may be exploited when considering using viruses as therapeutic tools to target cellular activities detrimental to our health, such as those displayed by cancer cells.

2 Oncolytic Viruses (OVs)

The ability of viruses to seek out their particular host cell, replicate and eventually destroy these cells has been exploited in cancer research for improved treatment of hard-to-treat cancers. Common human viruses can be genetically engineered in such a way that they seek out cancer cells via specific or enhanced receptor recognition, infect and destroy the cell and eventually attract immune cells to the site of infection. Cancer cells are therefore destroyed by the virus, and the immune system is primed to recognized cancer cell antigens, which are released into the microenvironment as part of the viral destruction process. Viruses that specifically target cancer cells in this way are known as oncolytic viruses (OVs). OVs are of great interest to cancer researchers or clinicians as they can infect cells, induce apoptosis or direct the immune response to the site of a tumour [9, 10]. This is especially of interest as it is known that cancer cells evade treatment by directly or indirectly suppressing the immune system. The use of OVs in such a context is known as a type of immunotherapy. Another attractive aspect of using OVs is that mutations in cancer cells lead to the generation of non-self structures that will re-educate the immune system to seek out and destroy other cancer cells, allowing the application of immunotherapy [11]. So far, the list of viruses that have been investigated for their potential as OVs include the adenoviruses, protoparvoviruses, vaccinia viruses, reoviruses and herpes simplex virus type 1 (HSV-1).

How can these viruses be harnessed for application as cancer treatments? Genetic engineering in natural viruses may make the virus tumour-selective. DNA viruses (excluding the poxviruses) and retroviruses may gain tumour selectivity by the insertion of specific promoters. These promoters can be integrated into the viral genome to induce gene expression of genes essential for virus replication. Another manipulation involves the targeting of gene transcription and translation. Cancer cells are characterized by defective interferon (IFN) signalling. By modification of OVs, proteins involved in the suppression of IFN signalling may be disabled. This means that in normal cells, replication is aborted. However, in a cancer cell, which has defective IFN signalling (excluding the poxviruses), translational shutdown mediated by IFN is prevented, and the virus can progress with replication [12]. Viruses can also be modified to inhibit proteins that promote apoptosis, which suppresses replication in normal healthy cells. Since cancer cells can avoid apoptosis via the acquisition of mutations, they are still susceptible to viral infection and eventual replication of the virus. Overexpression or abundance of specific surface receptors on the cancer cells facilitates virus entry [12, 13]. Tropisms of viruses for the receptors may be enhanced by genetic engineering for advanced targeting of the virus to the host cell [14].

Possibly the first attempts of immunotherapy were made by Egyptian physicians in 2600 BC with application of poultice to incisions. It was believed that promoting infections by such means would lead to the regression of tumours [15]. A potential effect of viral infection on cancer progression and the use of OVs was investigated in medical cases recorded in the 1800s when a 42-year-old woman and a 4-year-old boy suffering from leukaemia went into remission after becoming infected with influenza and chickenpox. Further similar observations of remission in two Hodgkin's disease patients after exposure to viral hepatitis in the 1940s led to the first clinical trials of immunotherapy [16]. Encouraged by the effects of hepatitis infection in the two patients, human trials were initiated in 1949 that involved 22 patients with the same disease. The trial was based on administering extracts from hepatitis virus contained within tissue or sera to the patients. Results from the trial showed that 7 patients improved for 1 month before they relapsed, 13 showed signs of hepatitis, and 1 died. This very first trial supplied the first evidence that it is possible to use certain OVs to target cancer cells under the right conditions [16, 17]. It was also demonstrated that tumour regression and temporary remission in younger patients suffering from immunosuppression is achievable [16].

The use of viruses in cancer treatment is now rapidly expanding, and there are high expectations to utilize OVs for routine application of such treatment options. However, thus far, only a single oncolytic viral therapy has been approved for use in the clinic. Talimogene laherparepvec (OncoVEX GM-CSF), known as T-VEC, was certified for use against advanced melanomas in 2015 [18]. Currently, there are no other approved viral treatments, although several potential therapies are in phase I/II clinical trials [19]. Cancer types in these trials cover some of the most aggressive cancers, including pancreatic cancer, melanoma, breast cancer, liver cancer, glioblastoma (GBM), prostate cancer and colorectal cancer. The limited success so far may be attributed to several factors. Physical barriers such as endothelial layers or the dense extracellular matrix (ECM) of solid tumours cause interstitial hypertension, which reduces the infiltrative potential of viruses. In addition, strong innate immune responses may clear the virus before it can reach its target site, and solid tumours promote the induction of an immunosuppressive environment that may counteract the ability of OVs to induce an immune response targeted at the tumour [20, 21]. To overcome the limitations highlighted by the first clinical trials, several approaches must be considered. Firstly, the choice of virus itself is important, as larger viruses are more suited for the insertion of additional therapeutic genes; however, they may be too large to overcome the physical barriers within the body. It has also been shown that enveloped viruses are less oncolytic and more likely to be eliminated by the immune response; thus, such viruses may be modified to enhance their oncolytic potential. Effective delivery methods of the OVs must be determined. These can be either intratumorally (the virus is directly injected into the tumour site) or systemic (intravenously or intraperitoneally). Intratumoural injection is the favoured route of admission in preclinical or clinical trials but may not be feasible for inaccessible or multifocal tumours. Systemic administration appears to be the ideal route as it allows the OVs to reach primary and secondary tumours and is a non-invasive and repeatable application [22, 23]. Recent trials have shown that bioavailability is not satisfactory as the viruses are exposed to rapid sequestration and destruction by the host immune system before they can reach their target. To prevent recognition by the immune response, virus particles may be covered by substances such as polyethylene glycol (PEG) polymers, which prevent recognition [24]. To overcome problems associated with immune responses, further developments support, for example, the use of decoy viruses that sequester pre-existing antibodies. Potentially, newly emerging viruses able to cross physical barriers such as the blood-brain barrier (BBB) include the currently circulating SARS-CoV-2 virus, which could be utilized in several ways to target difficult-to-reach body parts such as the brain, either directly or through conjugation with other anti-tumour therapeutic agents, in combination with other proteins that will allow increased specificity towards tumour.

3 Microtubules (MTs) and Dynamics of Cell Migration

The role of the MTs in cell migration has been discussed in other chapters of this book. It is known that MTs fulfil vital functions in cell migratory processes associated with development, wound healing and the immune response. Apart from the roles already mentioned, they are also vital for the formation of the spindle apparatus during mitosis and for intracellular transport. The MT structure is optimized to perform these various roles. They are tube-like protein structures and vital components of the cytoskeleton in eukaryotes, made from subunits called tubulins. These are small, globular proteins with different shapes and make up a superfamily of tubulins. There are six families: alpha, beta, gamma, delta, epsilon and zeta [25]; the tubulins of relevance to the MTs are the alpha and beta tubulins. Heterodimers are formed from the interaction of an alpha tubulin with a beta tubulin. The heterodimers are characterized by a plus and minus end. The alpha tubulin forms the minus end and the beta tubulin the plus end; this arrangement is important for the role of the MTs in cell migration, which is based on the dynamic instability of the MTs. Tubulin heterodimers align themselves in parallel and form long chains; these are called protofilaments. During growth of the chains, the dimers orient themselves so that there is a changing repeat of the alpha and beta tubulins. The minus and plus ends are now referred to as the ends of the protofilament. Growth and extension of the new dimers, as well as of the protofilament, always occur at the plus end of the chain. MTs are formed when 13 protofilaments are arranged laterally and in close proximity, leading to the formation of rigid and stable rods. MT diameters of 20–30 nm have been recorded. The MT rods are connected to the *m*icrotubule-*o*rganizing *c*entre (MTOC) via the minus end, whereas the plus end can grow towards the cell membrane or shrink. As such, some MTs are dynamically unstable; they grow and shrink by addition or loss from the plus end. This is due to the binding of GTP to the plus end of the MTs. If new tubulin dimers are added to the growing protofilaments, the GTP molecules localized at the end are hydrolysed to GDP. Normally, this is a slow turnover, so there is always a GTP cap at the growing end of the filament. This is necessary to maintain the stability of the MTs. However, if the concentration of new tubulin dimers in the vicinity of the growing MTs is too low, the addition of new dimers is slower than the hydrolysis of the ATP molecules in the already created dimers. This is called *catastrophe*. The GTP cap has been removed, leading to instability of the MT end, resulting in the loss of the dimers and MT shrinkage. This process is reversed when there are enough tubulin dimers that can be incorporated.

In this case, a *rescue* can form a new GTP cap, and the MTs can grow again until the next catastrophe occurs. The structure of MTs and their subcomponents are illustrated in Fig. 2 (for further detailed information on MT structure and function, please refer to textbooks such as Alberts' *Molecular Biology of the Cell*, 7th edition.)

As part of the migration process, MTs also act as tracts for motor proteins. Motor proteins are vital for intracellular transport, as they bind vesicles, organelles or biomacromolecules for transport throughout the cell. The motor proteins that have been thoroughly investigated are kinesins and dyneins. Kinesins and dyneins are characterized by similar striking structures; they have two motor domains by which they can attach to the MTs. They also possess two long, entwined tail domains, where the cargo is attached to. Some kinesins do not possess motor domains and do not bind cargo but instead bind two MTs. Kinesins move along the MTs towards the plus end and transport their cargo from the cell centre to the cell edge. Dyneins move in the opposite direction. To move, both use the 'hand-over-hand' mechanism; the two motor domains are placed interchangeably in front of each so that the motor protein can move along the MTs in a forward direction. As such, at least one of the motor domains is attached to the MTs. As part of this activity, both dyneins and kinesins require energy in the form of ATP. For each step in the motor domain, one ATP molecule is required and is replaced for the next step with a new molecule. Depending on whether ATP, its hydrolysed form or none is bound, the motor domain is connected tightly or weakly to the MTs, enabling the protein to move forward.

As mentioned previously, two types of MTs can be found in a cell: stable and dynamic. Stable MTs are responsible for the transport of vesicles and the stability of the cells, while dynamic MTs are needed for processes that require a rapid response to change, such as cell migration or mitosis. As such, dynamic instability of MTs also allows for the directed migration of cells, as previously discussed. It follows that activities targeting the dynamic nature of MTs in cell migration will affect cell migration.

For further, thorough reviews on MTs, please refer to [26–31]. Figure 2 illustrates the structural components of MTs and the transport of motor proteins along MTs.

4 Viruses and MT Interactions

Viruses are known to be highly efficient in utilizing a cell's machinery for replication, but we also now know that they utilize the MT cytoskeleton for travel within the invaded cell. This is due to the fact that molecules with a molecular weight greater than 500 kDa do not diffuse freely within a cell. In addition, structures in excess of 20 nm also require energy-dependent transport. For viruses with a large capsid or genome, directed transport must be achieved with the help of the actin filament via myosins moving along the filaments and/or the MTs via dynein and kinesin [32, 33]. In this chapter, we will focus on the interaction of viruses with the MT. For further examples on the interaction with the actin filament, see [8]. Dyneins and kinesins move their cargo along MT tracks as previously described; this allows viruses to either benefit from the existing transport system by 'hitchhiking' inside

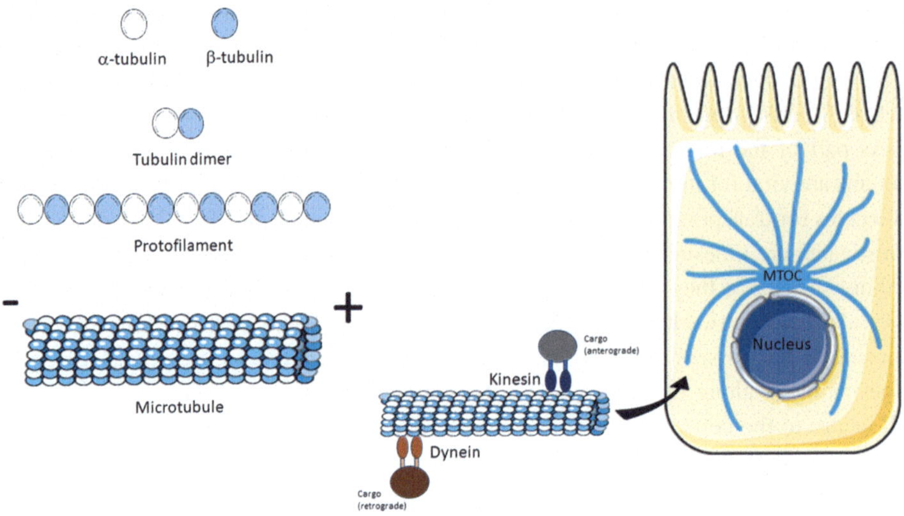

Fig. 2 Structural components of the microtubules (MTs). Two tubulins, alpha and beta, are the subunits or building blocks of the MTs. When an alpha tubulin combines with a beta tubulin, they form a tubulin dimer. A long chain of tubulin dimers makes up a protofilament. Protofilament arranged in a parallel manner and packed tightly together form an MT. The MT is characterized by a plus and minus end, which allows dynamic instability. MTs perform many functions within the cell; they act as cellular tracks for proteins such as the motor and cargo transport proteins dynein and kinesin. While dynein facilitates cargo transport in retrograde fashion towards the microtubule-organizing centre (MTOC), kinesins aid in anterograde transport towards the cell edge. Also shown is an overall view of MT arrangement within a cell, with its anchoring point, the MTOC

vesicles or organelles or bind to the motor proteins directly once they have entered the host cell. During viral replication, a virus first needs to reach its site of replication, which is either the cytosol or the nucleus. Research has uncovered that many viruses utilize and need the MTs to reach the nucleus; this is achieved either by transport inside endosomes or as cytosolic cargo [33].

Some viruses, such as Kaposi's sarcoma-associated herpesvirus, manipulate the MTs to stabilize them by activating associated signalling cascades involving the focal adhesion kinase (FAK) and PI3-kinase, leading to RHO activity on the MTs for travel within endosomes along the stabilized MTs to the nucleus [34]. Other viruses, such as adenoviruses or herpesviruses, can manipulate the direction of transport on the MTs by modulating the amount of plus-end-directed transport. Therefore, minus-end transport towards the nucleus becomes more frequent, allowing migration in a preferred direction. Interactions between viral components and dynein have been reported for certain viruses, allowing direct interaction and transport along the MTs [35]. Disruption of host adaptors such as dynactin, which link the cargo to dynein, has been shown to affect virus transport, such as towards the MTOC, which indicates that these viruses are able to utilize such adaptors rather than the motor proteins themselves to hitch a ride [8]. Finally, some viruses, such as herpesviruses or human immunodeficiency virus (HIV), make use of proteins located at the growing end of the MTs, the tip-tracking proteins such as EB1, CLIP-170 or the kinesin Kif4. By direct interaction with these tip-tracking proteins, the viruses induce MT stabilization, allowing transport towards the MTOC. Other mechanisms of MT stabilization are achieved by interference with Rac1 signalling or enhancing tyrosination or acetylation of the MTs on egress. The MTs are subjected to post-translational modifications such as acetylation or tyrosination to allow diversification and differential functionality among the MTs and interaction with specific proteins, including the tip-tracking proteins and motor proteins [36]. Further interaction with the cell cytoskeleton takes place during replication and virus assembly and is described in detail here [8, 37, 38]. The various strategies employed by viruses to move within cells in summarized in Fig. 3.

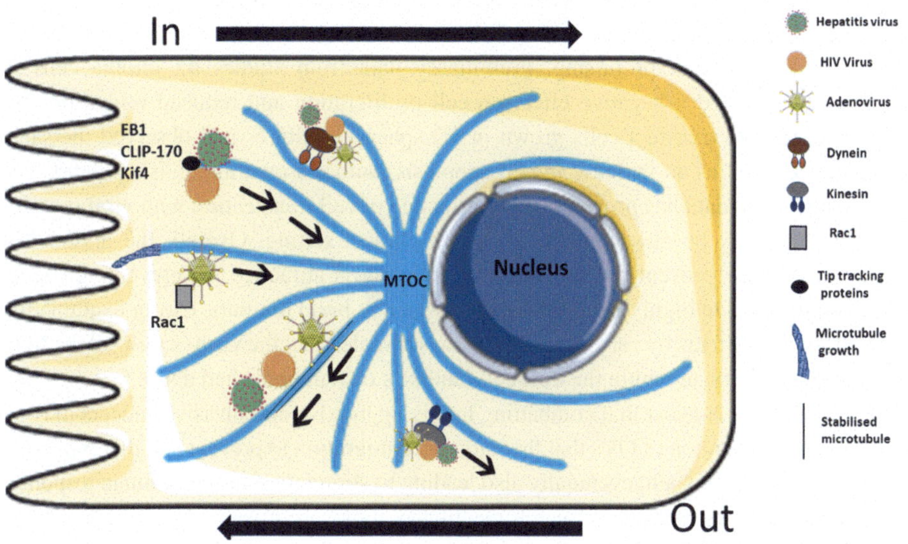

Fig. 3 Viruses and their interactions with the MTs. Viruses such as herpesvirus, adenovirus and HIV can utilize MTs for movement within the host cell. This is achieved by manipulating the dynamic nature of the MT. By binding to tip-tracking proteins such as EB1, Clip-70 and Kif1a, viruses can induce stabilization of the MTs. Interactions with Rac1 have also been reported to induce stabilization. Viruses can hitchhike on motor proteins to reach their site of replication and use motor proteins to move towards the cell edge for egress. MTs may be stabilized for virus exit by influencing tyrosination and acetylation status of the MT

5 Examples of OVs and the Study of Cell Migration

As mentioned earlier in this chapter, viruses are characterized by their inability to replicate on their own and their reliance on a host cell for replication to occur. They therefore need to enter their host, induce the production of their subunits to be assembled into newly made viruses and finally leave the host cell. Once inside the host cells, OVs have to also move within the host to reach their site of replication and orchestrate the production of their own genetic and structural units.

We have recently investigated the use of different viruses as OVs for the treatment of high-grade paediatric and adult brain tumours and observed that cell migration of the targeted cancer cells is affected during infection with the viruses. The OVs had been engineered to specifically target cancer cells for destruction. This research indicated that the OV HSV produced significant anti-migratory effects and eventually induced cell death in glioma cells, affecting migration and proliferation in brain tumour cells [39, 40].

In addition, our first studies with subunits of the bluetongue virus (BTV) transfected into glioma cells look promising. Preliminary data from studies on glioma cells in monolayers revealed a negative effect on cell proliferation and reduced migration. We next proceeded to transfect cells grown in 3D spheroid cultures and observed the same outcomes, as well as an increase in cell apoptosis (author's own unpublished data). The preliminary BTV data has proven that the principle of cell transfection with viral proteins is viable and may be also applied in a more clinical setting. Overall, the notion that these viruses may prevent cancer cells from moving is attractive for preventing cancer cell dissemination in highly aggressive cancers such as brain tumours. Closer examination of the described effects by various in vitro assays revealed that the viruses utilized the MTs as a means to circulate within the host cell (author's own unpublished data).

These findings suggest that combatting highly aggressive and invasive cancers may be aided by the application of OVs that first target cell migration to prevent cell dissemination and then induce cell death, eventually also leading to stimulation of the immune response. Here, two examples from the research mentioned above are given to describe the noted effects in detail.

5.1 Common Human Viruses as OVs and Their Effect on MTs

Our recent work investigated the potential of common human viruses to target cell migration in cancer, including HSV, reovirus and vaccinia virus. These viruses had been previously engineered to exclusively target cancer cells, including those derived from high-grade paediatric gliomas and diffuse intrinsic pontine glioma (DIPGs). Differences were observed among these viruses in terms of their effect on cell migration. The oncolytic HSV appeared to be the most promising OV tested, as it induced specific and pronounced anti-migratory and anti-invasive effects on cell migration, not as a by-product of its effect on proliferation. The effects observed by live cell imaging, scratch assays and 3D invasion

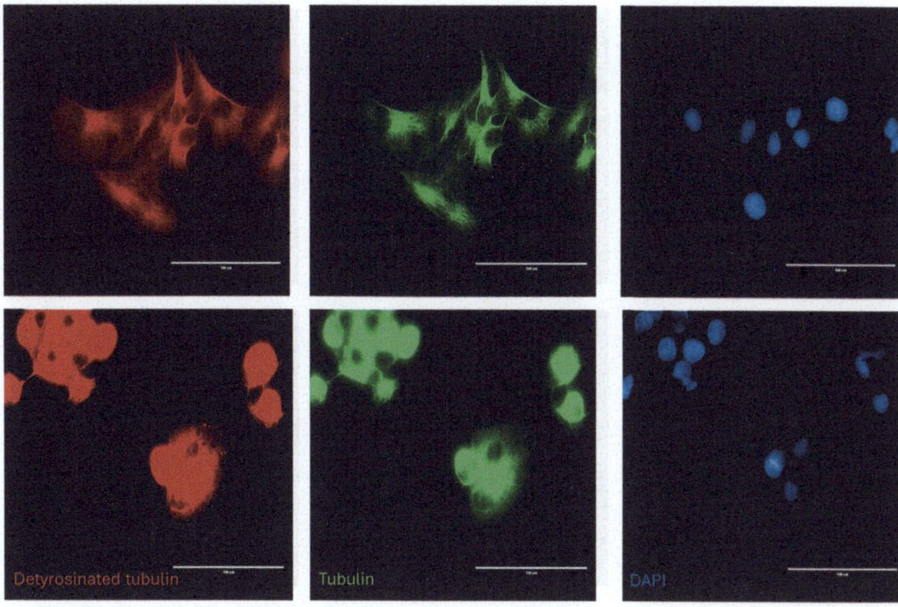

Fig. 4 HSV targets MT tyrosination for exit from the host cell. Immunofluorescent images of the paediatric glioma cell line SF188, either not infected (top row) or infected (bottom row) with HSV over 24 h. When the cells are probed for detyrosinated tubulins, the levels of detyrosinated tubulin have increased in the virus-treated cells, as indicated by stronger fluorescent levels. Fluorescent colours indicate the following: red, detyrosinated tubulin; green, tubulin; blue, DAPI (DNA marker). Scale bar = 50 μm (Image Julia Cockle)

assays were confirmed by immunofluorescence microscopy and Western blotting of the targeted cells. It was discovered that the OVs altered cytoskeletal dynamics of the cancer cells, leading to the stabilization of MTs via the accumulation of post-translational tubulin modifications [40]. The effects of viral infection on MT post-translational modifications are highlighted in Figs. 4 and 5. Here, deacetylation and detyrosination of tubulin were enhanced in cancer cells after infection with oncolytic HSV.

Interestingly, the treatment of the cells with the OVs also led to the inhibition of the glycogen synthase kinase 3 (GSK-3), an enzyme that modulates cell migration in this cancer type by regulating actin polymerization, focal adhesion maturation and MT dynamics. Consequently, localized clustering of the adenomatous polyposis coli protein (APC), which associates with the tips of the MTs and regulates MT dynamics, was inhibited, suggesting disruption of MT dynamics [39, 40]. This is shown in Fig. 6, where localized clustering of APC was suppressed in cancer cells after the addition of oncolytic HSV.

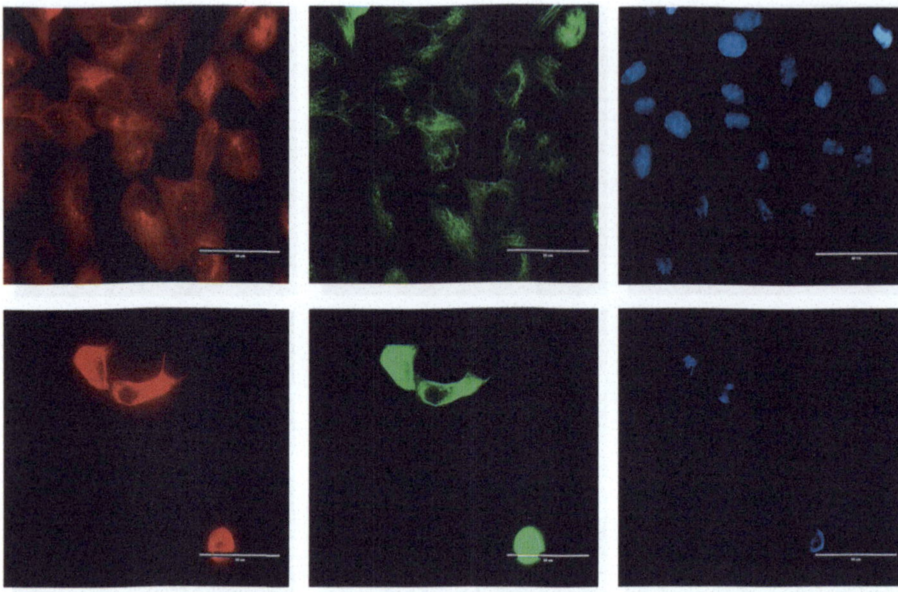

Fig. 5 HSV targets MT acetylation for exit from the host cell. Immunofluorescent images of the paediatric glioma cell line KNS42, either not infected (top row) or infected (bottom row), with HSV over 24 h. When the cells are probed for deacetylated tubulins, the levels of deacetylated tubulin have increased in the virus-treated cells, as indicated by stronger fluorescent levels. Fluorescent colours indicate the following: red, deacetylated tubulin; green, tubulin; blue, DAPI (DNA marker). Scale bar = 50 μm (Image Julia Cockle)

5.2 The Animal Virus Bluetongue Virus (BTV) and Its Effect on MT Dynamics

Supportive evidence comes from the literature regarding the use of viruses affecting animals. In their initial research investigating the effect of BTV subunits, Shaw et al. [41] discuss the possibility of using BTV to target GBM. This virus belongs to the family Reoviridae, genus *Orbivirus*; so far, 20 species in the genus have been identified. The non-enveloped virus is complex with a capsid and double-stranded (ds) RNA genome. It infects domestic animals, including sheep, goats and cattle, as well as wildlife [42]. The virus consists of seven structural proteins (VP1–VP7) and at least of four non-structural proteins (NS1–NS4) encoded by ten segments of dsRNA. The structural proteins are associated with the two-layer core of the virus, encapsulating the RNA polymerase and segmented genome [43]. The non-structural proteins play a role in promoting viral replication within the host and targeting the host's immunity. One of the most abundant non-structural proteins, NS2, is associated with the viral inclusion bodies (VIB), which are viral factories characterized by dense aggregates of viral particles that allow virus synthesis [43–45]. It has been reported that an animal virus such as BTV can be utilized to target cancer

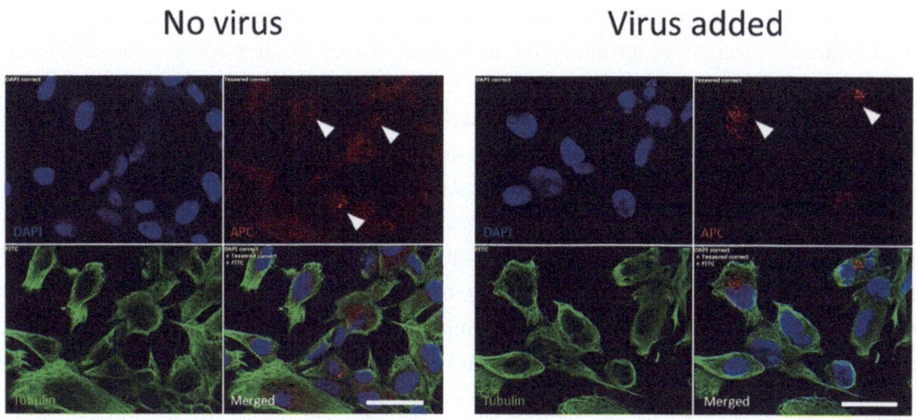

Fig. 6 Effect of virus infection on microtubule tip-trackers such as APC. HSV induces the relocation of the tip-tracking protein APC from the centrosome to other parts of the cell, as shown by immunofluorescence. In non-infected conditions, APC is found at the centrosome in the paediatric cell line SF188, indicated by bright single red spots. After infection with HSV for 24 h, the dots have disappeared and are replaced by scattered red patterns near the nucleus. Arrowheads indicate the change in appearance and location of APC. Fluorescent colours indicate the following: red, APC; green, tubulin; blue, DAPI (DNA marker); merged images overlaid. Scale bar = 20 μm (Image Julia Cockle)

cells. Whole BTV was used to successfully target renal cancer cells in vitro and in vivo by Wang et al. [46]. In xenograft models, intratumoural injection with BTV led to a significant reduction in tumour volume.

Studies on the non-structural BTV protein NS2 revealed that the virus can affect mammalian mitosis via the activity of NS2. GBM cells transfected with an NS2 construct exhibited mitotic defects, disrupting the formation of the mitotic spindle as observed by live cell imaging and immunofluorescence microscopy analysis. In addition, the interaction between the MTs and the chromosomes was also affected [41]. Research by Bhattacharya et al. [47] further showed the dependency of the BTV virus on interaction with the MT cytoskeleton. When the MT networks were disrupted by pharmacological intervention, an increase in intracellular BTV was observed with a concomitant decrease in released virus. It was concluded that BTV utilizes the MT network as a transport system for the export of the mature virus particle from the host cell. As described in previous chapters, the MTs play a vital role in cell migration. Together with the actin filament and the association with focal adhesions, the MTs allow polarization of the cell in the direction of migration [48]. If this activity is disturbed, this will have a follow-on effect on overall cell migration. In a recent study by the author, it was noted that targeting the glioma cell line U87 transfected with the NS2 BTV construct, versus mock-transfected cells, had an anti-migratory effect on the cancer cells. U87 cells transfected with the BTV NS2 protein were found to have a significantly lower accumulated distance (AD) compared to the mock-transfected cells (A. Brüning-Richardson and D. Abed, unpublished data, Figs. 7a, b). 'Accumulated distance' is an indicator of distance travelled after analysis of movies of migrating cells using Tracker tools [49]. As the MTs play a key role in cell motility, the observed reduction in AD may indicate a loss of MT function aiding polarization of the cells. The mechanical coordination of the cytoskeleton is compromised by disruption of MT activity, leading to the observed effect. Another striking observation was that a high proportion of the cells appeared stationary during the live cell imaging, highlighting the detrimental effect of transfection with the viral subunit. Interestingly, a second glioma cell line, U251, used in this study appeared not to be affected by the same treatment. This may be due to the heterogeneity of GBM. U87 and U251 have been thoroughly characterized, and differences in proliferation, migration and invasion have been reported. U87 has been shown to have greatly enhanced migratory capacities compared to U251. Research by Li et al. [50] concluded through a proteomic approach that the observed differences in phenotypes were due to differences in protein expression and their associated biological processes. Diao et al. [51] studied four different glioma cell lines, including U87 and U251, in a 3D microfabricated model consisting of hollow microchamber arrays containing a natural collagen interface. These authors found that the cell lines differed in terms of cell morphology, proliferation, migration and invasion. U251 was found to be the cell line that migrated the least in terms of cell speed, directionality and displacement. Such differences in protein expression and migratory activities may be the reason for the observed different responses to transfection with BTV NS2. Preliminary studies on 3D spheroids transfected with the BTV NS construct also indicate anti-migratory effects in U87 (Fig. 7c) (author's own unpublished data).

These observations also hold clear significance when considering BTV as a novel OV for clinical use. If BTV is able to inhibit GBM cell motility, it may allow for more effective treatment. Stopping cell migration of single cells away from the original tumour would enable treatments such as chemotherapy or radiotherapy to combat a larger portion of cells in the first round of treatment, reducing the risk of patient relapse.

6 Methods in Cell Migration Using Viruses

In the examples mentioned in this chapter, we utilized commonly used migration and invasion assays to determine the anti-migratory activity of OVs. For adaption for use with viruses, we carried out the following methods. Guidelines regarding the growing and handling of live viruses must be followed at the location where the experiments are conducted.

One assay that can be easily carried out is a scratch assay, as previously discussed.

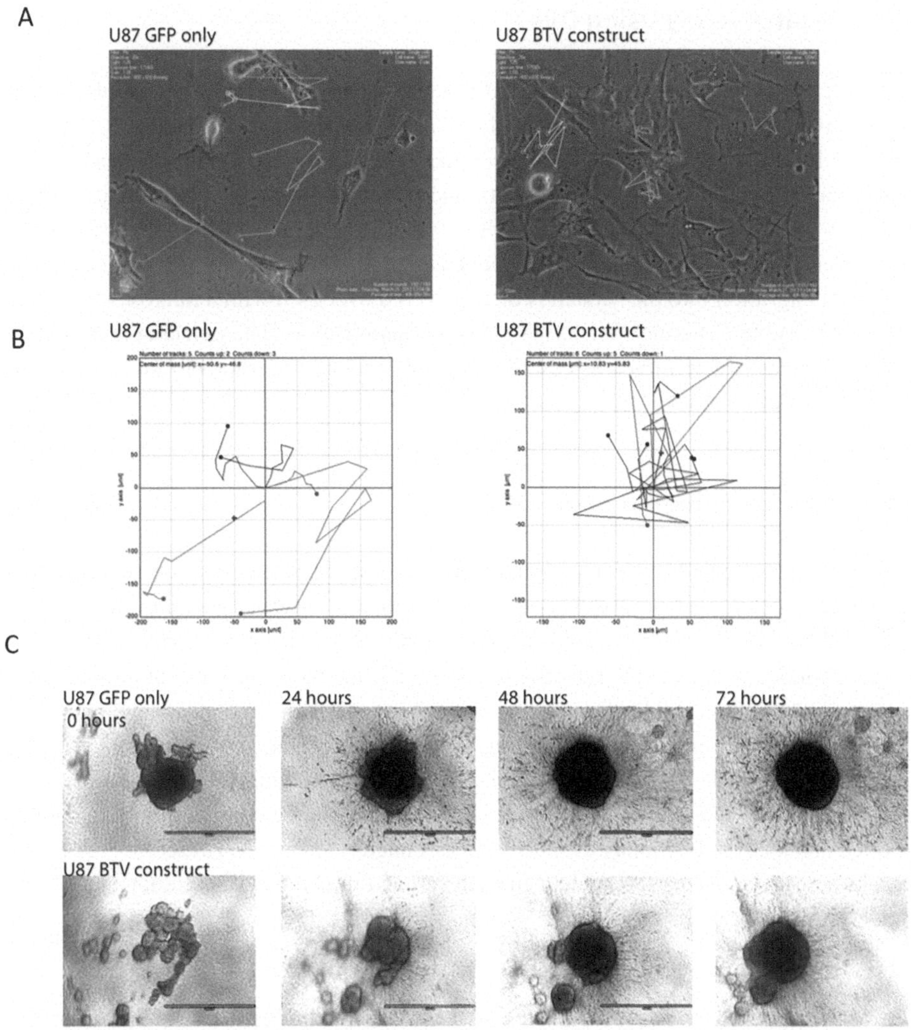

Fig. 7 The BTV non-structural protein NS2 targets cell migration in brain tumour cells. (**a**) Snapshots of the glioma cell line U87 after random cell migration. Traces of the migrating cells over time are indicated by different coloured lines. After transfection with the NS2 BTV protein, cells cover less distance than mock-transfected cells. (**b**) Rose plots indicate the effect of the treatment with the NS2 BTV constructs. Each line represents an individual single cell. Transfection with the NS2 BTV construct results in less distance travelled. (**c**) Migration of cancer cells is affected by NS2 BTV transfection in U87 spheroids migrating/invading into a collagen-based matrix. The control cells (top line) are able to migrate into the collagen over time, as indicated by single cells emanating away from the original spheroid. In contrast, spheroids transfected with the NS2 BTV construct are characterized by reduced migration, as indicated by the reduced distance covered by the single migrating cells (bottom). Scale bar = 1000 μm (Images: Anke Brüning-Richardson and Delnia Abed)

6.1 Scratch Assay Using OVs

The cancer cells in question, here paediatric glioma or DIPG cell lines, are plated at a predetermined cell density into 24-well plates and allowed to reach 80–90% confluence. After incubation for 24 h, a line is drawn on the underside of each plate. As described in the previous chapter, a gentle scratch is introduced, and the wells are washed with culture medium to remove detached cells and cell debris. Next, culture medium containing the virus under investigation is added at different plaque-forming units (pfu) ranging from 0.01 to 50. Cell migration into the scratch is then observed over 24 h with a microscope system such as the EVOS (Thermo Fisher, UK) at ×4 magnification. The percentage change in the scratch areas is determined as described in Part II Chap. "Methods to Investigate Cell Migration" of this book.

6.2 Live Cell Imaging Using OVs

Cells to be interrogated are counted and added to Ibidi (Ibidi, Germany) or equivalent specialized imaging dishes at sub-confluence and then allowed to attach to the imaging plate for 2 h at 37 °C. The medium is removed and replaced with the virus to be tested in culture medium, such as HSV, reovirus or vaccinia virus; the virus concentration used in our system is 10 pfu/cell. This figure will need to be predetermined depending on the virus used. The dishes are incubated in imaging systems such as the BioStation (Nikon, UK) culturing chamber, which is attached to the live imaging system. Imaging is carried out over 48 h at 3-min intervals; BioStation software is used to create movies that can then be exported and analysed by analysis tools such as ImageJ [52] and MTrackJ [53]. Tracking is achieved by identifying the nucleus in each cell and tracking cell migration at 150-min intervals.

6.3 Migration/Invasion Assays and Live Cell Imaging of 3D Spheroids Infected with Virus

The spheroid invasion assay, as mentioned in Part II Chap. "Methods to Investigate Cell Migration", is followed. After spheroid embedding in collagen and collagen polymerization, the spheroids are exposed to the virus by the addition of 8×10^2, 8×10^3, 8×10^4 or 4×10^5 pfu/well of virus in the appropriate culture medium. Note that these are predetermined numbers for the viruses used in our studies. In case of the migration assays, spheroids are imaged with a microscopic system such as the EVOS cell imaging system (Thermo Fisher, UK) at ×4 magnification. Spheroids are imaged at $t = 0$, 24, 48 and 72 h post-embedding, though this may vary according to the cell lines used. The images are then used to determine the migration index of the spheroids over time.

For live cell imaging, we use imaging systems such as the Incucyte Zoom at ×4 magnification. The spheroids are overlayed with medium containing virus at 8×10^4 pfu/well for HSV and vaccinia virus. For visualization, the viruses used also express GFP. Images are taken once every hour over 24 h. Movies are generated with the associated Incucyte™ software for data analysis with ImageJ and visualization of the GFP.

6.4 Immunofluorescence Analysis of the Effect of Virus Infection on Cell Migration

To assess the effect of viral infection on cell migration via targeting the MTs, we routinely use immunofluorescence. For this, we probe virus-infected cells attached to coverslips after fixing with 100% ice-cold methanol. The slides are fixed for 2 min at −20 °C and then washed three times with PBS. Blocking of the slides is achieved by using 0.05% skimmed milk powder in PBS for 5 min at room temperature. Meanwhile, primary antibodies raised against, for example, acetylated and detyrosinated tubulin as markers of disrupted MT dynamics are diluted in the blocking buffer, and the antibody solutions are then microfuged for 5 min at 13,000 rpm. Two hundred microlitres of antibody solution are added to pieces of parafilm floating on water in an incubation chamber. The coverslips are placed upside down onto the antibody solution and incubated for 1 h at room temperature. After incubation, the coverslips are removed and washed three time in PBS. Fluorescently labelled secondary antibodies to the species in which the first antibodies are generated are prepared in the blocking solution as described for the primary antibodies. Also added to this preparation is DAPI as the nuclear stain, followed by incubation as described before on the floating parafilm pieces. Care must be taken to cover the incubation chamber to prevent signal decay due to bleaching of the fluorescent labels. After another incubation of 1 h, the coverslips are washed three times in PBS and then mounted down in spots of Fluoromount-G or an equivalent mounting medium. The slides are then allowed to air-dry overnight covered by foil. For analysis of the effect of the virus on protein expression, fluorescent signals for each antibody are recorded for approximately 200 cells per condition and measured by determining the corrected total cell fluorescence (CTCF) as described by [54]. Additionally, changes in protein localization should be noted. This can be scored by counting the number of cells within one slide that show a particular localization of the proteins, e.g. tyrosinated or acetylated tubulin, in the cell populations.

Take-Home Message

In this chapter, the concept of using viruses for the treatment of cancers and their ability to target cell migration was introduced. Several key characteristics of viruses and their effects on their host cells were discussed:

- Viruses are infectious agents that cannot replicate on their own and rely on a host cell for replication.
- Viruses enter the host cell and utilize the MT network to move within the infected cell.
- OVs are viruses that specifically target cancer cells via special cancer cell receptors and virus interactions.
- OVs utilize host cell MT cytoskeleton to reach their site of replication within the cancer cell.
- OVs can be used to study the effects on MT dynamics through in vitro assays in 2D and 3D.

References

1. Moelling K, Broecker F. Viruses and evolution - viruses first? A personal perspective. Front Microbiol. 2019;10:523. https://doi.org/10.3389/fmicb.2019.00523. PMID: 30941110; PMCID: PMC6433886
2. Lander ES, Linton LM, Birren B, Nusbaum C, Zody MC, Baldwin J, Devon K, Dewar K, Doyle M, FitzHugh W, Funke R, Gage D, Harris K, Heaford A, Howland J, Kann L, Lehoczky J, LeVine R, McEwan P, McKernan K, Meldrim J, Mesirov JP, Miranda C, Morris W, Naylor J, Raymond C, Rosetti M, Santos R, Sheridan A, Sougnez C, Stange-Thomann Y, Stojanovic N, Subramanian A, Wyman D, Rogers J, Sulston J, Ainscough R, Beck S, Bentley D, Burton J, Clee C, Carter N, Coulson A, Deadman R, Deloukas P, Dunham A, Dunham I, Durbin R, French L, Grafham D, Gregory S, Hubbard T, Humphray S, Hunt A, Jones M, Lloyd C, McMurray A, Matthews L, Mercer S, Milne S, Mullikin JC, Mungall A, Plumb R, Ross M, Shownkeen R, Sims S, Waterston RH, Wilson RK, Hillier LW, McPherson JD, Marra MA, Mardis ER, Fulton LA, Chinwalla AT, Pepin KH, Gish WR, Chissoe SL, Wendl MC, Delehaunty KD, Miner TL, Delehaunty A, Kramer JB, Cook LL, Fulton RS, Johnson DL, Minx PJ, Clifton SW, Hawkins T, Branscomb E, Predki P, Richardson P, Wenning S, Slezak T, Doggett N, Cheng JF, Olsen A, Lucas S, Elkin C, Uberbacher E, Frazier M, Gibbs RA, Muzny DM, Scherer SE, Bouck JB, Sodergren EJ, Worley KC, Rives CM, Gorrell JH, Metzker ML, Naylor SL, Kucherlapati RS, Nelson DL, Weinstock GM, Sakaki Y, Fujiyama A, Hattori M, Yada T, Toyoda A, Itoh T, Kawagoe C, Watanabe H, Totoki Y, Taylor T, Weissenbach J, Heilig R, Saurin W, Artiguenave F, Brottier P, Bruls T, Pelletier E, Robert C, Wincker P, Smith DR, Doucette-Stamm L, Rubenfield M, Weinstock K, Lee HM, Dubois J, Rosenthal A, Platzer M, Nyakatura G, Taudien S, Rump A, Yang H, Yu J, Wang J, Huang G, Gu J, Hood L, Rowen L, Madan A, Qin S, Davis RW, Federspiel NA, Abola AP, Proctor MJ, Myers RM, Schmutz J, Dickson M, Grimwood J, Cox DR, Olson MV, Kaul R, Raymond C, Shimizu N, Kawasaki K, Minoshima S, Evans GA, Athanasiou M, Schultz R, Roe BA, Chen F, Pan H, Ramser J, Lehrach H, Reinhardt R, McCombie WR, de la

Bastide M, Dedhia N, Blöcker H, Hornischer K, Nordsiek G, Agarwala R, Aravind L, Bailey JA, Bateman A, Batzoglou S, Birney E, Bork P, Brown DG, Burge CB, Cerutti L, Chen HC, Church D, Clamp M, Copley RR, Doerks T, Eddy SR, Eichler EE, Furey TS, Galagan J, Gilbert JG, Harmon C, Hayashizaki Y, Haussler D, Hermjakob H, Hokamp K, Jang W, Johnson LS, Jones TA, Kasif S, Kaspryzk A, Kennedy S, Kent WJ, Kitts P, Koonin EV, Korf I, Kulp D, Lancet D, Lowe TM, McLysaght A, Mikkelsen T, Moran JV, Mulder N, Pollara VJ, Ponting CP, Schuler G, Schultz J, Slater G, Smit AF, Stupka E, Szustakowki J, Thierry-Mieg D, Thierry-Mieg J, Wagner L, Wallis J, Wheeler R, Williams A, Wolf YI, Wolfe KH, Yang SP, Yeh RF, Collins F, Guyer MS, Peterson J, Felsenfeld A, Wetterstrand KA, Patrinos A, Morgan MJ, de Jong P, Catanese JJ, Osoegawa K, Shizuya H, Choi S, Chen YJ, Szustakowki J, International Human Genome Sequencing Consortium. Initial sequencing and analysis of the human genome. Nature. 2001;409(6822):860–921. https://doi.org/10.1038/35057062. Erratum in: Nature 2001 Aug 2;412(6846):565. Erratum in: Nature 2001 Jun 7;411(6838):720. Szustakowki, J [corrected to Szustakowski, J]. PMID: 11237011

3. Chial H. DNA sequencing technologies key to the human genome project. Nat Educ. 2008;1(1):219.
4. Gates AJ, Morselli Gysi D, Kellis M, Barabási AL. A wealth of discovery built on the human genome project. Nature. 2021;590:212–5.
5. Enard D, Cai L, Gwennap C, Petrov DA. Viruses are a dominant driver of protein adaptation in mammals. Elife. 2016;5:e12469. https://doi.org/10.7554/eLife.12469. PMID: 27187613; PMCID: PMC4869911
6. https://www.sciencefocus.com/the-human-body/virus-human-evolution/
7. https://www.scientificamerican.com/article/are-viruses-alive-2004/
8. Radtke K, Döhner K, Sodeik B. Viral interactions with the cytoskeleton: a hitchhiker's guide to the cell. Cell Microbiol. 2006;8:387–400. https://doi.org/10.1111/j.1462-5822.2005.00679.
9. Cao GD, He XB, Sun Q, Chen S, Wan K, Xu X, Feng X, Li PP, Chen B, Xiong MM. The oncolytic virus in cancer diagnosis and treatment. Front Oncol. 2020;10:1786. https://doi.org/10.3389/fonc.2020.01786. PMID: 33014876; PMCID: PMC7509414
10. Martin NT, Bell JC. Oncolytic virus combination therapy: killing one bird with two stones. Mol Ther. 2018;26(6):1414–22. https://doi.org/10.1016/j.ymthe.2018.04.001. Epub 2018 Apr 5. PMID: 29703699; PMCID: PMC5986726
11. Hemminki O, Dos Santos JM, Hemminki A. Oncolytic viruses for cancer immunotherapy. J Hematol Oncol. 2020;13(1):84. https://doi.org/10.1186/s13045-020-00922-1. PMID: 32600470; PMCID: PMC7325106
12. Kaufman HL, Kohlhapp FJ, Zloza A. Oncolytic viruses: a new class of immunotherapy drugs. Nat Rev Drug Discov. 2015;14(9):642–62. https://doi.org/10.1038/nrd4663. Erratum in: Nat Rev Drug Discov. 2016 Aug 30;15(9):660. PMID: 26323545; PMCID: PMC7097180
13. Li J, Meng Q, Zhou X, Zhao H, Wang K, Niu H, Wang Y. Gospel of malignant Glioma: Oncolytic virus therapy. Gene. 2022;818:146217. https://doi.org/10.1016/j.gene.2022.146217. Epub 2022 Jan 28. PMID: 35093451
14. Gao Y, Wu Y, Huan T, Wang X, Xu J, Xu Q, Yu F, Shi H. The application of oncolytic viruses in cancer therapy. Biotechnol Lett. 2021;43(10):1945–54. https://doi.org/10.1007/s10529-021-03173-3. Epub 2021 Aug 26. PMID: 34448096
15. Jessy T. Immunity over inability: the spontaneous regression of cancer. J Nat Sci Biol Med. 2011;2(1):43–9. https://doi.org/10.4103/0976-9668.82318. PMID: 22470233; PMCID: PMC3312698
16. Kelly E, Russell SJ. History of oncolytic viruses: genesis to genetic engineering. Mol Ther. 2007;15(4):651–9. https://doi.org/10.1038/sj.mt.6300108. Epub 2007 Feb 13. PMID: 17299401

17. Hoster HA, Zanes RP Jr, von Haam E. Studies in Hodgkin's disease; the association of viral hepatitis and Hodgkin's disease; a preliminary report. Cancer Res. 1949;9(8):473–80.
18. Mondal M, Guo J, He P, Zhou D. Recent advances of oncolytic virus in cancer therapy. Hum Vaccin Immunother. 2020;16(10):2389–402. https://doi.org/10.1080/21645515.2020.1723363. Epub 2020 Feb 20. PMID: 32078405; PMCID: PMC7644205
19. Cook M, Chauhan A. Clinical application of oncolytic viruses: a systematic review. Int J Mol Sci. 2020;21(20):7505. https://doi.org/10.3390/ijms21207505. PMID: 33053757; PMCID: PMC7589713
20. Lawler SE, Speranza MC, Cho CF, Chiocca EA. Oncolytic viruses in cancer treatment: a review. JAMA Oncol. 2017;3(6):841–9. https://doi.org/10.1001/jamaoncol.2016.2064. PMID: 27441411
21. Zheng M, Huang J, Tong A, Yang H. Oncolytic viruses for cancer therapy: barriers and recent advances. Mol Ther Oncolytics. 2019;15:234–47. https://doi.org/10.1016/j.omto.2019.10.007. PMID: 31872046; PMCID: PMC6911943
22. Shi T, Song X, Wang Y, Liu F, Wei J. Combining oncolytic viruses with cancer immunotherapy: establishing a new generation of cancer treatment. Front Immunol. 2020;11:683. https://doi.org/10.3389/fimmu.2020.00683. PMID: 32411132; PMCID: PMC7198760
23. Ylösmäki E, Cerullo V. Design and application of oncolytic viruses for cancer immunotherapy. Curr Opin Biotechnol. 2020;65:25–36. https://doi.org/10.1016/j.copbio.2019.11.016. Epub 2019 Dec 23. PMID: 31874424. – Limitations
24. Santos Apolonio J, de Souza L, Gonçalves V, Cordeiro Santos ML, Silva Luz M, Silva Souza JV, Rocha Pinheiro SL, de Souza WR, Sande Loureiro M, de Melo FF. Oncolytic virus therapy in cancer: a current review. World J Virol. 2021;10(5):229–55. https://doi.org/10.5501/wjv.v10.i5.229. PMID: 34631474; PMCID: PMC8474975
25. Oakley BR. An abundance of tubulins. Trends Cell Biol. 2000;10(12):537–42. https://doi.org/10.1016/s0962-8924(00)01857-2. PMID: 11121746
26. Goodson HV, Jonasson EM. Microtubules and microtubule-associated proteins. Cold Spring Harb Perspect Biol. 2018;10(6):a022608. https://doi.org/10.1101/cshperspect.a022608. PMID: 29858272; PMCID: PMC5983186
27. Bera A, Gupta ML Jr. Microtubules in microorganisms: how tubulin isotypes contribute to diverse cytoskeletal functions. Front Cell Dev Biol. 2022;10:913809. https://doi.org/10.3389/fcell.2022.913809. PMID: 35865635; PMCID: PMC9294176
28. Janke C, Magiera MM. The tubulin code and its role in controlling microtubule properties and functions. Nat Rev Mol Cell Biol. 2020;21(6):307–26. https://doi.org/10.1038/s41580-020-0214-3. Epub 2020 Feb 27. PMID: 32107477
29. Gudimchuk NB, McIntosh JR. Regulation of microtubule dynamics, mechanics and function through the growing tip. Nat Rev Mol Cell Biol. 2021;22(12):777–95. https://doi.org/10.1038/s41580-021-00399-x. Epub 2021 Aug 18. PMID: 34408299
30. Wade RH. On and around microtubules: an overview. Mol Biotechnol. 2009;43(2):177–91. https://doi.org/10.1007/s12033-009-9193-5. Epub 2009 Jun 30. PMID: 19565362
31. Zwetsloot AJ, Tut G, Straube A. Measuring microtubule dynamics. Essays Biochem. 2018;62(6):725–35. https://doi.org/10.1042/EBC20180035. PMID: 30287587; PMCID: PMC6281472
32. Seksek O, Biwersi J, Verkman AS. Translational diffusion of macromolecule-sized solutes in cytoplasm and nucleus. J Cell Biol. 1997;138(1):131–42.
33. Stidwell RP, Greber UF. Intracellular virus trafficking reveals physiological characteristics of cytoskeleton. News Physiol Sci. 2000;15:67–71.

34. Sharma-Walia N, Naranatt PP, Krishnan HH, Zeng L, Chandran B. Kaposi's sarcoma associated herpesvirus envelope glycoprotein gB induces the integrin-dependent focal adhesion kinase-Src-PI-3K-RhoGTPase signal pathways and cytoskeletal rearrangements. J Virol. 2004;78:4207–23.
35. Ye GJ, Vaughan KT, Vallee RB, Roizman B. The herpes simplex virus 1 UL 34 protein interacts with a cytoplasmic dynein intermediate chain and targets nuclear membrane. J Virol. 2000;74:1355–63.
36. Wang IH, Burckhardt CJ, Yakimovich A, Greber UF. Imaging, tracking and computational analyses of virus entry and egress with the cytoskeleton. Viruses. 2018;10(4):166. https://doi.org/10.3390/v10040166. PMID: 29614729; PMCID: PMC5923460
37. Naghavi MH, Walsh D. Microtubule regulation and function during virus infection. J Virol. 2017;91(16):e00538–17. https://doi.org/10.1128/JVI.00538-17. PMID: 28615197; PMCID: PMC5533906
38. Walsh D, Naghavi MH. Exploitation of cytoskeletal networks during early viral infection. Trends Microbiol. 2019;27(1):39–50. https://doi.org/10.1016/j.tim.2018.06.008. Epub 2018 Jul 20, PMID: 30033343; PMCID: PMC6309480
39. Cockle JV, Brüning-Richardson A, Scott KJ, Thompson J, Kottke T, Morrison E, Ismail A, Carcaboso AM, Rose A, Selby P, Conner J, Picton S, Short S, Vile R, Melcher A, Ilett E. Oncolytic herpes simplex virus inhibits pediatric brain tumor migration and invasion. Mol Ther Oncolytics. 2017;5:75–86. https://doi.org/10.1016/j.omto.2017.04.002. PMID: 28547002; PMCID: PMC5435599
40. Cockle JV. Development of an anti-invasive therapeutic strategy for paediatric high grade glioma. 2013. https://etheses.whiterose.ac.uk/12515/1/CockleJV_Medicine_PhD_2016.pdf_
41. Shaw AE, Brüning-Richardson A, Morrison EE, Bond J, Simpson J, Ross-Smith N, Alpar O, Mertens PP, Monaghan P. Bluetongue virus infection induces aberrant mitosis in mammalian cells. Virol J. 2013;10:319. https://doi.org/10.1186/1743-422X-10-319. PMID: 24165208; PMCID: PMC3874736
42. https://www.pirbright.ac.uk/bluetongue
43. Rojas JM, Martín V, Sevilla N. Vaccination as a strategy to prevent bluetongue virus vertical transmission. Pathogens. 2021;10(11):1528. https://doi.org/10.3390/pathogens10111528. PMID: 34832683; PMCID: PMC8622840
44. Roy P. Bluetongue virus structure and assembly. Curr Opin Virol. 2017;24:115–23. https://doi.org/10.1016/j.coviro.2017.05.003. Epub 2017 Jun 10. PMID: 28609677
45. Mertens PP, Diprose J, Maan S, Singh KP, Attoui H, Samuel AR. Bluetongue virus replication, molecular and structural biology. Vet Ital. 2004;40(4):426–37. PMID: 20422565
46. Wang H, Song L, Zhang X, Zhang X, Zhou X. Bluetongue viruses act as novel oncolytic viruses to effectively inhibit human renal cancer cell growth in vitro and in vivo. Med Sci Monit. 2021;27:e930634. https://doi.org/10.12659/MSM.930634. PMID: 33507885; PMCID: PMC7852039
47. Bhattacharya B, Noad RJ, Roy P. Interaction between bluetongue virus outer capsid protein VP2 and vimentin is necessary for virus egress. Virol J. 2007;4:7. https://doi.org/10.1186/1743-422X-4-7. PMID: 17224050; PMCID: PMC1783847
48. Seetharaman S, Etienne-Manneville S. Cytoskeletal crosstalk in cell migration. Trends Cell Biol. 2020;30(9):720–35. https://doi.org/10.1016/j.tcb.2020.06.004. Epub 2020 Jul 13. PMID: 32674938
49. https://ibidi.com/content/306-data-analysis-of-chemotaxis-assays
50. Li H, Lei B, Xiang W, Wang H, Feng W, Liu Y, Qi S. Differences in protein expression between the U251 and U87 cell lines. Turk Neurosurg. 2017;27(6):894–903. https://doi.org/10.5137/1019-5149.JTN.17746-16.1. PMID: 27651343

51. Diao W, Tong X, Yang C, Zhang F, Bao C, Chen H, Liu L, Li M, Ye F, Fan Q, Wang J, Ou-Yang ZC. Behaviors of glioblastoma cells in in vitro microenvironments. Sci Rep. 2019;9(1):85. https://doi.org/10.1038/s41598-018-36347-7. PMID: 30643153; PMCID: PMC6331579
52. https://imagej.nih.gov/ij/
53. https://imagescience.org/meijering/software/mtrackj/manual/
54. https://theolb.readthedocs.io/en/latest/imaging/measuring-cell-fluorescence-using-imagej.html

In Vivo Models of Cell Migration

Kannan Govindaraj and Prasanna Padmanaban

What You Will Learn in This Chapter

The relevance of studies of cell migration *in vitro* needs to be validated by the study of cell migration *in vivo*. This chapter will give an update on *in vivo* studies in various animal systems such as worms, fruit flies, fish, frogs, chickens, and rodents, with a focus on technological developments to investigate cell migration *in vivo*, particularly in angiogenesis, wound healing, and cancer metastasis.

K. Govindaraj (✉)
Developmental Bioengineering, Technical Medical Centre, University of Twente, Enschede, The Netherlands

Department of Medicine, Brigham's and Women's Hospital, Harvard Medical School, Cambridge, MA, USA
e-mail: k.govindaraj@utwente.nl

P. Padmanaban (✉)
Vascularization Lab, Biomechanical Engineering Department, Technical Medical Centre, University of Twente, Enschede, The Netherlands

European Molecular Biology Laboratory, Barcelona, Spain
e-mail: prasanna.padmanaban@embl.es

A. Brüning-Richardson, S. Knipp (eds.), *Cell Migration in Development, Health and Disease*, Learning Materials in Biosciences,
https://doi.org/10.1007/978-3-031-64532-7_14

1 Introduction

Cell migration is prominent in many processes such as angiogenesis, cancer metastasis, and wound healing. In angiogenesis and cancer metastasis, a single motile cell or multicellular aggregates penetrate through the extracellular matrices of neighboring tissues, whereas in wound healing, collective migration of cells is governed by tight intercellular connections. The formation of new blood vessels is called *angiogenesis*, which is a process common to both health and disease. Sprouting and intussusceptive angiogenesis are the two main *in vivo* mechanisms by which new blood vessels are formed. Proliferation and directional migration of endothelial cells are integral processes of angiogenesis. Unraveling the mechanisms involved in endothelial cell migration is crucial to understanding the process of angiogenesis. Processes such as tissue regeneration, wound healing, and the growth of cancerous tumors require the formation of new blood vessels. Moreover, there is a compelling need for the successful formation of functional blood vessels *in vitro* for tissue engineering applications. However, the process of *in vivo* angiogenesis is not completely understood, which hampers the successful development of functional engineered tissues. For the past several years, much effort has been made to understand angiogenesis in rodents, zebrafish, and small animals. *In vivo* angiogenesis is best studied using fluorescent methods because they allow high-resolution imaging at the level of single cells.

Migration of cancer cells from the primary tumor to a new location or tissue is called *metastasis*. Cell signaling and the mechanisms orchestrating cancer metastasis are still not fully understood. Having control over cancer cell migration will greatly benefit the control of cancer spreading and treatment. To investigate the process of cancer cell metastasis, scientists use several model organisms, such as athymic mice and rodents, that are compatible with intravital microscopy (IVM) as well as bioluminescence imaging (BLI).

Epithelial cells play an important role in *wound healing*. Understanding the process of wound healing helps to improve clinical procedures in treating wounds in the patients with diabetes and treating operational wounds. Cellular processes such as how cells collectively move to the wounded site, the biochemical stimuli that drive cell migration involved, and cell-cell interaction at the wound site are not yet fully understood. Mouse models have been effectively used to understand the process of cell migration in wound healing. Methods such as IVM, BLI, and histological analysis have been successfully used to study wound healing processes.

As you have studied in the previous chapters, studying and visualizing cell migration *in vitro* is very straightforward. Cell migration on cell culture substrates can be directly visualized using conventional light and fluorescent microscopy. Noninvasive *in vivo* imaging has advanced our knowledge in cell signaling and function in health and disease. However, studying cell migration *in vivo* is more complicated and involves complex methods and infrastructure. Autofluorescence and physical motion of animals impose additional challenges in *in vivo* imaging. In this chapter, we provide an update on various *in vivo* models with a focus on technological developments to visualize and characterize complex mechanisms of angiogenesis, cancer metastasis, and wound healing.

2 Animal Models

2.1 *Caenorhabditis elegans* (Nematode)

Due to its small size, transparency, and the help of diverse genetic techniques such as forward genetics (random mutation screening), reverse genetics (target specific genes for RNA interference or CRISPR-Cas9 gene editing), transgenesis (foreign DNA introduction in the *C. elegans* genome), and reporter gene analysis (using green fluorescent protein to visualize expressed genes), *C. elegans* is an emerging model organism both in the context of regenerative medicine and fundamental research. This 1-mm-long nematode is widely used to demonstrate *in vivo* intercellular interactions and detect and screen genetic mutations from the single-cell to whole-organism level. Vascular endothelial growth factor (VEGF) and platelet-derived growth factor (PDGF) play an important role in angiogenesis processes. Tarsitano and others showed that the polyvinyl fluoride-1 (PVF-1) molecule present in *C. elegans* plays a crucial role in neural development and has similar biochemical properties to VEGF/PDGF in vertebrates. These studies promote the possibility of using *C. elegans* as a simple experimental model system to study growth factor effects in driving angiogenesis using a combination of *in vitro* and *in vivo* assays [1]. The authors first cloned and characterized the PVF-1 gene from *C. elegans*, which encodes a protein that is structurally similar to VEGF. They then expressed the PVF-1 protein in mammalian cells and demonstrated that it is able to bind to and activate VEGF receptors *in vitro*. To investigate the angiogenic activity of PVF-1 *in vivo*, the authors performed a series of assays using zebrafish embryos, which have a transparent body that allows for easy visualization of blood vessels. They injected PVF-1 protein into the yolk sac of zebrafish embryos and observed a dose-dependent increase in angiogenesis. The authors also generated transgenic *C. elegans* strains expressing PVF-1 and demonstrated that overexpression of PVF-1 in *C. elegans* leads to increased angiogenesis and branching of the gonadal vasculature. Finally, the authors investigated the role of PVF-1 in regulating angiogenesis *in vivo* using a mouse model. They injected PVF-1 protein into Matrigel plugs implanted in mice and observed a significant increase in blood vessel formation. Xu and Chisholm showed that intracellular wounding triggers the local production of mitochondrial reactive oxygen species (mtROS) in *C. elegans* that promotes actin-mediated repair by uptaking calcium ions. This manipulation of mtROS enhances cellular or tissue repair [2]. The authors first induced epidermal wounds in *C. elegans* and observed a rapid and transient increase in mtROS levels at the wound site. They then used genetic and pharmacological approaches to manipulate mtROS levels in *C. elegans* and observed that increased mtROS levels promote wound repair, while decreased mtROS levels impair wound repair. To further investigate the role of mtROS in wound repair, the authors examined the downstream effects of mtROS signaling. They found that mtROS signaling activates the p38 mitogen-activated protein kinase (MAPK) pathway, which promotes actin polymerization and cell migration at the wound site. The authors also identified a transcription factor, DVE-1, which is required for the upregulation of mitochondrial antioxidant genes in response to wounding, and demonstrated that DVE-1 is required for proper wound repair in *C. elegans*. Overall, the study demonstrates that epidermal wounding in *C. elegans* induces an mtROS burst that promotes tissue or wound repair by activating the p38 MAPK pathway and promoting actin polymerization and cell migration. This study provides insights into the molecular mechanisms underlying the regulation of wound repair by mtROS signaling. Recently, Kelley and colleagues have shown that the anchor cells, which are specialized cells found in the posterior part of the gonad arm in *C. elegans*, have a significant role in directing the movement of the developing vulval precursor cells (VPCs) of *C. elegans*. This organism contains only six genes, making it a fascinating model to investigate the process of basement membrane breaching in the absence of matrix metalloproteinases (MMPs), which are critical for cancer invasion [3].

2.2 *Drosophila melanogaster* (Fruit Fly)

Drosophila is an interesting model organism for live imaging to directly observe stem cell behavior in both development and disease progression. Oncogenic carboxy-terminal binding protein (CtBP) is well-known because of its overexpression in a wide variety of cancers and nonmalignant tumors. Interestingly, the *Drosophila* genome encodes a single CtBP, which makes it a good target for cancer research [4]. Wu and colleagues showed that depletion of CtBP triggers tumor invasion by using a combination of genetic, molecular, and imaging approaches. The authors first used a genetic screen to identify genes that modulate tumor invasion in *Drosophila*. They found that depletion of CtBP, which is overexpressed in many human cancers, increased the invasiveness of tumor cells *in vivo*. They performed gene expression analysis and found that depletion of CtBP led to upregulation of genes involved in epithelial-mesenchymal transition (EMT), a process that enables cancer cells to invade and metastasize. They also found that depletion of CtBP resulted in increased expression of the transcription factor Snail, a master regulator of EMT. The authors then used live imaging to visualize the behavior of tumor cells *in vivo* and found that depletion of CtBP led to increased cell motility and invasiveness, consistent with an EMT phenotype. Additionally, the same authors reported that CtBP plays an important role during *Drosophila* development in loss-of-cell polarity-triggered cell invasion in the wing discs and cell migration in thorax closure. Jun-Kit Hu and colleagues, in another study, showed that the intestine of *Drosophila* is an interesting and powerful model to study intestinal stem cell (ISC) functionality. They demonstrated that

ISCs exhibit migratory behavior after tissue damage and that this migration is dependent on actin-based lamellipodia. Moreover, non-canonical Wnt signaling promotes ISC-driven migration [5].

2.3 *Danio rerio* (Zebrafish)

Zebrafish have become a reliable model for studying angiogenesis and cancer research. Due to their fast developmental cycles, embryo transparency, and small size, they are interesting candidates for exploring the live dynamics of development and disease progression. Wang and colleagues demonstrated the role of long noncoding RNAs (lncRNAs) as significant players in regulating colorectal cancer (CRC) using zebrafish xenograft model. The authors established a xenograft model by injecting human CRC cells into zebrafish embryos and observing their growth and migration [6]. Then they used this model to investigate the effects of SNHG4 on CRC cell behavior. The authors found that overexpression of SNHG4 in CRC cells increased their proliferation and migration in the zebrafish model, while knockdown of SNHG4 had the opposite effect. They also observed changes in the expression of genes involved in cell proliferation, apoptosis, and migration, suggesting that SNHG4 regulates these processes in CRC cells. In another study, Martínez-Pena used zebrafish models to study the metastatic competency of breast cancer circulating tumor cells (CTCs). They showed that disseminated clusters of tumor cells survive better than single cells in zebrafish [7]. The authors labeled breast cancer cells with a fluorescent protein and injected them into the circulation of zebrafish embryos. They then used time-lapse imaging to visualize the behavior of the cancer cells as they entered the circulation and formed metastases in different tissues. Different techniques were applied to manipulate the size of these clusters, either by disrupting them or by artificially increasing their size. Gene expression analysis was performed on single cells and clusters, and it was determined that clusters had higher expression of genes involved in cell-cell adhesion, survival, and proliferation compared to single cells. Based on these results, the authors concluded that disseminated tumor cells clusters are better able to survive and grow in distant tissues than single cells. Shazeli Che Zain and colleagues used the zebrafish model to study wound healing. They tested palm oil as a botanical-based healing agent to treat dermal wounds and injuries created by biopsy punch in the posterior area to the gill in adult zebrafish [8]. Angiogenesis is a crucial developmental process as well as an important therapeutic target for conditions ranging from cancer to heart disease. Parab and colleagues studied VEGF-mediated vessel-selective patterning in zebrafish. They identified angiogenic molecular cues required for myelencephalic choroid plexus (mCP) vascular development in zebrafish [9].

2.4 *Xenopus laevis* (Frog)

Xenopus embryos are widely used in cancer and wound healing research. Recently, Tejeda-Muñoz and colleagues published a quick and easily reproducible protocol that allows scientists to simply inject DNA/mRNA of interest into the early embryonic stage of *Xenopus* embryos and dissociate cells within a short period of time to visualize cell migration, focal adhesion, and, in this case, Wnt signaling [10]. The authors visualized cell migration in *Xenopus* by using time-lapse confocal microscopy, expressing fluorescently labeled integrin beta-1 and focal adhesion kinase (FAK) proteins in the neural crest cells of *Xenopus* embryos. These proteins are known to play a crucial role in cell migration and adhesion. By analyzing the changes in the position and morphology of the labeled cells over time, the authors were able to visualize and quantify the migration of neural crest cells in response to Wnt signaling. Stanisstreet proposed an interesting hypothesis that wound healing in *Xenopus* embryos is affected by cell shape changes that are similar to cells that uptake calcium [11]. The author observes that during wound healing process, cells at the edge of the wound undergo significant changes in shape, including an increase in cell height and a decrease in cell width. These changes in cell shape are similar to the shape changes that occur in cells actively taking up calcium. This study suggests that the uptake of calcium may be involved in wound healing in *Xenopus* embryos. The author further investigates this hypothesis by treating the embryos with calcium channel blockers, which prevent the uptake of calcium into cells. The results show that the blockade of calcium uptake significantly impairs the ability of cells to migrate to the wound site and complete the wound healing process. Cancer tissues have similar characteristics to *Xenopus* embryonic tissue development, particularly the EMT. Tanaka and colleagues utilized *Xenopus* embryos to screen anticancer chemical compounds. After performing blind screening of more than 100 compounds, they showed that C-157 (an analog of podophyllotoxin) and D-572 (an indole alkaloid) disrupt interphase microtubules and prevent cancer cell invasion [12].

2.5 *Gallus gallus domesticus* (Chicken)

Chick chorioallantoic membrane (CAM) models are mainly used to study angiogenesis processes and *in vivo* testing of biomaterials that are applied in cancer and wound healing research. Huang and others demonstrated that chick embryos can be cultured over a long period outside the eggshell to study various mechanisms such as angiogenesis, vascular perfusion, and cell-biomaterial interactions to examine cancer cell migration and wound repair [13]. More recently, Padmanaban and colleagues characterized the blood flows within multiscale hierarchical blood vessels of developing chick embryos using laser speckle contrast imaging (LSCI) and sidestream dark field imaging [14]. These engineering-based, accessible *in vivo* models are essential for tissue engineers to efficiently study the complex processes occurring during development and to translate or fine-tune *in vitro* experiments. Pawlikowska and colleagues exploited the CAM assay as a preclinical platform to perform anti-metastatic drug testing. They implanted tumor cells into the CAM membrane to test the metastatic capacity. This study highlights that CAM is a high vascularized membrane that supports the growth of tumors, making it an ideal platform for studying the progression of cancer. This noninvasive, transparent CAM assay platform provided a rapid and inexpensive means of evaluating the efficacy of the drug before proceeding to *in vivo* studies [15]. In another study, Winter and colleagues demonstrated the stimulating effect of intensive LED light in pulse mode, both *in vitro* and *in vivo*, to investigate endothelial cell migration and network formation, which is essential in angiogenesis and wound healing processes [16].

2.6 Mice (Rodent)

Using rodents such as mice and rats is a well-established method for studying wound healing and cancer research due to its physiological relevance to humans. However, this type of *in vivo* model is technically challenging in terms of experimental design and imaging, as it requires expertise and specialized training to perform the tasks. Nicotinamide adenine dinucleotide (NAD) plays a major role in kidney dysfunction, cardiomyopathy, and metabolic disorders. Recently, Li and colleagues demonstrated that hyperglycemia reduces the biosynthesis of NAD+, which impairs corneal epithelial wound healing in diabetic mice. Type 1 diabetes mellitus (DM) was induced in mice by streptozotocin (an antibiotic that causes pancreatic islet β-cell destruction), and corneal epithelial wounds were created in both diabetic and nondiabetic mice by epithelial scraping. It was determined that the diabetic mice had impaired corneal epithelial wound healing compared to the nondiabetic mice. The impaired healing was attributed to reduced biosynthesis of NAD+ in the diabetic mice, which was caused by hyperglycemia. Increasing NAD+ and its precursors could be a potential therapeutic strategy for diabetic corneal complications [17].

Keratinocytes play a major role in wound closure. Holt and colleagues recently discovered PIEZO1, a mechanosensitive ion channel, in regulating keratinocyte migration during wound healing. Interestingly, this study demonstrated that the spatiotemporal dynamics of PIEZOI localization control keratinocyte migration during wound healing. They found that PIEZO1 is expressed in migrating keratinocytes during wound healing and that its localization is regulated by the activity of the RhoA signaling pathway, which is involved in cell migration. Using genetically engineered mice, they studied the role of PIEZO1 in wound healing. It was demonstrated that epidermal-specific PIEZO1 knockout mice exhibited faster wound closure, while gain-of-function mice displayed slower wound closure compared to controls. PIEZO1 is initially localized to the leading edge of migrating keratinocytes, but its localization shifts to the trailing edge as the cells migrate. This spatiotemporal regulation of PIEZO1 localization is important for controlling the directionality of keratinocyte migration during wound healing [18]. Immunotherapy plays an important role in reducing the spread of metastatic cancer cells.

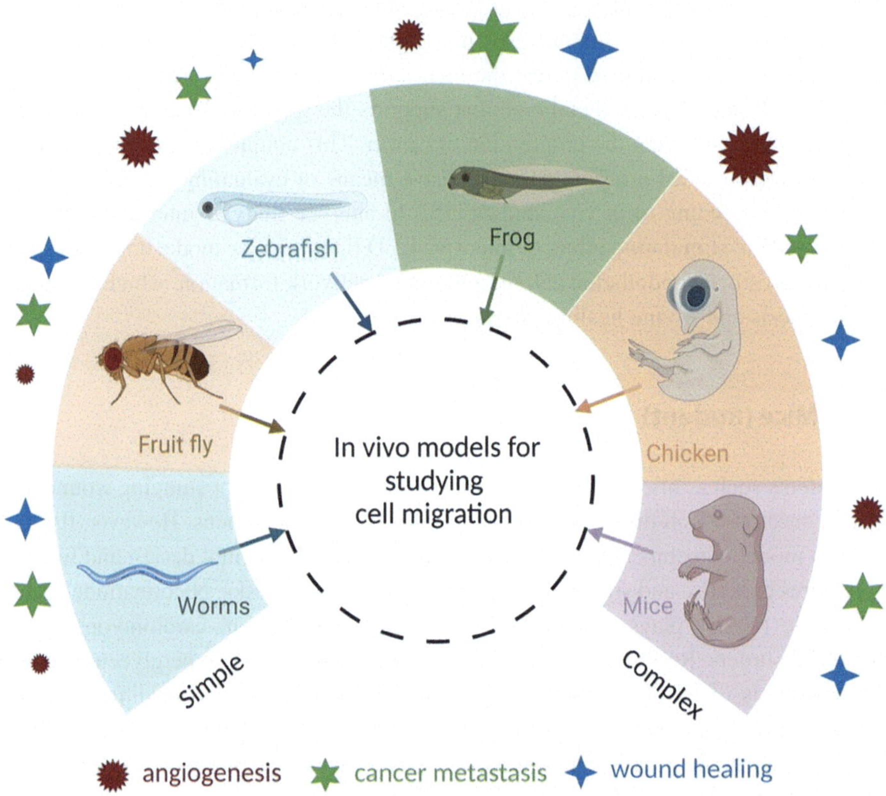

Fig. 1 *In vivo* models used for studying cell migration. Schematic illustrations are created using BioRender (agreement number NA24LMLPXX). The size of the colored star-shaped icons represents the potential of the respective model systems for studying cell migration in angiogenesis, cancer metastasis, and wound healing mechanisms

More recently, Kaczanowska and colleagues utilized genetically engineered myeloid cells (GEMys) to improve anti-tumor immune responses and inhibit cancer metastasis. The researchers used GEMys to overexpress the protein IFN-beta, which is known to activate anti-tumor immune responses. They then tested these cells in mouse models of metastatic cancer and found that they were able to reduce the growth of metastatic tumors and prolong survival. These GEMys reduced the number of immune-suppressive cells and increased the number of immune-stimulatory cells in the tumor microenvironment. These cells also induced the expression of genes involved in anti-tumor immunity, leading to improved immune responses against cancer cells. This study highlights the potential of GEMys as a therapeutic approach for treating metastatic cancer [19] (Fig. 1).

3 Imaging Techniques

3.1 Intravital Microscopy (IVM)

The discovery and cloning of green or red fluorescent proteins (GFP/RFP) revolutionized the field of cell biology. A gene of a protein of interest can be specifically paired with a gene of one of these fluorescent proteins and can be easily expressed inside mammalian cells without impairing their function. Cells stably expressing a fluorescently tagged protein of interest can be introduced into living animals, and their location and migration can be tracked by fluorescence microscopy methods such as confocal or multiphoton microscopy for several days, if not weeks. Confocal microscopy can acquire high-resolution images. Here, the fluorophore is excited with laser light, and within the constraints of the wavelength of light, confocal microscopy can achieve lateral image resolution up to 200 nm and axial resolution up to 500 nm. In one such early study of using IVM, Chishima et al. [20] injected ANIP 973 cells—stably expressing GFP—into mice and studied their retention in the lung after 7 days [20]. Over the past 20 years, researchers have used fluorescent proteins *in vivo* to study many cellular processes and disease mechanisms such as tumor metastasis, angiogenesis, host-pathogen interactions, etc. [21]. On the other hand, multiphoton microscopy simultaneously uses two or three longer-wavelength (deep red) photons to excite the fluorophore. Multiphoton excitation of the fluorophore enables deep-tissue imaging and can image up to 1.3 mm [22]. In one such study, the nuclei and cytoplasm of live human fibrosarcoma cancer cells were labeled with GFP and RFP, respectively, and injected into mice. The cell migration of the cancer cells through blood vessels was imaged using an *in vivo* small animal imaging fluorescence microscope [23]. Recently, Ilina et al. studied collective cell migration in mice using spheroid micro-implantation. This study demonstrated how collective cells transition to single-cell migration in breast cancer metastasis. In addition, collective invasion of tumor cells and cytoskeletal plasticity were characterized in detail [24]. IVM has been an important tool in studying *in vivo* blood vessel development (angiogenesis). Due to the close proximity of blood vessels, the mouse cornea is a readily suitable tissue for studying angiogenesis. Wang et al. used a mouse that specifically expresses enhanced yellow fluorescent protein (eYFP) in endothelial cells to study cell migration in the cornea. Using IVM, migration of endothelial cells, sprouting of blood vessels, and vascular remodeling processes have been imaged at the single-cell level resolution in the mouse cornea [25].

Advantages and limitations: Fluorescent cells can be easily tracked and visualized using IVM. However, due to the limitations of the wavelength of light, imaging depth is limited to a few hundred micrometers (<1000 μm, max). Therefore, cell migration close to the surface can only be imaged, and this method is not suitable for deep-tissue imaging.

3.2 Bioluminescence Imaging (BLI)

Luciferase reporters are yet another tool that is successfully used to locate cells and study cell migration in live animals. Luciferase is an enzyme that, upon reaction with its substrate luciferin, produces luminescence. The generated luminescence can be detected using luminometers. Like fluorescent proteins, luciferase enzymes can also be stably expressed in cell lines. These cells are then injected into animals. Prior to imaging, a substrate of luciferase, D-luciferin, is injected into the animal that had been previously injected with luciferase-expressing cells. Animal imaging systems capable of detecting luminescence signals can then be used to precisely locate the cells in the live animals. Using near-infrared wavelengths (1000–1700 nm, NIR-II) combined with recent technological developments, the fluorescence imaging depth has improved to a little more than 1 cm [26]. In addition to observing cell migration close to the skin, imaging of deep tissues such as the lymphatic system and targets within 3 cm from the skin is possible with BLI. Recently, using NIR-II BLI probes, the lymphatic systems of chickens and mice have been imaged [26]. Additionally, using NIR probes with two distinct wavelengths, Lu et al. have imaged tumor tissue and lymphatic vessels in mice [26]. Moreover, light-producing transgenic animals (LPTA) expressing luciferase reporters have been developed to study multiple cellular processes such as intracellular signal transduction, cancer metastasis, and metabolic monitoring. These LPTAs do not require the injection of cells expressing luciferase enzymes. Upon injection of the luciferin substrate, it binds to the luciferase enzyme produced by the target cells, which then produce luminescent light. These cells can be readily imaged by a BLI imager [27, 28]. In another study, using biocompatible nanoparticles, Chen et al. imaged 3.2-cm-thick pork tissue [29]. Moreover, these methods can achieve spatial resolution at the cellular level.

Advantages and limitations: Imaging depths are deeper than those achieved by fluorescence-based IVM. However, injecting the luciferase substrate is necessary for BLI. Even though imaging depth is higher than IVM, whole-animal imaging is still not possible with BLI.

3.3 Magnetic Resonance Imaging (MRI)

MRI uses magnetic fields and radio waves to image tissues and bones inside the body. It has also been used to study *in vivo* cell migration in live animals with high spatial resolution. However, the cells under investigation need to be treated with contrast agents, such as paramagnetic gadolinium ion complexes or superparamagnetic (iron oxide) magnetite particles. Cells pretreated with contrast agents are then injected into animals, and their *in vivo* migration can be tracked over weeks by subjecting the living animals to MRI. For example, highly metastatic glioma cells labeled with iron oxide nanoparticles were implemented in the caudate nucleus of mice, and their mobility was tracked using MRI for 3 weeks. Glioma cells migrated from the caudate nucleus to the anterior commissure

[30]. In another study, embryonic stem (ES) cells expressing GFP were loaded with MRI contrast agents and implanted in rat brains. ES cells that were implanted in the contralateral hemisphere migrated along the corpus callosum and were detected on the other side of the implantation site [31], which was confirmed by GFP fluorescence. Migration of multipotent neuroblasts (NB) has also been studied in rats. Contrast agent-treated NB cells that follow sharply defined rostral migratory streams migrate toward the subventricular zone and olfactory bulb [32]. These studies have specifically tracked cell migration speed in the brains of mice and rats using MRI.

Advantages and limitations: Deep-tissue imaging is possible. Any organ within the animal can be specifically labeled and imaged. However, treatment with contrast agents may harm the animal. Resolution at the single-cell level is not possible.

3.4 Positron Emission Tomography (PET) and Single-Photon Emission Computed Tomography (SPECT)

PET and SPECT employ radiolabeled tracers to image *in vivo* targets. Molecules or proteins that specifically bind to or interact with target cells or tissues are labeled with a radioactive atom. These radiolabeled tracers are then injected into the animal, and the *in vivo* location of these tracers is detected and imaged by PET or SPECT. Although PET and SPECT use the same principle of detecting radioactive tracers, they differ in the types of radiolabels used. SPECT uses gamma-ray emitting nuclides such as ^{111}In, ^{123}I, and ^{99m}Tc for radiolabeling the tracers, while PET uses ^{18}F-labeled tracers [33]. Cancer cells are known for uncontrolled rapid cell division, and they take up unusually high amounts of glucose to meet their energy demands. Such glucose molecules can be labeled with ^{18}F, and cancer metastasis can be tracked by PET after injecting ^{18}F-labeled glucose into live animals. However, even though there may be applications in cancer research, SPECT is mainly used to diagnose heart diseases such as blockage in the arteries by tracing gamma-ray emitting tracers.

Advantages and limitations: PET and SPECT offer whole-animal imaging with the possibility to image any location or tissue in live animals. However, the instrumentation requires sophisticated infrastructure for handling harmful radioactive materials.

3.5 Histological Studies

Unlike the other methods discussed above, histological studies are not suitable for living animals. However, cell migration can be studied in fixed animal tissues using simple brightfield or fluorescence microscopy. For example, if cancer metastasis needs to be studied in a particular tissue, a small portion of the tissue of interest is excised from the animal, fixed, and (for instance) embedded in paraffin. The paraffin-embedded tissue is then sectioned into thin slices, and the cells are stained with hematoxylin and eosin (H&E) [34]. In addition, the tissue sections can be stained for other intracellular targets using immunohistochemistry. Depending on the type of staining used, cells can be visualized using brightfield or fluorescence microscopy, or both. Modo et al. studied the temporal dynamics of neural cell migration in rat brains over 14 days. In this study, MHP36 cells were labelled with GRID chelate—a contrast agent for MRI and a fluorescent label for histological imaging—and were implanted in the contralateral hemisphere of a rat brain. After 14 days of implantation, histological analysis of sectioned rat brains showed that the MHP36 cells migrated to the peri-lesion area in the rat brain, as indicated by the fluorescence signal from the GRID labeling [1].

Advantages and limitations: Any part of tissue can be excised from the animal, stained with histochemical dyes or specific antibodies, and readily visualized under a microscope. However, preparing tissue sections and staining is a tedious process and requires sacrificing the animal before processing the tissue (Fig. 2 and Table 1).

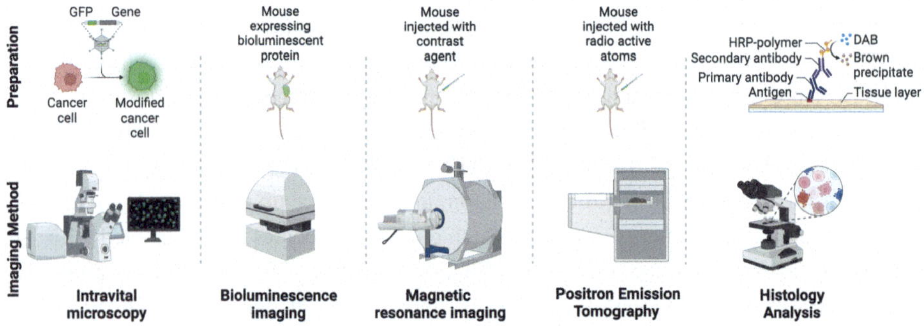

Fig. 2 Schematics illustrating sample preparation and imaging methods. In IVM, cells expressing a protein of interest tagged with a fluorescent protein are injected into the animal and imaged using a confocal microscope. In BLI, within an animal, target cells expressing the luciferase enzyme produce luminescent light upon reaction with its substrate luciferin, and these light-producing cells can be imaged using a BLI imager. In MRI, cells pretreated with a contrast agent are injected into the animal and imaged using an MRI scanner. In PET, cells pretreated with radioactive atoms are injected into an animal and imaged using a PET scanner. In histological analysis, cells in tissue sections are either stained with histochemical dyes or target proteins. Immunohistochemistry is used to image these stained cells using a light or a fluorescence microscope [35–38]

Table 1 Comparison of imaging technologies applied within *in vivo* models to study cell migration

Method	Imaging depth	Imaging probes	Advantages	Limitations
IVM	1.3 mm	Fluorescent proteins or molecules	• Compatible with live animals • Resolution at the single-cell level • Multicolor/target labeling • Cell migration can be tracked for several weeks	• Not suitable to visualize deep tissues • Autofluorescence may decrease signal-to-noise ratio • Cells expressing fluorescent protein must be injected into the animal prior to imaging
BLI	~3 cm	Luciferase enzyme and substrates	• Compatible with live animals • Single-cell level resolution is possible • Multiple cellular processes can be imaged simultaneously • Cell migration can be tracked for several weeks	• Whole-animal imaging is not possible • Cells expressing luciferase enzymes injected into the animal prior to imaging • Additionally, luciferase substrates need to be injected prior to imaging
MRI	Any depth	Contrast agents containing ionic magnetite particles	• Imaging any depth of a living animal is possible • Tissues of interest can be specifically labeled and imaged	• Complex and expensive instrumentation is required • Injection of contrast agent or radiolabeled atoms may harm the animal
PET/SPECT	Any depth	Radioactive tracers		
Histology	Any depth	Histochemical dyes or fluorescently labeled antibodies	• Cells at any tissue location can be visualized • After histology/immunostaining, tissue sections can be readily visualized using widefield/fluorescence microscopy	• Not suitable for live animals • Time-consuming and extensive processing of tissue sections

Take-Home Message

- Animal models are used for studying *in vivo* cell migration.
- There are advantages and limitations of current animal models for studying *in vivo* cell migration that need to be considered.
- A range of techniques and equipment are used for investigating *in vivo* cell migration, depending on the particular research question.
- There are advantages and limitations of current technologies for studying *in vivo* cell migration.

Acknowledgements Funding information: KG would like to acknowledge Marie Curie award from European Commission, Grant No. GAP-101109659.

P. P. would like to acknowledge the EIPOD4 fellowship program, funded by EMBL and Marie Sklodowska Curie Actions (grant agreement 847543).

References

1. Tarsitano M, De Falco S, Colonna V, McGhee JD, Persico MG. The C. elegans pvf-1 gene encodes a PDGF/VEGF-like factor able to bind mammalian VEGF receptors and to induce angiogenesis. FASEB J. 2006;20(2):227–33.
2. Xu S, Chisholm AD. C. elegans epidermal wounding induces a mitochondrial ROS burst that promotes wound repair. Dev Cell. 2014;31(1):48–60.
3. Kelley LC, Chi Q, Caceres R, Hastie E, Schindler AJ, Jiang Y, et al. Adaptive F-actin polymerization and localized ATP production drive basement membrane invasion in the absence of MMPs. Dev Cell. 2019;48(3):313–28 e8.
4. Wu C, Ding X, Li Z, Huang Y, Xu Q, Zou R, et al. CtBP modulates snail-mediated tumor invasion in drosophila. Cell Death Discov. 2021;7(1):202.
5. Hu DJ, Yun J, Elstrott J, Jasper H. Non-canonical Wnt signaling promotes directed migration of intestinal stem cells to sites of injury. Nat Commun. 2021;12(1):7150.
6. Wang J, Zhang XY, Xu DY. Zebrafish xenograft model for studying the function of lncRNA SNHG4 in the proliferation and migration of colorectal cancer. J Gastrointest Oncol. 2022;13(1):210–20.
7. Martinez-Pena I, Hurtado P, Carmona-Ule N, Abuin C, Davila-Ibanez AB, Sanchez L, et al. Dissecting breast cancer circulating tumor cells competence via modelling metastasis in zebrafish. Int J Mol Sci. 2021;22(17):9279.
8. Zain MSC, Edirisinghe SL, Kim C-H, De Zoysa M, Shaari K. Nanoemulsion of flavonoid-enriched oil palm (Elaeis guineensis Jacq.) leaf extract enhances wound healing in zebrafish. Phytomedicine Plus. 2021;1(4):100124.
9. Parab S, Quick RE, Matsuoka RL. Endothelial cell-type-specific molecular requirements for angiogenesis drive fenestrated vessel development in the brain. Elife. 2021;10:e64295.

10. Tejeda-Munoz N, Morselli M, Moriyama Y, Sheladiya P, Pellegrini M, De Robertis EM. Canonical Wnt signaling induces focal adhesion and integrin beta-1 endocytosis. iScience. 2022;25(4):104123.
11. Stanisstreet M. Calcium and wound healing in Xenopus early embryos. J Embryol Exp Morphol. 1982;67:195–205.
12. Tanaka M, Kuriyama S, Itoh G, Kohyama A, Iwabuchi Y, Shibata H, et al. Identification of anti-cancer chemical compounds using xenopus embryos. Cancer Sci. 2016;107(6):803–11.
13. Huang W, Itayama M, Arai F, Furukawa KS, Ushida T, Kawahara T. An angiogenesis platform using a cubic artificial eggshell with patterned blood vessels on chicken chorioallantoic membrane. PLoS One. 2017;12(4):e0175595.
14. Padmanaban P, Chizari A, Knop T, Zhang J, Trikalitis VD, Koopman B, et al. Assessment of flow within developing chicken vasculature and biofabricated vascularized tissues using multimodal imaging techniques. Sci Rep. 2021;11(1):18251.
15. Pawlikowska P, Tayoun T, Oulhen M, Faugeroux V, Rouffiac V, Aberlenc A, et al. Exploitation of the chick embryo chorioallantoic membrane (CAM) as a platform for anti-metastatic drug testing. Sci Rep. 2020;10(1):16876.
16. Winter R, Dungel P, Reischies FMJ, Rohringer S, Slezak P, Smolle C, et al. Photobiomodulation (PBM) promotes angiogenesis in-vitro and in chick embryo chorioallantoic membrane model. Sci Rep. 2018;8(1):17080.
17. Li Y, Li J, Zhao C, Yang L, Qi X, Wang X, et al. Hyperglycemia-reduced NAD(+) biosynthesis impairs corneal epithelial wound healing in diabetic mice. Metabolism. 2021;114:154402.
18. Holt JR, Zeng WZ, Evans EL, Woo SH, Ma S, Abuwarda H, et al. Spatiotemporal dynamics of PIEZO1 localization controls keratinocyte migration during wound healing. Elife. 2021;10:e65415.
19. Kaczanowska S, Beury DW, Gopalan V, Tycko AK, Qin H, Clements ME, et al. Genetically engineered myeloid cells rebalance the core immune suppression program in metastasis. Cell. 2021;184(8):2033–52.e21.
20. Chishima T, Yang M, Miyagi Y, Li L, Tan Y, Baranov E, et al. Governing step of metastasis visualized in vitro. Proc Natl Acad Sci USA. 1997;94(21):11573–6.
21. Hoffman RM. In vivo cell biology of cancer cells visualized with fluorescent proteins. Curr Top Dev Biol. 2005;70:121–44.
22. Horton NG, Wang K, Kobat D, Clark CG, Wise FW, Schaffer CB, et al. In vivo three-photon microscopy of subcortical structures within an intact mouse brain. Nat Photonics. 2013;7(3):205–9.
23. Hoffman RM, Yang M. Subcellular imaging in the live mouse. Nat Protoc. 2006;1(2):775–82.
24. Ilina O, Campanello L, Gritsenko PG, Vullings M, Wang C, Bult P, et al. Intravital microscopy of collective invasion plasticity in breast cancer. Dis Model Mech. 2018;11(9):dmm034330.
25. Wang Y, Jin Y, Lavina B, Jakobsson L. Characterization of multi-cellular dynamics of angiogenesis and vascular remodelling by intravital imaging of the wounded mouse cornea. Sci Rep. 2018;8(1):10672.
26. Lu L, Li B, Ding S, Fan Y, Wang S, Sun C, et al. NIR-II bioluminescence for in vivo high contrast imaging and in situ ATP-mediated metastases tracing. Nat Commun. 2020;11(1):4192.
27. Zagozdzon AM, O'Leary P, Callanan JJ, Crown J, Gallagher WM, Zagozdzon R. Generation of a new bioluminescent model for visualisation of mammary tumour development in transgenic mice. BMC Cancer. 2012;12:209.
28. Zhang N, Lyons S, Lim E, Lassota P. A spontaneous acinar cell carcinoma model for monitoring progression of pancreatic lesions and response to treatment through noninvasive bioluminescence imaging. Clin Cancer Res. 2009;15(15):4915–24.

29. Chen G, Shen J, Ohulchanskyy TY, Patel NJ, Kutikov A, Li Z, et al. (α-NaYbF4:Tm(3+))/CaF2 core/shell nanoparticles with efficient near-infrared to near-infrared upconversion for high-contrast deep tissue bioimaging. ACS Nano. 2012;6(9):8280–7.
30. Zhang F, Xie J, Liu G, He Y, Lu G, Chen X. In vivo MRI tracking of cell invasion and migration in a rat glioma model. Mol Imaging Biol. 2011;13(4):695–701.
31. Hoehn M, Kustermann E, Blunk J, Wiedermann D, Trapp T, Wecker S, et al. Monitoring of implanted stem cell migration in vivo: a highly resolved in vivo magnetic resonance imaging investigation of experimental stroke in rat. Proc Natl Acad Sci USA. 2002;99(25):16267–72.
32. Nieman BJ, Shyu JY, Rodriguez JJ, Garcia AD, Joyner AL, Turnbull DH. In vivo MRI of neural cell migration dynamics in the mouse brain. Neuroimage. 2010;50(2):456–64.
33. James ML, Gambhir SS. A molecular imaging primer: modalities, imaging agents, and applications. Physiol Rev. 2012;92(2):897–965.
34. Mangir N, Raza A, Haycock JW, Chapple C, Macneil S. An improved in vivo methodology to visualise tumour induced changes in vasculature using the chick chorionic allantoic membrane assay. In Vivo. 2018;32(3):461–72.
35. Turk M, Biernaskie J, Mahoney DJ, Jenne CN. Intravital microscopy techniques to image wound healing in mouse skin. In: Heit B, editor. Fluorescent microscopy. Methods in molecular biology, vol. 2440. New York, NY: Humana; 2022. https://doi.org/10.1007/978-1-0716-2051-9_10.
36. Hamblin MR, O'Donnell DA, Murthy N, Contag CH, Hasan T. Rapid control of wound infections by targeted photodynamic therapy monitored by in vivo bioluminescence imaging. Photochem Photobiol. 2007;75(1):51–7.
37. Vyver M, Boodhoo K, Frazier T, Hamel K, Kopcewicz M, et al. Histology scoring system for murine cutaneous wounds. Stem Cells Dev. 2021;30:1141–52.
38. Modo M, Mellodew K, Cash D, Fraser SE, Meade TJ, Price J, Williams SC. Mapping transplanted stem cell migration after a stroke: a serial, in vivo magnetic resonance imaging study. Neuroimage. 2004;21:311–7.

Mathematical and Computational Modelling of Immune Cell Responses in Cell Migration

Temitope O. Benson

What You Will Learn in This Chapter

Cell migration is a fundamental mechanism in all living organisms, including plants and animals. It is essential for embryogenesis, tissue morphogenesis, and human health, but also plays an important role in the development and promotion of diseases. The focus of this chapter is on immune cell responses in human cell migration and the way we use mathematical and computational models to investigate and understand immune cell migratory activities. We will examine the migratory strategies used by immune cells to perform their functions and how these aid in defense mechanisms against pathogens and diseases focusing on one of the deadliest diseases, cancer.

Cancer is a disease in which abnormal cells divide uncontrollably and destroy body tissues, leading to tumor formation and eventually spreading to other parts of the body, in a process known as metastasis. We will discuss how the immune system is activated when cancer cells are detected and how cancer cells avoid and respond to migrating immune cells. We will review some experimental approaches and cell culture techniques that have been used to understand how cancer cells spread and the

(continued)

T. O. Benson (✉)
Institute for Computational and Data Sciences, University at Buffalo, State University of New York, New York, USA
e-mail: tbenson2@buffalo.edu

A. Brüning-Richardson, S. Knipp (eds.), *Cell Migration in Development, Health and Disease*, Learning Materials in Biosciences,
https://doi.org/10.1007/978-3-031-64532-7_15

interaction of normal and cancerous cells with cells of the immune response. This review will provide a background understanding of how mathematical models are formulated. After the review, we dive deeper into the principles of mathematical and computational models used to investigate the migration of immune cells in cell migration. Finally, we will discuss their continued refinement, limitations, and significant progress made in the field of mathematical and computational modelling.

1 Introduction

Cell movement is a complicated and diversified process, as both simple and multicellular organisms rely on cell migration for their development. Cell migration is involved in tissue formation during embryonic development, wound healing, immunological responses, homeostasis, and cancer metastasis. Several factors and processes are responsible for cell migration and movement. Mechanical and chemical cues cause cells to migrate in response to external signals. Failure or errors in cell migration can lead to various pathological problems including mental illness and impairment, vascular disease, tumorigenesis, and cancer invasion [1, 2]. Cells may proceed as individual cells or as part of a collective of cells. Cell migration is characterized by distinct migratory strategies, ranging from the polarization of migrating cells, extension of their protrusions, and formation of cellular adhesion between the cells to drive them forward or inhibit their movements. Individual cell migration is categorized as either amoeboid or mesenchymal migration, with amoeboid movement characterized by rounded, ellipsoid entities that move without focal adhesion or cell attachment, but with the support of a propulsive bleb. Strong focal attachment to the extracellular matrix (ECM), cytoskeletal contractility, and elongated spindle-like cell bodies characterize mesenchymal motility. The movement of a cellular cohort across the ECM while maintaining functional cell–cell connections is known as collective cell migration [3]. The ECM is a noncellular environment that exists in all tissues of multicellular organisms. Water, carbohydrates, and proteins are the main components of the ECM. Collagen, elastin, fibronectin, and laminin are fibrous proteins found in the ECM, with collagen being the most prevalent and structurally vital protein in the ECM. Cell migration studies have been conducted using various methods, including in vitro models, in vivo models, in silico (or computational), and mathematical models. Because of the complexities of in vivo cell migration, an in vitro experiment cannot fully replicate in vivo events. As a result, in silico and mathematical models have been employed to simulate the observed outcome in in vivo assays [4].

In this chapter, we will focus on the mechanisms involved in in silico and mathematical modelling of cell migration and how a mathematical model is used to predict the migratory activities of immune cells. In addition, we will explore various methods that have been used to better understand immune cell response to cell migration, particularly immune cell responses to migrating cancer cells. Some pathogens can utilize the migratory cellular machinery in living organisms, thereby suppressing, or promoting cell migration and culminating in potentially disastrous outcomes. As a result, knowledge of the processes driving cell migration is critical for developing cancer treatments and therapies.

2 Overview of Human Immune Systems

The human immune system is a complex network of cells, tissues, and organs that work together to protect the body against infections, diseases, and foreign invaders. It operates in a series of steps to perform its functions. These steps can be grouped into four categories: recognition of foreign invaders, activation of immune cells to eliminate invaders, response of immune cells based on their deployment, and memory retention of previous invaders [5]. There are two major types of the immune system: the innate immune system and the adaptive immune system. The innate immune cells, such as neutrophils, are the first line of defense against infections. Neutrophils are recruited to the site of infection or tissue damage by chemokines, which are small signalling molecules secreted by damaged cells, infected cells, or other immune cells. Neutrophils follow a chemokine gradient, which guides them over to the site of infection. Once they arrive at the site of infection, neutrophils engulf and destroy the pathogens or damaged cells [6–10].

Macrophages are another type of innate immune cells that play a crucial role in cell migration. Macrophages are large cells that engulf and destroy pathogens, present antigens to T-cells, and secrete cytokines. Similar to neutrophils, macrophages are recruited to the site of infection or tissue damage by chemokines. Unlike neutrophils, macrophages are long-lived immune cells that can remain in the tissue for an extended period, ranging from months to years. Dendritic cells, on the other hand, are specialized antigen-presenting cells that can activate T-cells. Like neutrophils and macrophages, dendritic cells are recruited to the site of infection by chemokines. Once they arrive at the site, dendritic cells capture and process antigens from pathogens or damaged cells. They migrate to the lymph nodes where they present antigens to T-cells, leading to the activation of adaptive immune responses [11].

In the adaptive immune system, T- and B-cells migrate to the lymph nodes and other lymphoid tissue for antigen presentation by dendritic cells. Once activated in this way, T-cells proliferate and differentiate into effector cells that can migrate to the site of infection or tissue damage to perform their functions such as eliminating infected cells or secreting cytokines. B-cells also play a crucial role in cell migration and produce antibodies that can recognize and neutralize pathogens. B-cells are also activated after antigen presentation in the lymph nodes. Once activated, B-cells proliferate and differentiate into plasma cells, which can migrate to the site of infection to secrete antibodies. Cell migration is a crucial process in both the innate and adaptive immune systems allowing immune cells to reach their targets efficiently in response to a potential threat. Chemokines play a crucial role in guiding immune cells to their targets. Chemokines bind to transmembrane G protein-coupled receptors that trigger intracellular signalling that drives cell polarization, adhesion, and migration [12–16].

2.1 Cell Migration in Innate and Adaptive Immunity

The innate and adaptive immune responses both rely on cell migration to function effectively. Innate immune cells, including neutrophils, macrophages, and dendritic cells, are the first line of defense against pathogens. Once activated, the cells utilize chemotaxis and phagocytosis: a process of ingesting bacteria's, to eliminate the pathogen [17]. Cell migration as part of the innate immune response is primarily mediated by chemokines, which are small signalling molecules that attract immune cells to sites of infection or inflammation. Chemokines are produced by a variety of cells, including infected or damaged cells, and act by binding to receptors on the surface of immune cells. This binding triggers a signalling cascade that induces changes in the cytoskeleton of the cell, allowing it to move toward the source of the chemokine. Once immune cells arrive at the site of infection, they use various mechanisms to eliminate the pathogen. Neutrophils, for example, release toxic substances that eliminate bacteria, whereas macrophages and dendritic cells engulf and break down invading microorganisms. In contrast to the innate immune response, the adaptive immune response is a more specialized and targeted response that relies on the actions of T and B lymphocytes [17]. These cells can recognize and respond to specific antigens, allowing them to target and eliminate certain pathogens. Cell migration in the adaptive immune response is mediated by a different set of chemokines and cytokines, which are produced by Antigen-Presenting Cells (APCs), such as dendritic cells and macrophages. These cells present antigens to T lymphocytes, which then differentiate into effector cells capable of targeting and eliminating pathogens. Once activated, effector T-lymphocytes undergo a complex process of migration and chemotaxis to reach the infection site. This migration is mediated by a variety of signalling molecules, including chemokines and integrins, which allow the T-cells to navigate through tissues and across endothelial barriers. Cell migration in both innate and adaptive immune responses enables immune cells to detect, target, and eliminate pathogens and damaged tissues. Understanding the mechanisms and processes by which cancer cells and immune cells interact and respond to one another is critical for developing effective strategies for treating infectious and non-infectious diseases [18–21]. In the next section, we will explore the interactions between cancer cells and immune cells.

3 Cancer and Immune Responses

Cancer arises from the abnormal growth and spread of cells in the body. Although there are different types of cancer, they all share a common origin: mutations in the genetic material of cells. These mutations can occur spontaneously or are caused by external factors such as exposure to radiation or certain chemicals. At the most basic level, cancer is caused by mutations in the DNA sequence of cells. In addition to DNA mutations, cancer can also be caused by changes in the epigenetic information that control gene expression. Epigenetic changes can alter the way genes are turned on or off, leading to abnormal cell growth, division, and migration [22–25].

Blood cell cancer and solid tumors are two main forms of cancers. Solid tumors are further divided into different types based on the cells that give rise to them. Carcinomas are epithelial-derived cancer cells that start in the tissue lining the inner or outer surface of the body. Leukemia, which arises from blood cells and is frequently detected in the bone marrow, and lymphoma, which arises from the immune system, are the two types of blood cancer cells [5, 26, 27].

There is a robust defensive mechanism that aids in the battle against diseases, such as cancer. However, cancer, a very complex disease, often eludes and evades immune cell responses [18]. In response to stress, physiological changes are initiated to support the immune system, aiming to address the challenges posed by stressors. However, these alterations can potentially disrupt the normal regulation of the immune response, consequently fostering the invasion of cancer cells and enhancing the survival of pathogens through immune dysregulation. The chronic activation of these stress responses, which include the hypothalamic–pituitary–adrenal axis and the sympathetic–adrenal–medullary axis, continuously produces glucocorticoids and catecholamines. Recent findings indicate that endogenous opioids play a role in diminishing the cytotoxicity of natural killer (NK) cells. The loss or mutation of p53 in cancer has profound effects on the recruitment and functionality of myeloid and T-cells, fostering immune evasion and propelling cancer progression. Notably, p53 exhibits functional significance in immune cells, yielding diverse outcomes that can either hinder or support tumor development. Various stress signals, including genotoxic damage, oncogene activation, nutrient deprivation, and hypoxia, can activate p53. A more comprehensive understanding of how p53 modulates the immune response to cancer holds the promise of developing targeted therapies and promoting enduring antitumor responses in patients [28].

3.1 Lung Cancer and Immune Response

Lung cancer is one of the leading causes of death worldwide [29]. It is characterized by the unchecked proliferation of abnormal cells in the lungs, which can disperse to other body areas and obstruct the typical operation of the lungs. The causes and available treatments for the various forms of lung cancer, such as small-cell lung cancer and non-small cell lung cancer, can vary. The immune system reaction to cancer cells is a significant factor in lung cancer [30]. The migration of immune cells plays a crucial role in the context of lung cancer, which stands as one of the prevalent forms of cancer worldwide. The formidable challenge posed by lung cancer to the host's immune defenses arises from its characteristics as a solid tumor, characterized by minimal antigenicity and a heterogeneous nature. Within the tumor-infiltrating lymphocytes (TIL), regulatory T-cells (Tregs or T reg cells), previously recognized as suppressor T-cells, can be distinguished by the expression of the fork-head box P3 (Foxp3) protein. In this intricate interplay, CTLA4 (cytotoxic T-lymphocyte antigen 4) emerges as a key player. Constantly present on Tregs, CTLA4 serves to inhibit T-cell activation. Notably, the expression of CTLA4 has been found to be elevated in lymphocytes derived from individuals afflicted with lung cancer. Recent breakthroughs indicate that antibodies suppressing CTLA4 have shown promise in assisting patients grappling with lung cancer. A critical aspect of the lung cancer microenvironment involves the reciprocal secretion of cytokines by both cancer cells and immune cells. These cytokines exert suppressive effects on the immune system within the cancer microenvironment, further underscoring the intricate dynamics at play in the immune response against lung cancer [31–33].

The intricate immune response system in lung cancer is significantly influenced by the dynamic activities of Transforming Growth Factor (TGF) and interleukin-10 (IL-10). TGF, with its pleiotropic functions, holds a prominent role in orchestrating the activities of the T helper 17 cells (Th17) population. This underscores the multifaceted impact of immune cell migration, especially of Th17 cells, in the context of lung cancer. Within the landscape of lung cancer, Tumor-Associated Macrophages (TAM) emerge as myeloid suppressor cells exhibiting the phenotype of M2 macrophages. Notably, these TAMs contribute to the immune response through certain regulatory features, including the production of IL-10 and TGF. The migration of these macrophages and their interactions within the tumor microenvironment further accentuates the complexity of immune cell dynamics in lung cancer. Adding another layer to this intricate network are Myeloid-Derived Suppressor Cells (MDSCs). These cells play a pivotal role in regulating cytotoxic T-cell function, and their activities are intricately interwoven with the specific conditions prevailing in the tumor environment. The contribution of immune cell migration, particularly that of MDSCs, underscores the nuanced and context-dependent nature of the immune response in lung cancer [11].

3.2 Breast Cancer and Immune Response

Breast cancer (BC) is the second most common cancer in women after skin cancer [34, 35]. There are various forms of breast cancer, and each type has a different prognosis and course of therapy. The function of the immune system in the onset and spread of the disease is a component of breast cancer that has recently received much attention. The immune system can either encourage or hinder the growth and metastasis of breast cancers. On the one hand, the immune system can identify and combat cancer cells, aiding in the slowing or halting of tumor growth. However, the immune system can also support the ability of cancer cells to escape detection and grow out of control, which promotes the development and spread of the illness [36–38]. Chronic inflammation is one way in which the immune system contributes to breast cancer. Inflammation is a reaction of the immune system to tissue damage or infection. Chronic inflammation can cause tissue damage and aid in the growth of cancer, whereas acute inflammation is a typical and required reaction to defend the body [31–33, 39].

The association between chronic inflammation, particularly in postmenopausal women, and the elevated risk of breast cancer underscores the intricate interplay between systemic immune responses and carcinogenesis. Of relevance is the role of immune cells within the tumor microenvironment (TME), as their presence profoundly influences the trajectory of breast cancer. The process by which immune cells ingress into the tumor microenvironment stands as a critical facet of this dynamic interaction. Initially attracted by chemotactic signals released by the tumor, immune cells undergo a series of orchestrated steps. These include adhesion to the vascular endothelium, facilitated by adhesion molecules such as selectins and integrins, followed by transmigration through the endothelial layer. Upon entry into the tumor tissue, immune cells navigate through the intricate ECM and stromal components, responding to microenvironmental signals that shape their function within the TME. Furthermore, the immune system assumes a pivotal role as the target of diverse pharmaceutical interventions designed to combat breast cancer. Noteworthy among these therapeutic strategies are cancer vaccines, which stimulate the immune system to recognize and mount a response against cancer cells. Additionally, immune checkpoint inhibitors represent a class of medications that disable specific proteins on immune cells, thereby unleashing their capacity to target and eliminate cancer cells. In elucidating the multifaceted relationship between chronic inflammation, immune cell dynamics in the TME, and therapeutic interventions targeting the immune system, a comprehensive understanding emerges, offering insights crucial to advancing both preventive and therapeutic paradigms in breast cancer management [31–33, 39].

Breast cancer, acknowledged as the most prevalent and formidable malignant tumor afflicting women, manifests a complex trajectory influenced by intricate tumor–host immunological interactions. The onset, progression, metastasis, and prognosis of breast cancer exhibit considerable variability among individuals, reflecting the nuanced interplay between the tumor and the host immune system. Within this context, diverse routes contribute to breast cancer (BC) pathogenesis, with a substantial focus on immune evasion mechanisms. The capability of tumor cells to modulate surface antigens and manipulate the surrounding microenvironment empowers them to elude the body's immune response, designed to target aberrant components and foreign entities. Notably, the sophisticated interplay involving the immune checkpoint inhibitors, specifically the PD-L1/PD-1 ligand–receptor pairs and CTLA-4, has been explored and harnessed in the realm of immunotherapies. The burgeoning landscape of checkpoint inhibitor research and clinical trials holds promise not only for therapeutic interventions but also for the development of advanced preclinical detection and diagnostic tools in breast cancer. As tumors evolve, they deploy an array of immune evasion strategies, demonstrating varying degrees of proficiency in concealing themselves from immune surveillance. Contrary to its historical characterization as an immunologically silent neoplasm, breast cancer has become a focal point for immunotherapeutic exploration, capitalizing on an enhanced understanding of tumor immune surveillance. While immunotherapy has exhibited efficacy in certain cancers like melanoma, bladder cancer, and renal cell carcinoma, its success in breast cancer has been limited. The immunological silencing of breast cancer arises from mechanisms that restrict immune recognition and induce significant immunosuppression. In summation, the immune system emerges as a dual protagonist in the narrative of breast cancer, capable of both promoting and hindering the growth and dissemination of cancer cells [31–33]. A comprehensive comprehension of these intricate immunological dynamics is pivotal for advancing therapeutic strategies and refining our approach to breast cancer management.

3.3 Prostate Cancer and Immune Response

Prostate cancer, affecting the walnut-shaped prostate gland in the male reproductive system, is a multifaceted disease wherein immune cell responses play a crucial role in both its initiation and progression. The interplay between the immune system and prostate cancer involves intricate mechanisms that extend beyond mere surveillance, evasion, and modulation. The immune system's surveillance mechanisms actively identify and eliminate cancerous cells within the prostate. T-cells and natural killer cells, integral components of the immune response, continuously patrol the microenvironment, recognizing and neutralizing aberrant cells before they undergo uncontrolled proliferation. This surveillance is instrumental in preventing the development of prostate cancer by maintaining a vigilant immune response against potential malignancies [40–42].

Prostate cancer cells, however, often employ sophisticated strategies to evade immune detection and destruction. They can secrete immunosuppressive chemicals, creating a microenvironment that hampers the function of immune cells. Additionally, cancer cells may downregulate surface proteins, making them less recognizable to the immune system. These evasion tactics contribute to the ability of prostate cancer cells to proliferate unchecked and disseminate to distant sites. The modulation of the immune system in the context of prostate cancer is a dynamic process. Specific cancer cells can either activate or suppress immune responses. For instance, certain prostate cancer cells activate the immune system, inducing the release of cytokines and immune cells that exert anti-cancer effects. On the contrary, other cancer cells may release immunosuppressive signals, inhibiting the immune system's ability to mount an effective response against the tumor. This intricate balance of activation and suppression contributes significantly to the heterogeneity and clinical variability observed in prostate cancer [43–45]. The prostate cancer microenvironment plays a pivotal role in dictating immune cell behavior. It is a complex milieu of cancer cells, stromal cells, and various signalling molecules. The migration of immune cells within this microenvironment is orchestrated by chemotactic signals released by cancer cells and other components. The spatial distribution of immune cells within the tumor microenvironment influences their efficacy in recognizing and eliminating cancer cells. Understanding the detailed dynamics of immune cell responses in prostate cancer is imperative for the development of targeted therapeutic interventions. As research advances, unraveling the intricate crosstalk between the immune system and prostate cancer holds promise for refining treatment strategies and improving outcomes for affected individuals [46]. In conclusion, prostate cancer formation and progression are complicated processes that involve the immune system.

3.4 Leukemia and Immune Response

Leukemia is a cancer that affects the bone marrow and the blood. It is characterized by aberrant white blood cell (WBC) production, which is a crucial component of the immune system. Leukemia presents in various forms, each distinguished by the kind of white blood cell it affects and the rate at which the illness advances. While some forms of leukemia are slow growing and may not show signs for many years, others are more aggressive and spread quickly. Leukemia is assumed to be caused by a combination of genetic and environmental factors; however, its specific etiology is unknown. Owing to their family history, past exposure to specific chemicals, or other medical issues, some persons may be more susceptible to developing leukemia [47]. Solid tumor immune responses are well understood, and our understanding of the immune evasion pathways used by these cancers has advanced dramatically. Immune activation and tolerance mechanisms are less visible in hematological malignancies, which develop and spread differently than solid tumors.

Leukemia and lymphoma originate in immune response compartments, leading to either inadequate immune activation (immunological ignorance) or substantial hindrance of anti-tumor immune responses (immune evasion) [48, 49]. Although solid and hematologic malignant cancers share many immune evasion routes, leukemia and lymphoma employ various "unique" immune evasion mechanisms. Classical Hodgkin's lymphoma is an immunogenic hematologic malignancy that promotes immune evasion by activating programmed cell death ligand 1 through a unique genetic pathway [31, 32, 39]. Although T-cell priming against solid cancer antigens is prevalent, it is frequently accompanied by functional impairment in the tumor context. In disseminated leukemia and lymphoma, T-cell tolerance is controlled at the level of "tolerogenic" antigen-presenting cells, i.e., capable of producing immunological tolerance, resulting in inadequate priming upon initial antigen contact. Immunosuppression is the primary cause of morbidity and death in individuals with Chronic Lymphocytic Leukemia (CLL), a hematological disease characterized by uncontrolled B-cell proliferation and substantial immunological dysfunction. The innate and adaptive immune response effectors are defective, favoring the establishment of a tolerant milieu that is conducive to disease spread [50–52]. The role of the innate immune response in relapse or blast crisis in chronic myeloid leukemia, as defined by the expression of the oncogene tyrosine kinase BCR-ABL, has been investigated. Acute Myeloid Leukemia (AML) is a systemic illness, and its routes to subvert the host immune response may differ significantly from those used by solid tumors. Much remains unknown regarding the immune escape routes created by AML,

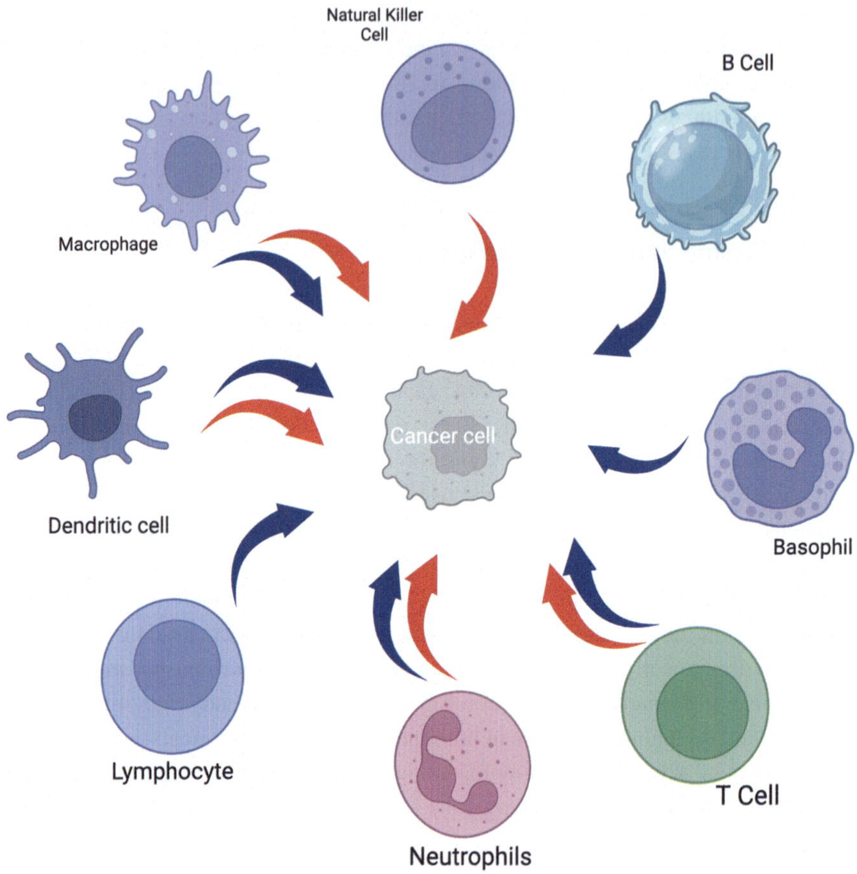

Fig. 1 Schematic representation of the interactions of Immune Cells and Cancer Cells. Dendritic cells (DC), natural killer (NK) cells, T-Cell, and macrophage. Immune cells infiltrate to the tumor and kill tumor cells by different mechanisms: phagocytosis, or tumor-specific cytotoxicity. T-cells, macrophages, and neutrophils secrete pro-tumoral cytokines that promote tumor cell survival and proliferation and induce angiogenesis or lymphangiogenesis. They also support tumor cell metastasis and suppress anti-tumoral immune cell recruitment and function in both primary and metastatic sites. Blue: Promotion, Red: Inhibition. Created with Biorender.com. https://help.biorender.com/hc/en-gb/articles/17605511350685-Citing-BioRender

as well as whether attempting to inhibit tolerance could impact the disease's fate [46, 53]. Immunological and inflammatory responses alter the milieu of the bone marrow, causing differentiation, and eventually malignancy (Fig. 1).

3.5 In Vitro Migration Model of Immune Responses to Cancer

The technique of in vitro modelling is frequently employed to analyze biological processes in a laboratory setting using cell cultures. Through this method, scientists can investigate cell functioning, gene expression, and the impact of various substances on cells without resorting to live animals as test subjects. To simulate, explore, and evaluate immune response in terms of cell migration, researchers have employed various tools and techniques, including laboratory studies and computational simulations. COMPIT is an in vitro platform that analyses and predicts the human immune response to pharmaceutical substances. The technique is unique because it can assess the immune response induced by medication-damaged cells [54, 55]. Researchers have explored the intricate in vivo immune response to cell migrations through diverse in vitro models. The process of immune in vitro modelling holds promise as a valuable tool for addressing fundamental biological questions, informing the development of therapies and vaccines, and providing novel insights into the progression of diseases. The process involves replicating aspects of the immune response in a controlled laboratory environment using cultured cells. By studying cell migrations and interactions in vitro, researchers gain a better understanding of the underlying mechanisms that govern immune responses. This knowledge contributes to advancements in therapeutic approaches, vaccine design, and a deeper comprehension of disease processes. The insights derived from immune in vitro modelling empower scientists to make informed decisions in the pursuit of innovative medical interventions and improved patient outcomes.

The comprehensive study of immune cell migration is essential in modelling two critical components of immunity: primary immune tissues and peripheral tissues housing immune elements [56]. Human Respiratory Syncytial Virus (HRSV) infections encompass a spectrum of symptoms, ranging from mild infections to severe bronchiolitis with life-threatening consequences. The pathophysiology of severe illness involves an exaggerated immune response leading to immunopathology. In addressing this, the authors of [57] present a method for elucidating the response of human immune cells to HRSV infection, employing a human in vitro model where live HRSV is introduced into peripheral blood mononuclear cells to human respiratory syncytial virus infection. Various advanced techniques have been explored for studying immune cell migration, including ex vivo cultures that maintain native tissue structure, microfluidic devices providing physiologically relevant fluid flow and environmental control, and engineered tissues offering precise control of the 3D microenvironment and biophysical cues [58, 59]. While existing models primarily focus on disease modelling, there is a recognized need for additional primary immune tissue models to enhance our understanding. Researchers anticipate that the proliferation of patient-specific models will shed light on the variations in immunity among individuals, facilitating the rapid comprehension and treatment of emerging diseases based on their unique characteristics. This emphasis on the study of immune cell migration is pivotal for advancing our understanding of immune responses and developing targeted strategies for addressing infectious diseases. Immunotherapy has undergone significant advancements in the last two decades, emerging as a cutting-edge and effective approach to cancer treatment. However, the intricacies of the mechanisms underlying novel therapies often remain unclear, prognostic markers exhibit limited efficacy, and preclinical testing lacks consistency. While in vivo experimental models and traditional two-dimensional culture techniques may not fully capture the complexity of tumor/immune system interactions, investigating these interactions is crucial.

This advancement is particularly relevant to the study of cell migration, especially immune cell migration, offering a more representative model for understanding the dynamics of tumor–immune interactions. Various advanced technologies, exemplified by 3D cultures, contribute to the progress of immunotherapy. This includes the development of novel medicines, the identification of biologics with potential therapeutic efficacy, and the screening of individuals suitable for immunotherapy treatments. The influence on cell migration, a critical aspect of immune responses against tumors, underscores the significance of these advancements in unraveling the complexities of cancer–immune interactions. As researchers delve deeper into these 3D models and related technologies, there is a growing potential to enhance our understanding of immune responses and optimize the effectiveness of immunotherapeutic interventions [60].

4 Mathematical and Computational Modelling

The application of mathematical concepts and methods to represent and analyze biological data is commonly referred to as mathematical modelling. This involves formulating equations, functions, or algorithms to elucidate relationships between variables. Mathematical modelling finds widespread use across diverse disciplines, including Engineering, Economics, Biology, and Physics, to comprehensively grasp and predict intricate systems and behaviors. Essential for problem solving and decision-making, mathematical modelling facilitates the generation of simulations and forecasts. These models can then be rigorously examined and validated using data derived from experimental observations. The utilization of mathematical principles to simulate and interpret biological phenomena is a defining characteristic of mathematical modelling, underscoring its significance in advancing our understanding and predictions in various scientific domains. To construct a mathematical model, a set of equations is formulated to represent each pertinent component. This process is initiated by identifying the factors that exert the most substantial influence on the investigated scenario. A notable advantage of mathematical modelling lies in its capacity to make predictions about real-world situations. By employing mathematical equations to illustrate the interactions among variables, one can anticipate how a system will respond under diverse conditions. For example, professionals like engineers or economists might utilize a mathematical model to project the repercussions of alterations in interest rates on the stock market or evaluate the resilience of a new bridge to natural disasters. Mathematical models, spanning various forms, prove versatile in depicting a wide array of systems.

Nonlinear models are frequently used to describe systems with variable rates of change, whereas linear models often represent systems with constant rates of change. Stochastic models are used to depict systems containing components of chance, while deterministic models are mathematical models that represent systems where the outcome is completely determined by the initial conditions and input parameters. A deterministic model is one in which parameters and sets of prior values uniquely define each set of variable states. In contrast, a stochastic model is one in which probability distributions characterize variable states.

Even though mathematical modelling has many benefits, it is crucial to remember that all models are abstractions of reality. All models have limits, and none of them can accurately capture all the intricate aspects that affect a system in the actual world. Therefore, it is crucial to use caution and be aware of a mathematical model's limitations when interpreting its findings. Mathematics serves as the fundamental underpinning for scientific inquiry, offering a systematic approach to addressing various challenges encountered by humanity. The articulation of models necessitates a specific language of communication. Mathematical modelling, employing this specialized mathematical language, serves as a means to depict the behavior of tangible entities and events. This approach proves advantageous, particularly when addressing real-world problems that pose challenges for experimental execution, potentially due to constraints such as computational expense, time limitations, and safety considerations [61, 62].

A mathematical model of infectious diseases is used to analyze and understand the dynamics of diseases and transmission mechanisms, make predictions, and provide insight into real-world disease phenomena. There are different mathematical model types and methods used in modelling these approaches. There are deterministic and stochastic methods, discrete and continuum (continuous), and static and dynamic, to mention a few. A discrete model treats an object as an agent or discrete object, whereas the continuous model considers objects in a continuous state, such as fluid flow. A dynamic model accounts for changes in the system's state over time, whereas a static model estimates the system in equilibrium and is thus time invariant [63]. Differential equations or difference equations are commonly used to represent dynamic models. The modelling and analysis of complex systems using computer methods and simulations are known as computational modelling. The ability of computational modelling to handle vast amounts of data is a significant advantage. Traditional analytical techniques frequently include manually evaluating data and drawing conclusions from that study. This can take a lot of time and is subject to human mistakes. On the other hand, computational modelling enables the automation of this procedure, enabling a more effective and precise process with no handler bias. The ability of computational modelling to simulate complicated systems that are challenging or impossible to investigate directly is another advantage. The requirement for precise data input presents one of the main challenges of computational modelling. The caliber of the data utilized to build the model significantly impacts its correctness. Incomplete or erroneous data will probably cause the model to produce unreliable findings. This implies that the information used in a computational model must be carefully gathered and

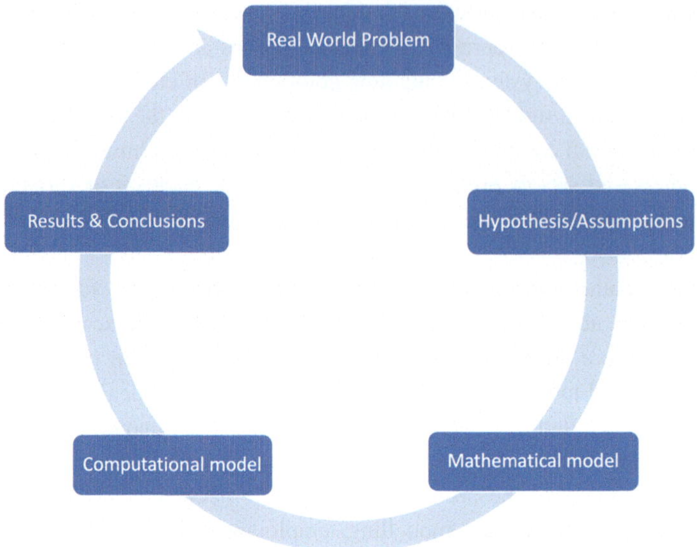

Fig. 2 Mathematical modelling life cycle. Mathematical Modeling Life Cycle: A cyclical process in which real-life problems are translated into mathematical language, solved within a symbolic system, and the solutions tested back within the real-life system

verified. Despite these difficulties, computational modelling has emerged as a crucial tool in numerous industries. It can fundamentally alter how we comprehend intricate systems and enable more precise forecasting and decision-making. Computational modelling will probably remain a vital tool for research and analysis for many years to come as a result [62] (Fig. 2).

4.1 Mathematical Model of Cancer Growth, Migration, and Immune Response

Historically, mathematical models of cancer have been categorized as continuous, discrete, or a combination of the two, depending on the physical description of the fundamental biological components (hybrid). However, continuous formulations enable more effective modelling of tissue-scale processes. Discrete models are good at capturing individual cells and their interconnections. Ordinary Differential Equation (ODE) formulations are used for systems where time is the only independent variable, whereas Partial Differential Equation (PDE) formulations are used for systems where space is present in more than one dimension. Developing a comprehensive, realistic mathematical model for every situation of the host tissue and environment is quite unlikely. Various mathematical models have been constructed to explain diverse events in the host organism at various stages [64]. A wide range of mathematical models from the basic to the complex formulation exists.

Despite the models' complexity, we still do not completely understand the processes that control immune responses and associated migratory activities. The strategy behind immune–target interactions has been studied using various mathematical models. The target could be any viruses, any infectious organism, including bacteria, or immunogenic tumor cells. The effector cell, a type of immune component, engages the target to stop it from causing additional harm. However, various interactions and processes occur in tumor cells, leading the effector to be overcome. The well-developed tools of phase plane analysis allow for a virtually comprehensive analytical discussion of the two-dimensional model, something that high-dimensional systems often cannot do [63].

Therapeutic vaccines are in active development to revolutionize the approach to combat, prevent, and treat a diverse range of illnesses. This innovative category of vaccines holds promise for not only traditional preventive measures but also as therapeutic interventions. Unlike conventional vaccines primarily designed for preventive immunization, therapeutic vaccines target existing diseases, aiming to modulate the immune response and mitigate the impact of the illness. The versatility of therapeutic vaccines offers a novel avenue in the realm of medical interventions, potentially transforming the landscape of disease management and treatment strategies. However, the process of immune response gets more difficult because of numerous unsolved concerns regarding how the immune system interacts with tumor cells and which immune system components play significant roles in responding to immunotherapy. In an analytical context, mathematical models are employed to handle most of these issues. The authors of [65] created a mathematical model of tumor–immune interactions to provide insight into the different functions of natural killer (NK) and CD8+ T-cells in inhibiting other tumor cell lines in mice and humans. Large granular lymphocytes known as NK cells are devoid of T or B-cell lineage markers. Regardless of prior exposure, they are a sort of innate immunity that, among other things, can identify and eliminate tumor cells. By eliminating abnormal cells before they can proliferate, enlarge, or migrate, NK cells are expected to play a crucial part in preventing clinical diagnosis by helping to defend the body against various diseases and infections, particularly viral infections and cancer. While NK cells are not involved in diagnosing diseases, they are part of the immune system that helps prevent the development of clinical symptoms and progression of diseases.

There are presently no self-regulatory terms in the equations or downregulation of an active immune response in the model presented in this chapter. While most current models do not address all immune system functions, they do, despite their simplicity, fit most of the actual evidence. These experimental and simulated results, when combined with the cell lysis data and parameter sensitivity analysis, demonstrate the impact of CD8+ T-cell activation on the outcome. The model's findings suggest that focusing on increasing CD8+ T-cell activity may be necessary (but not sufficient) for tumor regression. Indeed, cytotoxicity tests have shown that the patient-specific effectiveness of the CD8+ T-cell response and the likelihood of a patient favorably responding to immunotherapy therapies have a direct positive relationship [42–44]. Anti-tumor activity of the immune system is increasingly recognized as critical for establishing a long-lasting and effective response to cancer development and invasion, as well as avoiding recurrence following resection or therapy. As our understanding of tumor–immune cell interactions has expanded, experimental research has been supplemented by mathematical modelling to measure and predict these interactions. A mathematical model was constructed and tested to examine the impact of macrophage abundance, influx, and capacity to destroy tumor cells in avascular tumors preferentially. Following that, the models looked at the ability of macrophages that have been designed to target tumor cells and discovered that such techniques are intuitive and sensitive to the tumor and therapeutic factors [48].

4.2 Computational Tools for Investigating Cell Migration of the Immune Cell

The comprehensive investigation of multicellular systems necessitates a focused examination of cell dynamics, encompassing processes such as movement, growth, division, interaction, and apoptosis. This approach is imperative for addressing the inherent complexities within these systems and elucidating their underlying mechanisms. Numerous interacting cells respond to and influence their microenvironment, resulting in tissue-scale dynamics. A perfect "virtual laboratory" model for modelling both the biochemical microenvironment (the "stage") and numerous mechanically and biochemically interacting cells (the "players" on the stage) is required for such multicellular systems. There have been several tools designed to overcome this issue [66]. We will focus on two of the most important and extensively used tools for tissue engineering, infectious disease modelling, cancer modelling, and immune response modelling, which are PhysiCell and CompuCell3D.

PhysiCell is an open-source physics-based 3D multicellular system simulator. The open-source PhysiCell platform provides the stage and actors for studying multiple interacting cells in dynamic tissue microenvironments. A multi-substrate bio-transport solution relates cell phenotype to various diffusing substrates and signalling factors. It includes physiologically driven sub-models for cell cycle, apoptosis, necrosis, solid and fluid volume changes, mechanics, and motility that work right out of the box. Because it has minimal dependencies, C++ code is easy to maintain and deploy across platforms [45, 67].

CompuCell3D is a three-dimensional C++ software problem-solving environment that uses morphogenesis models to replicate biocomplexity concerns. Cell clustering, growth, division, death, adhesion, and volume and surface area limitations can all be represented using the Cellular Potts Model (CPM), which also includes partial differential equation solvers for modelling external chemical field reaction-diffusion and cell type automata for differentiation. By merging these models, CompuCell3D can mimic cellular reactions to external chemical fields like secretion or reabsorption, as well as responses like chemotaxis and haptotaxis [68]. Both tools have been extensively used to model immune response to tumor growth and cancer cell migration [62, 69], where models have been developed. Technologies have been combined to provide a computational mechanism for high-throughput hypothesis testing and, eventually, optimization (one for assisting in creating multiscale agent-based models and the other for enhancing model exploration and optimization. They are also utilized to imitate other infectious illnesses, including COVID-19, in addition to modelling and simulating immunological responses to cancer. They are also used to create a multiscale, multicellular, spatiotemporal model of a generic virus invading epithelial tissue, a simplified cellular immune response, and viral and immunological-induced tissue damage. Finally, they have been utilized to create modular components that can be readily expanded and changed collaboratively to reflect specific viral infections, tissue types, and immune responses [62].

4.3 Mathematical and Computational Modelling of Cell Migration in Immune Cells

In the realm of biomedical research, the study of immune cell dynamics, specifically their migration and responses, is of paramount importance. To elucidate these intricate processes, mathematical modelling emerges as a sophisticated and indispensable tool, relying on both ordinary differential equations (ODE) and partial differential equations (PDE). Foundational to this approach is the work of Kermack and McKendrick, exemplified by the Susceptible, Infected, and Recovered (SIR) model, which utilizes ODE to systematically track the dynamics of individuals during an epidemic. The primary aim of such models is to mitigate the reproductive number, representing the number of secondary cases arising from a single infected individual. In recent years, the evolution of mathematical techniques has propelled research to consider both spatial and temporal scales. This necessitates the formulation of intricate PDE models to simulate the nuanced behavior of immune cells, particularly in the context of cancer. Unlike ODEs, solving PDEs analytically is a formidable challenge, necessitating the deployment of computational and numerical approximations. The integration of ODE and PDE models in mathematical frameworks provides a holistic exploration of immune cell behaviors. This analytical approach contributes not only to the theoretical understanding of immune responses but also holds promise for informing practical strategies in disease control and therapeutic interventions. By leveraging mathematical principles, researchers gain valuable insights into the migration patterns and responses of immune cells, advancing our comprehension of these dynamic processes within the intricate landscape of cancer biology [70].

The immune responses triggered during the invasion of foreign organisms, viruses, antigens, or tumor cells alert the host body to the impending danger it is exposed to. The mathematical model investigates the interactions between the host, its surroundings, and intruder [48, 71, 72].

The spread and transmission of a specific disease within a population are represented using mathematical equations in the mathematical modelling of infectious diseases. This may encompass variables like the disease's transmission rate, how susceptible people are

Table 1 Some mathematical and computational techniques and their data format

Model	Type of data	Number of parameters			References
Ordinary differential equations	Individual or average	2	Or	More	[63, 77, 78]
Partial differential equation	Individual or average	2	Or	More	[50, 62, 68, 69]
Cellular Automata	Individual	1	Or	More	[66]
Cellular Potts	Individual or average	1	Or	More	[68]
Center based	Individual or average	1	Or	More	[67]

to it, and the efficacy of treatments like vaccination or isolation measures. Differential equations, which describe how the number of people in different stages of the disease (for example, infected and recovered), changes over time, are a typical method for modelling infectious diseases. These formulas can be used to forecast the disease's spread and pinpoint prospective treatments that could aid in containing or controlling it.

Agent-based modelling, which simulates the interactions and movements of individual "agents" that represent members of a population to forecast the spread of infectious diseases, is another method that may be used in the mathematical modelling of infectious diseases [73, 74]. Monte Carlo simulations use random sampling from a distribution of potential outcomes to generate various scenarios. Mathematical modelling of infectious diseases can help guide public health policy and decision-making by offering useful insights into the transmission and management of infections [75, 76]. In Table 1, we find a summary of mathematical models that are used to model cancer and immune cell migration.

Take-Home Message

1. Cell migration is a very important and essential mechanism in development, health, and disease.
2. Immune cell helps keep our body safe by fighting off pathogens and diseases.
3. Innate and adaptive immune systems respond quickly and slowly respectively, while adaptive immune systems are much more effective.
4. In vitro models are very important for understanding cancer and immune cell response. Nevertheless, due to the absence of specific structures, some processes may not accurately portray the intricate interactions that are necessary to model in vivo interactions.
5. Mathematical and Computational models help us explore different strategies to understand cell migration in immune response.
6. More research is still needed to better understand cancer cell migration and immune cell response.

References

1. Horwitz R, Webb D. Cell migration. Curr Biol. 2003;13(19):R756–9.
2. Vicente-Manzanares M, Webb DJ, Rick A, Horwitz. Cell migration at a glance. J Cell Sci. 2005;118(21):4917–9.
3. Trepat X, Chen Z, Jacobson K. Cell migration. Compr Physiol. 2012;2(4):2369.
4. Friedl P, Sahai E, Weiss S, Yamada KM. New dimensions in cell migration. Nat Rev Mol Cell Biol. 2012;13(11):743–7.
5. Sompayrac LM. How the immune system works. New York: Wiley; 2022.
6. Rosales C. Neutrophil: a cell with many roles in inflammation or several cell types? Front Physiol. 2018;9:113.
7. Liew PX, Kubes P. The neutrophil's role during health and disease. Physiol Rev. 2019;99(2):1223–48.
8. Kolaczkowska E, Kubes P. Neutrophil recruitment and function in health and inflammation. Nat Rev Immunol. 2013;13(3):159–75.
9. Wang J. Neutrophils in tissue injury and repair. Cell Tissue Res. 2018;371:531–9.
10. De Oliveira S, Rosowski EE, Huttenlocher A. Neutrophil migration in infection and wound repair: going forward in reverse. Nat Rev Immunol. 2016;16(6):378–91.
11. Hirayama D, Iida T, Nakase H. The phagocytic function of macrophage-enforcing innate immunity and tissue homeostasis. Int J Mol Sci. 2017;19(1):92.
12. Ozga AJ, Chow MT, Luster AD. Chemokines and the immune response to cancer. Immunity. 2021;54(5):859–74.
13. Kohli K, Pillarisetty VG, Kim TS. Key chemokines direct migration of immune cells in solid tumors. Cancer Gene Therapy. 2022;29(1):10–21.
14. Griffith JW, Sokol CL, Luster AD. Chemokines and chemokine receptors: positioning cells for host defense and immunity. Annu Rev Immunol. 2014;32:659–702.
15. Franciszkiewicz K, Boissonnas A, Boutet M, Combadiere C, Mami-Chouaib F. Role of chemokines and chemokine receptors in shaping the effector phase of the antitumor immune response. Cancer Res. 2012;72(24):6325–32.
16. Liu Y, Zeng G. Cancer and innate immune system interactions: translational potentials for cancer immunotherapy. J Immunother. 2012;35(4):299.
17. Vesperini D, Montalvo G, Bin Q, Lautenschlager F. Characterization of immune cell migration using microfabrication. Biophys Rev. 2021;13:185–202.
18. Chaplin DD. Overview of the immune response. J Allergy Clin Immunol. 2010;125(2):S3–S23.
19. Marshall JS, Warrington R, Watson W, Kim HL. An introduction to immunology and immunopathology. Allergy Asthma Clin Immunol. 2018;14(2):1–10.
20. Kim SK, Cho SW. The evasion mechanisms of cancer immunity and drug intervention in the tumor microenvironment. Front Pharmacol. 2022;13:868695.
21. Alberts B, Johnson A, Lewis J, Raff M, Roberts K, Walter P. Cancer as a microevolutionary process. In: Molecular biology of the cell. 4th ed. New York: Garland Science; 2002.
22. Laconi E, Marongiu F, DeGregori J. Cancer as a disease of old age: changing mutational and microenvironmental landscapes. Br J Cancer. 2020;122(7):943–52.
23. Ostroverkhova D, Przytycka TM, Panchenko AR. Cancer driver mutations: predictions and reality. Trends Mol Med. 2023;29(7):554–66.
24. Baghban R, Roshangar L, Jahanban-Esfahlan R, Seidi K, Ebrahimi-Kalan A, Jaymand M, Kolahian S, Javaheri T, Zare P. Tumor microenvironment complexity and therapeutic implications at a glance. Cell Commun Signal. 2020;18:1–19.
25. Foon KA, Schroff RW, Gale RP. Surface markers on leukemia and lymphoma cells: recent advances. Blood. 1982;60(1):1–19.

26. Sabattini E, Bacci F, Sagramoso C, Pileri SA. Who classification of tumours of haematopoietic and lymphoid tissues in 2008: an overview. Pathologica. 2010;102(3):83–7.
27. Buchbinder EI, Desai A. CTLA-4 and PD-1 pathways: similarities, differences, and implications of their inhibition. Am J Clin Oncol. 2016;39(1):98.
28. Blagih J, Buck MD, Vousden KH. p53, cancer and the immune response. J Cell Sci. 2020;133(5):jcs237453.
29. Thandra KC, Barsouk A, Saginala K, Aluru JS, Barsouk A. Epidemiology of lung cancer. Contemp Oncol (Pozn). 2021;25(1):45–52.
30. Domagala-Kulawik J. The role of the immune system in non-small cell lung carcinoma and potential for therapeutic intervention. Transl Lung Cancer Res. 2015;4(2):177.
31. Rotte A. Combination of CTLA-4 and PD-1 blockers for treatment of cancer. J Exp Clin Cancer Res. 2019;38:1–12.
32. Julius Clemence R, Hafalla CC, Couper KN, Grau GE, Laurent Renia J, de Souza B, Riley EM. The CTLA-4 and PD-1/PD-L1 inhibitory pathways independently regulate host resistance to plasmodium-induced acute immune pathology. PLoS Pathog. 2012;8(2):e1002504.
33. Zheng G, Hemminki A, Forsti A, Sundquist J, Sundquist K, Hemminki K. Second primary cancer after female breast cancer: familial risks and cause of death. Cancer Med. 2019;8(1):400–7.
34. Mellemkjær L, Friis S, Olsen JH, Scélo G, Hemminki K, Tracey E, Andersen A, Brewster DH, Pukkala E, McBride ML, et al. Risk of second cancer among women with breast cancer. Int J Cancer. 2006;118(9):2285–92.
35. Nikolov I, Kostev K, Kalder M. Incidence of other cancer diagnoses in women with breast cancer: A retrospective cohort study with 42,248 women. Breast Cancer Res Treat. 2022;195(1):75–82.
36. Ali HR, Provenzano E, Dawson S-J, Blows FM, Liu B, Shah M, Earl HM, Poole CJ, Hiller L, Dunn JA, et al. Association between CD8+ T-cell infiltration and breast cancer survival in 12,439 patients. Ann Oncol. 2014;25(8):1536–43.
37. Plitas G, Konopacki C, Kenmin W, Bos PD, Morrow M, Putintseva EV, Chudakov DM, Rudensky AY. Regulatory T cells exhibit distinct features in human breast cancer. Immunity. 2016;45(5):1122–34.
38. Ott PA, Stephen Hodi F, Robert C. CTLA-4 and PD-1/PD-L1 blockade: new immunotherapeutic modalities with durable clinical benefit in melanoma patients. Clin Cancer Res. 2013;19(19):5300–9.
39. Gilligan T, Kantoff PW. Chemotherapy for prostate cancer. Urology. 2002;60(3):94–100.
40. Nader R, El Amm J, Aragon-Ching JB. Role of chemotherapy in prostate cancer. Asian J Androl. 2018;20(3):221.
41. Pienta KJ, Smith DC. Advances in prostate cancer chemotherapy: a new era begins. CA Cancer J Clin. 2005;55(5):300–18.
42. Bou-Dargham MJ, Sha L, Sang Q-XA, Zhang J. Immune landscape of human prostate cancer: immune evasion mechanisms and biomarkers for personalized immunotherapy. BMC Cancer. 2020;20:1–10.
43. Akkın S, Varan G, Bilensoy E. A review on cancer immunotherapy and applications of nanotechnology to chemoimmunotherapy of different cancers. Molecules. 2021;26(11):3382.
44. Zhang P, Xiong X, Rolfo C, Xuexiang D, Zhang Y, Yang H, Russo A, Devenport M, Zhou P, Liu Y, et al. Mechanism- and immune landscape-based ranking of therapeutic responsiveness of 22 major human cancers to next generation anti-ctla-4 antibodies. Cancers. 2020;12(2):284.
45. Demaria S, Golden EB, Formenti SC. Role of local radiation therapy in cancer immunotherapy. JAMA Oncol. 2015;1(9):1325–32.

46. Teague RM, Kline J. Immune evasion in acute myeloid leukemia: current concepts and future directions. J Immunother Cancer. 2013;1(1):1–11.
47. Bakker E, Qattan M, Mutti L, Demonacos C, Krstic-Demonacos M. The role of microenvironment and immunity in drug response in leukemia. Biochim Biophys Acta. 2016;1863(3):414–26.
48. Quarteroni A. Mathematical models in science and engineering. Notices AMS. 2009;56(1):10–9.
49. Oberkampf WL, Trucano TG, Hirsch C. Verification, validation, and predictive capability in computational engineering and physics. Appl Mech Rev. 2004;57(5):345–84.
50. Curran EK, Godfrey J, Kline J. Mechanisms of immune tolerance in leukemia and lymphoma. Trends Immunol. 2017;38(7):513–25.
51. Arruga F, Gyau BB, Iannello A, Vitale N, Vaisitti T, Deaglio S. Immune response dysfunction in chronic lymphocytic leukemia: dissecting molecular mechanisms and microenvironmental conditions. Int J Mol Sci. 2020;21(5):1825.
52. Moreno C, Muñoz C, Terol MJ, Hernández-Rivas J-Á, Villanueva M. Restoration of the immune function as a complementary strategy to treat chronic lymphocytic leukemia effectively. J Exp Clin Cancer Res. 2021;40(1):1–16.
53. Huang W, Liu B, Eklund EA. Investigating the role of the innate immune response in relapse or blast crisis in chronic myeloid leukemia. Leukemia. 2020;34(9):2364–74.
54. Ricon-Becker I, Haldar R, Simon MS, Gutman M, Cole SW, Ben-Eliyahu S, Zmora O. Effect of perioperative cox-2 and beta-adrenergic inhibition on 5-year disease-free-survival in colorectal cancer: a pilot randomized controlled colorectal metastasis prevention trial (compit). Eur J Surg Oncol. 2023;49(3):655–61.
55. Hirsch C, Schildknecht S. In vitro research reproducibility: Keeping up high standards. Front Pharmacol. 2019;10:1484.
56. Hanahan D, Weinberg RA. Hallmarks of cancer: the next generation. Cell. 2011;144(5):646–74.
57. Vissers M, Habets MN, Ahout IML, Jans J, de Jonge MI, Diavatopoulos DA, Ferwerda G. An in vitro model to study immune responses of human peripheral blood mononuclear cells to human respiratory syncytial virus infection. J Vis Exp. 2013;82:e50766.
58. Bonner MG, Gudapati H, Mou X, Musah S. Microfluidic systems for modeling human development. Development. 2022;149(3):dev199463.
59. Hammel JH, Cook SR, Belanger MC, Munson JM, Pompano RR. Modeling immunity in vitro: slices, chips, and engineered tissues. Annu Rev Biomed Eng. 2021;23:461–91.
60. Mengus C, Muraro MG, Mele V, Amicarella F, Manfredonia C, Foglietta F, Muenst S, Soysal SD, Iezzi G, Spagnoli GC. In vitro modeling of tumor–immune system interaction. ACS Biomater Sci Eng. 2017;4(2):314–23.
61. Mahlbacher GE, Reihmer KC, Frieboes HB. Mathematical modeling of tumor-immune cell interactions. J Theor Biol. 2019;469:47–60.
62. Sego TJ, Aponte-Serrano JO, Gianlupi JF, Heaps SR, Breithaupt K, Brusch L, Osborne JM, Quardokus EM, Glazier JA. A modular framework for multiscale multicellular spatial modeling of viral infection, immune response and drug therapy timing and efficacy in epithelial tissues: a multiscale model of viral infection in epithelial tissues. bioRxiv. 2020;16(12):e1008451. https://doi.org/10.1101/2020.04.27.064139.
63. Adeniyi MO, Oke SI, Ekum MI, Benson T, Adewole MO. Assessing the impact of public compliance on the use of non-pharmaceutical intervention with cost-effectiveness analysis on the transmission dynamics of COVID-19: insight from mathematical modeling. In: Modeling, control and drug development for COVID-19 outbreak prevention. New York: Springer; 2022. p. 579–618.
64. Mayer H, Zaenker KS, An Der U, Heiden. A basic mathematical model of the immune response. Chaos Interdiscip J Nonlinear Sci. 1995;5(1):155–61.

65. de Pillis LG, Radunskaya AE, Wiseman CL. A validated mathematical model of cell-mediated immune response to tumor growth. Cancer Res. 2005;65(17):7950–8.
66. Metzcar J, Wang Y, Heiland R, Macklin P. A review of cell-based computational modeling in cancer biology. JCO Clin cancer Inform. 2019;2:1–13.
67. Ghaffarizadeh A, Heiland R, Friedman SH, Mumenthaler SM, Macklin P. PhysiCell: an open source physics-based cell simulator for 3-D multicellular systems. PLoS Comput Biol. 2018;14(2):e1005991.
68. Swat MH, Thomas GL, Belmonte JM, Shirinifard A, Hmeljak D, Glazier JA. Multi-scale modeling of tissues using CompuCell3D. In: Methods in cell biology, vol. 110. New York: Elsevier; 2012. p. 325–66.
69. Ozik J, Collier N, Wozniak JM, Macal C, Cockrell C, Friedman SH, Ghaffarizadeh A, Heiland R, An G, Macklin P. High-throughput cancer hypothesis testing with an integrated PhysiCell-EMEWS workflow. BMC Bioinformatics. 2018;19(18):81–97.
70. Brauer F. The Kermack–Mckendrick epidemic model revisited. Math Biosci. 2005;198(2):119–31.
71. Kretzschmar M, Wallinga J. Mathematical models in infectious disease epidemiology. In: Modern infectious disease epidemiology: concepts, methods, mathematical models, and public health. New York: Springer; 2010. p. 209–21.
72. Ojo MM, Benson TO, Peter OJ, Goufo EFD. Nonlinear optimal control strategies for a mathematical model of covid-19 and influenza co-infection. Physica A. 2022;607:128173.
73. Poleszczuk J, Macklin P, Enderling H. Agent-based modeling of cancer stem cell driven solid tumor growth. In: Stem cell heterogeneity: methods and protocols. New York: Springer; 2016. p. 335–46.
74. Cuevas E. An agent-based model to evaluate the covid-19 transmission risks in facilities. Comput Biol Med. 2020;121:103827.
75. Owoyemi A, Osuchukwu JI, Azubuike C, Ikpe RK, Nwachukwu BC, Akinde CB, Biokoro GW, Ajose AB, Nwokoma EI, Mfon NE, et al. Digital solutions for community and primary health workers: lessons from implementations in Africa. Front Digit Health. 2022;4:876957.
76. Owoyemi A, Nnaemeka E, Benson TO, Ikpe R, Nwachukwu B, Isedowo T. Machine learning recommendation system for health insurance decision making in Nigeria. arXiv preprint arXiv:2305.10708, 2023.
77. Ojo MM, Gbadamosi B, Benson TO, Adebimpe O, Georgina AL. Modeling the dynamics of Lassa fever in Nigeria. J Egypt Math Soc. 2021;29(1):1–19.
78. Ojo MM, Benson TO, Shittu AR, Goufo EFD. Optimal control and cost-effectiveness analysis for the dynamic modeling of Lassa fever. J Math Comput Sci. 2022;12:136.

SPRINGER NATURE

GPSR Compliance

The European Union's (EU) General Product Safety Regulation (GPSR) is a set of rules that requires consumer products to be safe and our obligations to ensure this.

If you have any concerns about our products, you can contact us on ProductSafety@springernature.com

In case Publisher is established outside the EU, the EU authorized representative is:

Springer Nature Customer Service Center GmbH
Europaplatz 3
69115 Heidelberg, Germany

The manufacturer's authorised representative in the EU is Springer Nature Customer Service Centre GmbH, Europaplatz 3, 69115 Heidelberg, Germany. If you have any concerns regarding our products, please contact ProductSafety@springernature.com

Printed and bound by CPI Group (UK) Ltd, Croydon, CR0 4YY
31/07/2026
02182992-0001